Convexification and Global Optimization in Continuous and Mixed-Integer Nonlinear Programming

Nonconvex Optimization and Its Applications

Volume 65

Managing Editor:
Panos Pardalos

Advisory Board:
J.R. Birge
Northwestern University, U.S.A.

Ding-Zhu Du
University of Minnesota, U.S.A.

C. A. Floudas
Princeton University, U.S.A.

J. Mockus
Lithuanian Academy of Sciences, Lithuania

H. D. Sherali
Virginia Polytechnic Institute and State University, U.S.A.

G. Stavroulakis
Technical University Braunschweig, Germany

The titles published in this series are listed at the end of this volume.

Convexification and Global Optimization in Continuous and Mixed-Integer Nonlinear Programming

Theory, Algorithms, Software, and Applications

by

Mohit Tawarmalani
Purdue University,
West Lafayette, IN, U.S.A.

and

Nikolaos V. Sahinidis
University of Illinois,
Urbana, IL, U.S.A.

KLUWER ACADEMIC PUBLISHERS
DORDRECHT / BOSTON / LONDON

A C.I.P. Catalogue record for this book is available from the Library of Congress.

ISBN 1-4020-1031-1

Published by Kluwer Academic Publishers,
P.O. Box 17, 3300 AA Dordrecht, The Netherlands.

Sold and distributed in North, Central and South America
by Kluwer Academic Publishers,
101 Philip Drive, Norwell, MA 02061, U.S.A.

In all other countries, sold and distributed
by Kluwer Academic Publishers,
P.O. Box 322, 3300 AH Dordrecht, The Netherlands.

Printed on acid-free paper

All Rights Reserved
© 2002 Kluwer Academic Publishers
No part of this work may be reproduced, stored in a retrieval system, or transmitted
in any form or by any means, electronic, mechanical, photocopying, microfilming, recording
or otherwise, without written permission from the Publisher, with the exception
of any material supplied specifically for the purpose of being entered
and executed on a computer system, for exclusive use by the purchaser of the work.

Printed in the Netherlands.

To our parents,

Girdhar and Usha Tawarmalani,

and

Vasilios and Aphrodite Sahinidis.

Contents

Preface	**xiii**
Acknowledgments	**xvii**
List of Figures	**xix**
List of Tables	**xxiii**

1 Introduction — 1
- 1.1 The Mixed-Integer Nonlinear Program 5
- 1.2 Branch-and-Bound 6
- 1.3 Illustrative Example 8
  - 1.3.1 A Separable Relaxation 9
  - 1.3.2 Tighter Relaxation 12
  - 1.3.3 Optimality-Based Range Reduction 12
  - 1.3.4 Drawing Inferences from Constraints 16
  - 1.3.5 Branching on the Incumbent 17
- 1.4 Outline of this Book 18

2 Convex Extensions — 25
- 2.1 Introduction 26
- 2.2 Convex Extensions of l.s.c. Functions 29
- 2.3 Multilinear Functions 40
- 2.4 Analysis of Convex Underestimators of x/y 43
  - 2.4.1 Convex Envelope of x/y 44
  - 2.4.2 Closed-Form Expression of Convex Envelope 45
  - 2.4.3 Theoretical Comparison of Underestimators 47
  - 2.4.4 Numerical Example 52

		2.4.5	Concave Envelope of x/y	56

        2.4.5   Concave Envelope of x/y 56
        2.4.6   Relaxing the Positivity Requirement 57
        2.4.7   Semidefinite Relaxation of x/y 60
  2.5 Generalizations and Applications 62
        2.5.1   Envelopes of $(ax+by)/(cx+dy)$ 64
        2.5.2   Convex Envelope of $f(x)y^2$ 65
        2.5.3   Convex Envelope of $f(x)/y$ 69
        2.5.4   Summation of Functions 69

3 Product Disaggregation   71
  3.1 Introduction 72
  3.2 Preliminaries 75
  3.3 Reformulations of a Rational Function 77
  3.4 Tightness of the Reformulation Scheme 81
  3.5 Special Instances of the Reformulation 91
  3.6 Examples of the Reformulation Scheme 93
        3.6.1   Example 1: Hock & Schittkowski (1981) 94
        3.6.2   Example 2: Nuclear Reactor Reload Pattern Design . 95
        3.6.3   Example 3: Catalyst Mixing for Packed Bed Reactor 97
  3.7 Reformulations of Hyperbolic Programs 100
  3.8 Upper Bounding of 0−1 Hyperbolic Programs 105
  3.9 A Branch-and-Bound Algorithm 108
  3.10 Cardinality Constrained Hyperbolic Programs 110
  3.11 Computational Results for CCH Programs 111
        3.11.1 Comparison of Bounds 112
        3.11.2 Performance of the Proposed Algorithm 112
        3.11.3 p-Choice Facility Location 115

4 Relaxations of Factorable Programs   125
  4.1 Nonlinear Relaxation Construction 125
        4.1.1   Concavoconvex Functions 130
  4.2 Polyhedral Outer-Approximation 132

5 Domain Reduction   147
  5.1 Preliminaries 147
        5.1.1   Legendre-Fenchel Transform 148
        5.1.2   Lagrangian Relaxation 152
  5.2 An Iterative Algorithm for Domain Reduction 153

5.3		Theoretical Framework: Abstract Minimization 154
5.4		Application to Traditional Models 160
5.5		Geometric Intuition 163
5.6		Domain Reduction Problem: Motivation 163
5.7		Relation to Earlier Works 164
	5.7.1	Bounds Via Monotone Complementarity 177
	5.7.2	Tightening using Reduced Costs 178
	5.7.3	Linearity-based Tightening 179
5.8		Probing 181
5.9		Learning Reduction Procedure 183

6 Node Partitioning   189

6.1		Introduction 189
6.2		Partitioning Factorable Programs 190
	6.2.1	Branching Variable Selection 190
	6.2.2	Branching Point Selection 194
6.3		Finiteness Issues 196
	6.3.1	Stochastic Integer Programs 197
	6.3.2	The Question of Finiteness 198
	6.3.3	Key to Finiteness 199
	6.3.4	Lower Bounding Problem 200
	6.3.5	Upper Bounding 202
	6.3.6	Branching Scheme 203
	6.3.7	Finiteness Proof 205
	6.3.8	Enhancements 205
	6.3.9	Extension to Mixed-Integer Recourse 207
	6.3.10	Computational Results for Stochastic Programs . . . 207

7 Implementation   213

7.1		Design Philosophy 213
7.2		Programming Languages and Portability 215
7.3		Supported Optimization Solvers 216
7.4		Data Storage and Associated Algorithms 216
	7.4.1	Management of Work-Array 216
	7.4.2	List of Open Nodes 217
	7.4.3	Module Storage: Factorable Programming 218
7.5		Evaluating Derivatives 219
7.6		Algorithmic Enhancements 221

		7.6.1	Multiple Solutions 221
		7.6.2	Local Upper Bounds 222
		7.6.3	Postponement 222
		7.6.4	Finite Branching Schemes 223
	7.7	Debugging Facilities 224
	7.8	BARON Interface 224

8 Refrigerant Design Problem   229
    8.1 Introduction 229
    8.2 Problem Statement 230
    8.3 Previous Work 231
    8.4 Optimization Formulation 232
        8.4.1 Modeling Physical Properties 235
        8.4.2 Modeling Structural Constraints 239
    8.5 Multiple Solutions 249
    8.6 Computational Results 249

9 The Pooling Problem   253
    9.1 Introduction 254
    9.2 The p- and q-Formulations 256
        9.2.1 The p-Formulation 256
        9.2.2 The q-Formulation 261
    9.3 The pq-Formulation 264
        9.3.1 Properties of the pq-Formulation 266
        9.3.2 Lagrangian Relaxations 273
    9.4 Global Optimization of the Pooling Problem 276
        9.4.1 Branching Strategy 278
        9.4.2 Computational Experience 279

10 Miscellaneous Problems   285
    10.1 Separable Concave Quadratic Programs 285
    10.2 Indefinite Quadratic Programs 289
    10.3 Linear Multiplicative Programs 293
    10.4 Generalized Linear Multiplicative Programs 297
    10.5 Univariate Polynomial Programs 298
    10.6 Miscellaneous Benchmark Problems 298
    10.7 Selected Mixed-Integer Nonlinear Programs 305
        10.7.1 Design of Just-in-Time Flowshops 305

| | | 10.7.2 | The Gupta-Ravindran Benchmarks . . . . 311 |

11 GAMS/BARON: A Tutorial   313
- 11.1 Introduction 314
- 11.2 Types of Problems GAMS/BARON Can Solve 315
  - 11.2.1 Factorable Nonlinear Programming: MIP, NLP, and MINLP 315
  - 11.2.2 Special Cases of BARON's Factorable Nonlinear Programming Solver 316
- 11.3 Software and Hardware Requirements 320
- 11.4 Model Requirements 320
  - 11.4.1 Variable and Expression Bounds 320
  - 11.4.2 Allowable Nonlinear Functions 321
- 11.5 How to Run GAMS/BARON 321
- 11.6 System Output 322
  - 11.6.1 System Log 322
  - 11.6.2 Termination Messages, Model and Solver Status . . . 324
- 11.7 Algorithmic and System Options 325
- 11.8 Application to Multiplicative Programs 325
  - 11.8.1 LMPs of Type 1 326
  - 11.8.2 Controlling Local Search Requirements 329
  - 11.8.3 Reducing Memory Requirements via Branching Options 331
  - 11.8.4 Controlling Memory Requirements via Probing . . . 333
  - 11.8.5 Effects of Reformulation 334
  - 11.8.6 LMPs of Type 2 335
  - 11.8.7 Controlling Time Spent on Preprocessing LPs 339
  - 11.8.8 LMPs of Type 3 342
  - 11.8.9 Comparison with Local Search 347
- 11.9 Application to Pooling Problems 356
  - 11.9.1 Controlling Time Spent in Preprocessing 364
  - 11.9.2 Reducing Memory Requirements 368
  - 11.9.3 Controlling the Size of the Search Tree 368
  - 11.9.4 Controlling Local Search Time During Navigation . . 371
  - 11.9.5 Reduced Branching Space 371
  - 11.9.6 Pooling Problem Computations 372
- 11.10 Problems from `globallib` and `minlplib` 376
- 11.11 Local Landscape Analyzer 380

11.12 Finding the K Best or All Feasible Solutions 383
    11.12.1 Motivation and Alternative Approaches 383
    11.12.2 Finding All Solutions to Combinatorial Optimization Problems 385
    11.12.3 Refrigerant Design Problem 391
    11.12.4 Finding All Solutions to Systems of Nonlinear Equations 394

A GAMS Models for Pooling Problems   403
  A.1 Problems Adhya 1, 2, 3, and 4 403
  A.2 Problems Bental 4 and 5 411
  A.3 Problems Foulds 2, 3, 4, and 5 416
  A.4 Problems Haverly 1, 2, and 3 428
  A.5 Problem RT 2 431

Bibliography   435

Index   463

Author Index   469

Preface

Interest in constrained optimization originated with the simple linear programming model since it was practical and perhaps the only computationally tractable model at the time. Constrained linear optimization models were soon adopted in numerous application areas and are perhaps the most widely used mathematical models in operations research and management science at the time of this writing. Modelers have, however, found the assumption of linearity to be overly restrictive in expressing the real-world phenomena and problems in economics, finance, business, communication, engineering design, computational biology, and other areas that frequently demand the use of nonlinear expressions and discrete variables in optimization models. Both of these extensions of the linear programming model are $\mathcal{NP}$-hard, thus representing very challenging problems. On the brighter side, recent advances in algorithmic and computing technology make it possible to revisit these problems with the hope of solving practically relevant problems in reasonable amounts of computational time.

Initial attempts at solving nonlinear programs concentrated on the development of local optimization methods guaranteeing globality under the assumption of convexity. On the other hand, the integer programming literature has concentrated on the development of methods that ensure global optima. The aim of this book is to marry the advancements in solving nonlinear and integer programming models and to develop new results in the more general framework of mixed-integer nonlinear programs (MINLPs) with the goal of devising practically efficient global optimization algorithms for MINLPs.

We embarked on the journey of developing an efficient global optimization algorithm for MINLPs in the early 1990s when we realized that there was no software that could even solve small-sized MINLPs to global optimality despite indications in the literature that such an algorithm could be easily constructed. Our initial attempts, however, found us struggling with many gaps in the literature in the specifications of such a global optimization algorithm. Therefore, about ten years ago, we decided to concentrate on special classes of mixed-integer nonlinear programs, including separable concave minimization problems and applications in capacity expansion of chemical processes. In the process, we developed the first branch-and-bound framework for MINLPs, the Branch-And-Reduce Optimization Navigator (BARON). Its initial purpose was to facilitate design and experimentation with global optimization algorithms. Drawing from this initial experience, in the last six years, we have concentrated on the automatic solution of a general class of MINLPs.

This book documents many of the theoretical advancements that have enabled us to develop BARON to the extent that it now makes it possible for the first time to solve many practically relevant problems in reasonable amounts of computational time in a completely automated manner. Theoretical and algorithmic developments that brought about this situation included:

- A constructive technique for characterizing convex envelopes of nonlinear functions (Chapter 2).

- Many strategies for reformulating mixed-integer nonlinear programs that enable efficient solution. For example, in Chapter 3, we show that "product disaggregation" (distributing the product over the sum) leads to tighter linear programming relaxations, much like variable disaggregation does in mixed-integer linear programming.

- Novel relaxations of nonlinear and mixed-integer nonlinear programs (Chapter 4) that are entirely linear and enable the use of robust and established linear programming techniques in solving MINLPs.

- A new theoretical framework for range reduction (Chapter 5) that helped us identify connections with Lagrangian outer-approximation and develop a unified treatment of existing and several new domain reduction techniques from the integer programming and constraint programming literatures.

- Techniques to traverse more efficiently the branch-and-bound tree, including:

    - the algorithm of Section 7.6.1 that finds all feasible solutions of

PREFACE

systems of nonlinear equations as well as combinatorial optimization problems through enumeration of a single search tree;

- the postponement strategy of Section 7.6.3;
- the branching scheme of Section 7.6.4 that guarantees finite termination for classes of problems for which previous algorithms were either convergent only in limit (*i.e.*, infinite) or resorted to explicit enumeration.

We demonstrate through computational experience that our implementation of these techniques in BARON can now routinely solve problems previously not amenable to standard optimization techniques. In particular:

- In Section 3.6.2, we present a small but difficult instance of a nuclear reactor pattern design problem that was solved for the first time to global optimality using our algorithms.

- In Chapter 8, we completely characterize the feasible space of a refrigerant design problem proposed 15 years ago revealing all the 29 candidate refrigerants that meet the design specifications.

- In Chapters 3, 9, 10, and 11, we provide new solutions and/or improved computational results compared to earlier approaches on various benchmark problems in stochastic decision making, pooling and blending problems in the petrochemical industry, a restaurant location problem, engineering design problems, and a large set of benchmark nonlinear and mixed-integer nonlinear programs.

In writing this book we had three aims. First, to provide a very comprehensive account of material previously available only in journals. Second, to offer a unified and cohesive treatment of a wealth of ideas at the operations research and computer science interface. Third, to present (in over half of the book) new material, including new algorithms, a detailed description of the implementation, extensive computational results, and many geometric interpretations and illustrations of the concepts throughout the book.

We expect that the readership of this book will vary significantly due to the rich mathematical structure and significant potential for applications of mixed-integer nonlinear programming. Students and researchers who wish to focus on convex analysis and its applications in MINLP should find Chapters 2 through 5 and Chapter 9 of interest. Chapters 3, 6, 7, 10, and 11 will appeal to readers interested in implementation and computational issues. Finally, the material in Sections 3.6.2, 3.6.3, and 3.11.3 as well as Chapters 8, 9, 10, and 11 cover modeling and applications of mixed-integer nonlinear programming.

We hope that this book will be used in graduate level courses in nonlinear optimization, integer programming, global optimization, convex analysis, applied mathematics, and engineering design. We also hope that the book will kindle the interest of graduate students, researchers, and practitioners in global optimization algorithms for mixed-integer nonlinear programming and that in the coming years we will witness works that bring forth improvements and applications of the algorithms proposed herein. To facilitate developments in these directions, we plan to maintain detailed descriptions of many of the models used in this book as well as other related information at: http://web.ics.purdue.edu/~mtawarma/minlpbook/.

Mohit Tawarmalani	Nikolaos V. Sahinidis
West Lafayette	Champaign
Indiana	Illinois

Acknowledgments

We would like to acknowledge many people who, knowingly or unknowingly, have helped us in our endeavor to write this book.

It has been our good fortune to have had a number of students and labmates with whom we had beneficial technical discussions and collaborations over the years. Chapter 3 and Section 6.3 resulted from joint work with Shabbir Ahmed (Tawarmalani, Ahmed & Sahinidis 2002a, Tawarmalani, Ahmed & Sahinidis 2002b). Parts of Chapter 8 resulted form joint work with Minrui Yu (Sahinidis, Tawarmalani & Yu 2002), while Section 10.7.1 resulted from joint work with Ramon Gutierrez (Gutierrez & Sahinidis 1996). We benefited from discussions with Nilanjan Adhya on earlier work related to the subject of Chapter 9. We are also indebted to Vinay Ghildyal, Ming Long Liu, Hong S. Ryoo, Joseph P. Shectman, and Russ J. Vander Wiel who developed and tested optimization algorithms that laid the ground work for obtaining some of the results described in this book. Kevin C. Furman, Anastasia Vaia, and Yannis Voudouris provided useful feedback while using BARON.

We wish to thank Professors Placid M. Ferreira, Lynn McLinden, Udatta S. Palekar, and Pravin Vaidya for their helpful guidance and suggestions while serving as members of Mohit Tawarmalani's dissertation committee. Special thanks are due to Lynn McLinden for his course, *Conjugate Analysis and Optimization*, which helped both of us understand a lot of the background material needed for the convexification part of this work.

Arne Drud, Alex Meeraus, and Mike Saunders deserve thanks for encouraging us over the past decade to continue the development of BARON. We are also thankful to Michael Overton and Matthew Wilkins. Michael proposed and Matt helped with the development of the mixed-integer semidefinite programming module of BARON when NVS was on sabbatical at New York University. We would also like to thank Hanif D. Sherali for his comments on the material of Chapter 9, and Arne J. Pearlstein for his suggestions related to Section 11.12.4.

We are truly indebted to Alex Meeraus and the folks of *The GAMS Development Corporation* who were instrumental in helping us improve the software implementation and making it available under GAMS. Alex, in particular, offered many critical comments on the book that helped us improve its quality and value to readers.

Thanks are also due to the optimization community that tested BARON and provided extensive feedback and an enormous set of test problems. Between October 1999 and August 2002, 164 researchers and users of optimization techniques submitted over 6100 problems to the BARON Web portal (an average of over six problems a day). We have learned a great deal and improved the algorithms and software considerably as a result of these submissions. Special thanks in this regard go to Dimitri Alevras, Leonardo Barreto, Paul Barton, Michael Bussieck, Indraneel Das, Steven Dirske, James Falk, Michael Ferris, Bob Fourer, Ignacio Grossmann, Alois Keusch, Leon Lasdon, Sangbum Lee, Alan Manne, Alex Meeraus, Domenico Mignone, Hans Mittelmann, Arnold Neumaier, Charles Ng, and Linus Schrage.

The editorial staff of Kluwer Academic Publishers, especially John Martindale and Angela Quilici, deserve thanks for their careful attention to the needs of our manuscript during the publication process. The book has also benefited a great deal from careful readings and suggestions from YoungJung Chang, Kevin C. Furman, Lindsay Gardner, Yanjun Li, Arne J. Pearlstein, Armin Pruessner, Louis Miguel Rios, Anastasia Vaia, and Wei Xie. We are particularly grateful to YoungJung and Lindsay.

The results presented in this book were developed through partial financial support from a number of sources. MT would like to acknowledge a Mechanical and Industrial Engineering Alumni Teaching Fellowship, and a Computational Science and Engineering Fellowship, both from the University of Illinois at Urbana-Champaign. NVS would like to acknowledge support received from the National Science Foundation (awards DMI 95-02722, BES 98-73586, ECS 00-98770, DMI 01-15166, and CTS 01-24751), a Lucent Technologies Industrial Ecology Fellowship, and the Department of Chemical and Biomolecular Engineering of the University of Illinois.

List of Figures

1.1	A multimodal objective function	2
1.2	A nonlinear feasible region	2
1.3	A convex objective function under integer constraints	3
1.4	The principles of branch-and-bound	7
1.5	Feasible region for illustrative example	8
1.6	Separable relaxation	9
1.7	Branch-and-bound tree for separable relaxation	11
1.8	Separable relaxations generated for illustrative example	13
1.9	Factorable relaxation for illustrative example	14
1.10	Range reduction using marginals	15
1.11	Range reduction using probing	16
1.12	Branch-and-reduce algorithm	19
2.1	A nonconvex function and a convex outer-approximation over an interval	26
2.2	A nonconvex function and its convex and concave envelopes over an interval	27
2.3	A concave function and its convex envelope over an interval	27
2.4	A nonconvex function and its convex envelope over an interval	27
2.5	A nonconvex function and a polyhedral convex envelope	28
2.6	Convex extension constructibility	31
2.7	Convex extensions of x^2 over $[0,1]$ restricted to $\{0,1\}$	34
2.8	Generating set of the epigraph of $-x^2$	38
2.9	Generating set of xy over a rectangle	41
2.10	Ratio: x/y	53
2.11	$x/y - z_f$	54
2.12	$x/y - z_c$	54
2.13	$z_c - z_f$	54

2.14	$z_c - z_g$ (restricted region for clarity)	55
2.15	$x/y - z_f$ (optimum at x^*, not x^U)	55
2.16	$\dfrac{-xy + xy^L + xy^U}{y^L y^U}$	57
2.17	x/y when $0 \in [x^L, x^U]$	58
2.18	$x^{0.8} y^2$	67
2.19	$\log_{10}(9x+1)y^2$	68
2.20	Convex envelope of $x^{0.8} y^2$	68
3.1	Comparison of reformulations: Ri → Rj ⇔ $V_{\text{LRi}} \geq V_{\text{LRj}}$.	105
4.1	Convex envelope for concavoconvex functions	131
4.2	Outer-approximation of convex functions	134
4.3	Interval bisection in sandwich algorithms	136
4.4	Slope bisection in sandwich algorithms	136
4.5	Maximum error rule in sandwich algorithms	137
4.6	Chord rule in sandwich algorithms	137
4.7	Angle bisection in sandwich algorithms	138
4.8	Maximum projective error rule in sandwich algorithms	138
4.9	Proof of Theorem 4.3	140
4.10	Analysis of outer-approximation for projective rule	143
4.11	Area removed by the supporting line	145
5.1	Affine minorants of f and f^*	148
5.2	The cosmic space	150
5.3	x^2 and conjugate $x^2/4$	165
5.4	Interpreting f^* in cosmic space	166
5.5	$\partial f(x) = \emptyset$ for a closed convex function	166
5.6	Perturbation problem and Lagrangian duality	167
5.7	Algorithm to lower bound M^P	168
5.8	Algorithm to solve R_c^A	169
5.9	Domain reduction problem	170
5.10	Hard problem	171
5.11	Easy problem	172
5.12	Bounds with initial approximations	173
5.13	Bounds with improved approximations	174
5.14	The Lagrangian subproblem	175
5.15	Closely related lower bounding procedures	176

LIST OF FIGURES

5.16	Equivalence with standard Lagrangian lower bounding	186
5.17	Branching constraint is not active	187
6.1	Objective function of (Ex)	199
6.2	Procedure to calculate ϵ	201
6.3	Branching scheme for 2SSIP	204
6.4	Mixed-integer second stage	208
7.1	Parser input for insulated tank design	226
7.2	BARON output for problem in Figure 7.1	227
7.3	BARON on the Web	228
8.1	Automotive refrigeration cycle	233
9.1	The pooling problem	255
9.2	An example of the p-formulation	259
9.3	An example of the q-formulation	263
9.4	An example of the pq-formulation	265
11.1	GAMS/CONOPT2 to GAMS/BARON objective function ratios for LMP1 problems	349
11.2	GAMS/MINOS to GAMS/BARON objective function ratios for LMP1 problems	350
11.3	Objective function ratios of the best of GAMS/CONOPT2 and GAMS/MINOS to GAMS/BARON for LMP1 problems	351
11.4	GAMS/BARON to GAMS/CONOPT2 solution time ratios for LMP1 problems	352
11.5	GAMS/BARON to GAMS/MINOS solution time ratios for LMP1 problems	353
11.6	GAMS/BARON to GAMS/CONOPT2/GAMS/MINOS solution time ratios for LMP1 problems	354
11.7	Time ratios of global to local solvers for LMP2 problems	354
11.8	GAMS/BARON presolver solution to global objective function ratios for LMP1 problems	356
11.9	GAMS/CONOPT2 versus GAMS/BARON solutions	377
11.10	GAMS/MINOS versus GAMS/BARON solutions	378
11.11	Local search versus GAMS/BARON solutions	379
11.12	Landscape analysis with stochastic search	382
11.13	Landscape analysis with stochastic search and range reduction	383

11.14 Number of binaries required for integer cuts in `icut.gms` .. 390
11.15 CPLEX nodes to solve `icut.gms` MIPs 390
11.16 CPLEX CPU seconds to solve `icut.gms` MIPs 391
11.17 Number of binaries required for integer cuts in `icut.gms` with enlarged search space 392
11.18 CPLEX nodes to solve `icut.gms` MIPs with enlarged search space 392
11.19 CPLEX CPU seconds to solve `icut.gms` MIPs with enlarged search space 393
11.20 Nodes per solution found for the robot problem 398
11.21 CPU times per solution found for the robot problem 398

List of Tables

2.1	Underestimators of x/y over $[0.1, 4] \times [0.1, 0.5]$	53
3.1	Size of the reformulations	104
3.2	Data for Example 3 in Saipe (1975)	111
3.3	Comparison of various bounds ($m = 5$)	113
3.4	Comparison of various bounds ($m = 10$)	117
3.5	Comparison of various bounds ($m = 15$)	118
3.6	Comparison of various bounds ($m = 20$)	119
3.7	Computational results for CCH ($m = 5$)	120
3.8	Computational results for CCH ($m = 5$)	121
3.9	Computational results for CCH ($m = 10$)	122
3.10	Computational results for CCH ($m = 20$)	123
6.1	Computational results for Test Set 1	209
6.2	Comparative performance for Test Set 2	209
6.3	Computational results for Test Set 2	210
6.4	Sizes of Test Set 3	210
6.5	Computational results for Test Set 3	211
6.6	Nodes in branch-and-bound tree	211
7.1	BARON modules and their capabilities	215
7.2	Evaluation procedure for products	220
8.1	Functional groups considered	234
8.2	Parameters in property estimations	238
8.3	Index notation for structural groups	242
8.4	Set of solutions to refrigerant design problem (Part I)	251
8.5	Set of solutions to refrigerant design problem (Part II)	252

9.1	Indices, variables, and parameters in the p-formulation	257
9.2	Characteristics of benchmark pooling problems	281
9.3	Comparison of bounds for benchmark pooling problems	282
9.4	Comparative computational results for pooling problems	283
10.1	Computational results for small SCQPs (Part I)	287
10.2	Computational results for small SCQPs (Part II)	288
10.3	Comparative computational results for large SCQPs	290
10.4	Computational results for indefinite quadratic programs (Part I)	291
10.5	Computational results for indefinite quadratic programs (Part II)	292
10.6	Computational results for LMPs	294
10.7	Computational results for LMPs of Type 1	295
10.8	Computational results for LMPs of Type 2	296
10.9	Computational results for LMPs of Type 3	296
10.10	Computational results for GLMPs	297
10.11	Computational results on univariate polynomial programs	298
10.12	Brief description of miscellaneous benchmark problems (Part I)	299
10.13	Brief description of miscellaneous benchmark problems (Part II)	300
10.14	Sources of miscellaneous benchmark problems (Part I)	301
10.15	Sources of miscellaneous benchmark problems (Part II)	302
10.16	Computational results on miscellaneous benchmarks (Part I)	303
10.17	Computational results on miscellaneous benchmarks (Part II)	304
10.18	JIT system design problem data for base case	308
10.19	JIT designs using DPSM	309
10.20	Globally optimal results for JIT system design	309
10.21	Comparison of heuristic and exact solutions for JIT system design	310
10.22	Selected MINLP problems from Gupta and Ravindran (1985)	312
11.1	Computational requirements of branch-and-bound for LMP1 with default GAMS/BARON options	330
11.2	Branch-and-bound requirements for LMP1 with `baron.opt` options	331

11.3	Branch-and-bound requirements for LMP1 with `baron.op2` options	333
11.4	Branch-and-bound requirements for LMP1 with `baron.op3` options	335
11.5	Branch-and-bound requirements for reformulated LMP1 with `baron.op3` options	336
11.6	Computational requirements of branch-and-bound for LMP2 with default GAMS/BARON options	340
11.7	Computational requirements of branch-and-bound for LMP2 with `baron.op4` options	341
11.8	Computational requirements of branch-and-bound for LMP2 with `baron.op5` options	343
11.9	Computational requirements of branch-and-bound for LMP3 with default GAMS/BARON options	347
11.10	Computational requirements of branch-and-bound for LMP3 with `baron.op6` options	348
11.11	CPU requirements for LMP1 with `baron.op7` options	357
11.12	Indices, variables, and parameters in the pq-formulation	358
11.13	BARON iterations (nodes) using different option settings	376
11.14	BARON CPU time (seconds) using different option settings	377
11.15	CPU time requirements and objective function values for GAMS/CONOPT2, GAMS/MINOS, and GAMS/BARON	378

Chapter 1

Introduction

Research in optimization attracted attention when significant advances were made in linear programming—the optimization of a linear objective over a linear constraint set—in the late 1940s. The focus of the optimization literature continued to remain in the domain of linearity for the next couple of decades and devoted itself to advancing the field of linear programming and its subclasses (Dantzig 1963, Ahuja, Magnanti & Orlin 1993). Motivated by applications, developments in nonlinear programming algorithms followed quickly and concerned themselves mostly with local optimization guaranteeing globality under certain convexity assumptions (cf. Minoux 1986, Bazaraa, Sherali & Shetty 1993). However, problems in many areas, including engineering design, logistics, manufacturing, and the chemical and biological sciences are often modeled via nonconvex formulations and exhibit multiple local optima.

Figures 1.1, 1.2, and 1.3 illustrate some of the challenges associated with solving nonconvex optimization problems. A multimodal objective function is shown in Figure 1.1. In the presence of multimodal functions, classical nonlinear programming techniques, such as steepest descent, terminate with solutions whose quality strongly depends on the starting point used. It is not uncommon for standard approaches to fail to identify a global optimum in the presence of multiple local optima. Even when the objective function is convex, nonlinearities in the constraint set may give rise to local optima as illustrated in Figure 1.2. Finally, integer requirements lead to nonconvexities (Figure 1.3) and present a significant challenge even in the absence of additional nonconvexities in the objective or constraint functions.

The potential gains that may be obtained through global optimization

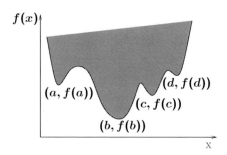

Figure 1.1: A multimodal objective function

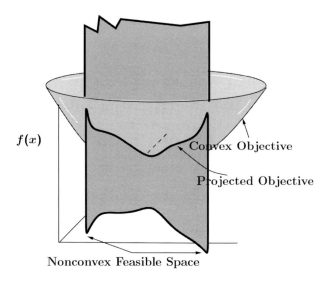

Figure 1.2: A nonlinear feasible region

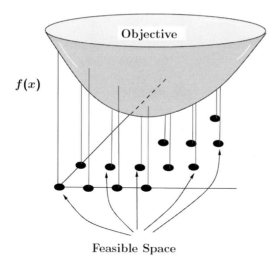

Figure 1.3: A convex objective function under integer constraints

of nonconvex problems motivated a stream of recent efforts by the mathematical programming community in this direction (Pardalos 1994-2002a, Pardalos 1994-2002b). Global optimization algorithms can be classified as either deterministic or stochastic. The deterministic approaches include branch-and-bound (Falk & Soland 1969, McCormick 1972, McCormick 1976, Hansen 1979, Gupta & Ravindran 1985, Hansen, Jaumard & Lu 1991, Hansen 1992, Sherali & Alameddine 1992, Nabar & Schrage 1992, Ben-Tal, Eiger & Gershovitz 1994, Ryoo & Sahinidis 1996, Kearfott 1996, Wolfe 1996, Vaidyanathan & El-Halwagi 1996, Smith & Pantelides 1996, Epperly & Swaney 1996, Borchers & Mitchell 1997, Lee & Mitchell 1997, Adjiman, Dallwig, Floudas & Neumaier 1998, Tawarmalani & Sahinidis 1999b), outer-approximation (Hoffman 1981, Duran & Grossmann 1986, Horst, Thoai & Tuy 1989), cutting planes (Tuy 1964, Hillestad & Jacobsen 1980, Tuy, Thieu & Thai 1985), and decomposition (Tuy 1985, Tuy 1987, Visweswaran & Floudas 1993). Stochastic methods include random search (Zhang & Wang 1993, Wu & Chow 1995, Zabinsky 1998), genetic algorithms (Wu & Chow 1995, Cheung, Langevin & Delmaire 1997), clustering algorithms (Rinnooy Kan & Timmer 1987a), and multi-level single linkage methods (Rinnooy Kan & Timmer 1987b, Li & Chou 1994). Recent reviews of such approaches are provided by Törn & Zilinskas (1989), Schoen (1991), Horst & Tuy (1996), and Sherali & Adams (1999).

While the field of integer *linear* programming has progressed at a considerable pace (cf. Schrijver 1986, Nemhauser & Wolsey 1988, Parker & Rardin 1988), solution strategies explicitly addressing mixed-integer *nonlinear* programs have appeared rather sporadically in the literature. The first published algorithms in this field mostly dealt with problems that are convex when integrality restrictions are dropped and/or contained nonlinearities restricted to quadratic functions (McBride & Yormark 1980, Lazimy 1982, Gupta & Ravindran 1985, Lazimy 1985, Duran & Grossmann 1986, Borchers & Mitchell 1994, Bienstock 1996, Borchers & Mitchell 1997, Lee & Mitchell 1997). Their implementations, whenever reported, were restrictive in scope and focused attention on special subclasses of mixed-integer nonlinear programming.

A few recent works have demonstrated that the application of deterministic branch-and-bound algorithms to the global optimization of MINLPs is promising (Ryoo & Sahinidis 1995, Smith & Pantelides 1996, Floudas 1999, Sherali & Wang 2001). Interestingly, elements of this approach appeared in the pioneering work of Beale and co-workers (Beale & Tomlin 1970, Beale & Forrest 1976, Beale 1979) on special ordered sets, integer programming, and global optimization. The latter works implemented an *a priori* discretization of nonlinear functions of separable programs. Modern approaches address more general factorable programs and rely on an *evolutionary* refinement of the search space, thus offering rigorous convergence guarantees.

This book presents work that advances the theory and application of the branch-and-bound method for globally optimizing mixed-integer nonlinear programs. Convergence of this algorithm is well-established as long as the partitioning and bounding schemes obey certain properties (cf. Horst & Tuy 1996). This leaves significant room in tailoring branch-and-bound for the problem class at hand. At the time of this writing, few general strategies for relaxing and partitioning mixed-integer nonlinear programs are known. The goal of this research is to develop appropriate range reduction tools, relaxation techniques, and branching strategies to demonstrate that branch-and-bound is now an effective algorithm for solving practically relevant MINLPs.

We begin in Section 1.1 by describing the basic mathematical problem addressed in this book. In Section 1.2 we provide an informal description of the branch-and-bound algorithm. An example is used in Section 1.3 to

1.1 The Mixed-Integer Nonlinear Program

Consider the following mixed-integer nonlinear program:

(P)
$$\begin{aligned} \min\ & f(x,y) \\ \text{s.t.}\ & g(x,y) \leq 0 \\ & x \in X \subseteq \mathbb{Z}^p \\ & y \in Y \subseteq \mathbb{R}^n \end{aligned}$$

where $f : (X,Y) \mapsto \mathbb{R}$ and $g : (X,Y) \mapsto \mathbb{R}^m$. The "relaxation" or "relaxed problem" is another optimization problem, R, whose solution provides a lower bound on the optimal objective function value of P. The relaxation is constructed by enlarging the feasible region and/or underestimating the objective function $f(x,y)$ as follows:

(R)
$$\begin{aligned} \min\ & \bar{f}(\bar{x},\bar{y}) \\ \text{s.t.}\ & \bar{g}(\bar{x},\bar{y}) \leq 0 \\ & \bar{x} \in \bar{X} \subseteq \mathbb{Z}^{\bar{p}} \\ & \bar{y} \in \bar{Y} \subseteq \mathbb{R}^{\bar{n}} \end{aligned}$$

where $\bar{f} : (\bar{X},\bar{Y}) \mapsto \mathbb{R}$ and $\bar{g} : (\bar{X},\bar{Y}) \mapsto \mathbb{R}^{\bar{m}}$ and for all $(x,y) \in (X,Y)$ such that $g(x,y) \leq 0$ there exists an $\bar{x}$ and a $\bar{y}$ such that $(\bar{x},\bar{y}) \in (\bar{X},\bar{Y})$, $\bar{g}(\bar{x},\bar{y}) \leq 0$ and $\bar{f}(\bar{x},\bar{y}) \leq f(x,y)$. In general, $\bar{x}$ and $\bar{y}$ need not correspond one-to-one to x and y. We are mostly interested in relaxations where $\bar{p} = 0$ and $\bar{f}$ and $\bar{g}$ are convex over $(\bar{X},\bar{Y})$ since such relaxations are tractable using conventional mathematical programming techniques.

Since P is its own relaxation, it is trivial to construct a relaxation R. Rather, the number of relaxations of P is infinite as long as there exists at least one point in the feasible domain of P where f is finite-valued. Following Weirstrass' theorem, there is in fact a convex relaxation with the same optimal objective function value as P provided that the feasible region of P is compact and f is a lower semicontinuous function. For practical reasons, the only restriction we place on the relaxation R is an abstract one, namely that it be much easier to solve than P itself.

1.2 Branch-and-Bound

Initially conceived as an algorithm to solve combinatorial optimization problems (Land & Doig 1960, Dakin 1965), branch-and-bound has since evolved into a method for solving more general multi-extremal problems in mathematical programming (Falk & Soland 1969, Horst & Tuy 1996).

As its name suggests, branch-and-bound solves problem P by constructing and solving its relaxation R—thereby lower bounding P's optimal objective function value—over successively refined partitions derived through branching in the feasible space.

The fundamental ideas behind this approach are illustrated in Figure 1.4, where we consider the minimization of a univariate function that exhibits two local minima. First, a relaxation, R, is constructed for the problem at hand (P). Once the relaxation is solved, a lower bound, L, on the optimal objective function value of P is obtained (Figure 1.4.a). Local minimization and other upper bounding heuristics are then employed to derive an upper bound U for the problem (Figure 1.4.b). The global minimum must therefore lie between L and U. If $U-L$ is sufficiently small, the procedure terminates with the current upper bounding solution. Otherwise, the feasible region is subdivided into partition elements. These partition elements are appended to a list of partition elements (open nodes) which need to be explored further in order to locate an optimal solution and verify its globality. A selection rule is applied to choose one partition element from the list of open nodes and a new relaxation is constructed taking advantage of its now reduced size. The branch-and-bound process is typically depicted as a tree, where the nodes and branches correspond to bounding and partitioning respectively (Figure 1.4.d). The branch-and-bound process also makes use of domain reduction techniques to fathom and contract search regions, when it is provable that by doing so the algorithm will still terminate with a global optimum (Figure 1.4.c).

A bounding scheme is called *consistent* if any unfathomed partition element can be further refined and, for any sequence of infinitely decreasing partition elements, the upper and lower bounds converge in the limit. A selection scheme is called *bound-improving* if a partition element with the current lowest bound is chosen after a finite number of steps for further investigation. It was shown by Horst & Tuy (1996) that, under these mild conditions, the branch-and-bound algorithm is convergent when it is applied to continuous optimization problems.

1.2. BRANCH-AND-BOUND

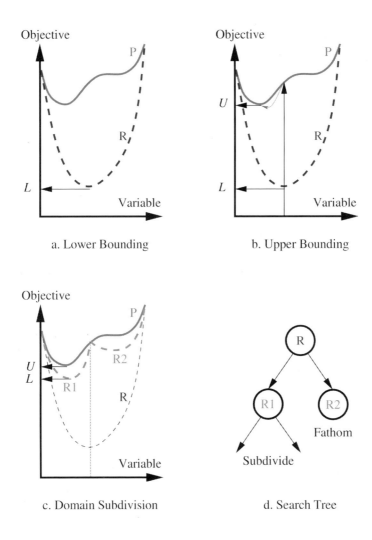

Figure 1.4: The principles of branch-and-bound

1.3 Illustrative Example

Our presentation of the branch-and-bound algorithm in the previous section has been rather simplistic, especially because we illustrated it on the problem of minimizing a univariate function over a convex set (interval). However, the extension to multivariate problems over nonconvex constraint sets is fairly straightforward and demonstrated here through an example. We also demonstrate the effect of acceleration techniques on the branch-and-bound algorithm.

The branch-and-bound algorithm for continuous global optimization is demonstrated on the following problem:

$$\text{(P)} \quad z_{\text{opt}} = \min \ f(x_1, x_2) = -x_1 - x_2$$
$$\text{s.t.} \ \ x_1 x_2 \leq 4$$
$$0 \leq x_1 \leq 6$$
$$0 \leq x_2 \leq 4$$

See Figure 1.5 for a graphical illustration of the feasible space of the problem. The problem exhibits two local minima at points A and B with objective function values of -5 and $-6\frac{2}{3}$, respectively. Point B is thus the global minimum we seek through the branch-and-bound algorithm.

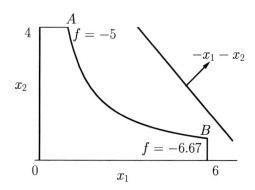

Figure 1.5: Feasible region for illustrative example

1.3.1 A Separable Relaxation

The feasible region of P is nonconvex and the objective function is linear (convex). We construct a convex relaxation of P by outer-approximating its feasible set with a convex set by using a separable reformulation scheme which employs the following algebraic identity:

$$x_1 x_2 = \frac{1}{2}\left\{(x_1 + x_2)^2 - x_1^2 - x_2^2\right\}$$

Amongst the three square terms on the right-hand-side of the above equation, $(x_1 + x_2)^2$ is convex while $-x_1^2$ and $-x_2^2$ are concave. We relax $-x_1^2$ and $-x_2^2$ by linear underestimators using the given bounds for x_1 ($[0,6]$) and x_2 ($[0,4]$). The following relaxation of P results:

(R) $\quad z_{opt} \geq \min \quad -x_1 - x_2$
$\quad\quad\quad\quad\quad\text{s.t.} \quad (x_1 + x_2)^2 - 6x_1 - 4x_2 \leq 8$
$\quad\quad\quad\quad\quad\quad\quad\quad 0 \leq x_1 \leq 6$
$\quad\quad\quad\quad\quad\quad\quad\quad 0 \leq x_2 \leq 4.$

This relaxation is depicted in Figure 1.6.

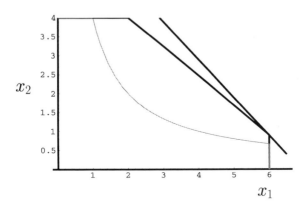

Figure 1.6: Separable relaxation

An alternate, and in fact tighter, relaxation for P can be derived using factorable programming techniques (McCormick 1976), which make use of the convex envelope of $x_1 x_2$ over the rectangle $[x_1^L, x_1^U] \times [x_2^L, x_2^U]$ as given

below:
$$xy \geq \max \left\langle \begin{array}{c} x^U y + y^U x - x^U y^U \\ x^L y + y^L x - x^L y^L \end{array} \right\rangle.$$

The factorable programming relaxation produces the convex hull for P and using it the branch-and-bound algorithm solves P at the root node. We detail the construction and use of factorable relaxation for this example later in this chapter. However, for illustrating the partitioning step and depicting the effect of acceleration techniques, here we use the separable relaxation.

Root Node: The optimal solution for R is $(x_1, x_2) \approx (6, 0.89)$ with an objective function value of -6.89. When a local search is performed on P starting at the relaxation solution $(6, 0.89)$, the point B, $(6, 2/3)$, is obtained with an objective function value of $-6\frac{2}{3}$. Thus, we have shown that:

$$z_{\text{opt}} \in \left[-6.89, -6\frac{2}{3}\right].$$

Branching Variable: The optimal solution for R has x_1 at its upper bound. Let us say we pick any $x_1^B \in [0, 6]$ and create two partition elements such that $x_1 \leq x_1^B$ in the left part and $x_1 \geq x_1^B$ in the right part. Then, the solution of the root node relaxation, *i.e.*, $(6, 0.89)$, stays feasible in the right branch yielding the same lower bound as derived at the root node. No improvement is observed because the linear underestimator of $-x_1^2$ is exact when x_1 is at its upper bound. As a result, no violation of the nonconvex functions at this solution point can be attributed to x_1. The violation can be attributed solely to the error introduced by the underestimation of $-x_2^2$. Therefore, we choose x_2 as the branching variable, and $x_2 = 2$ as the branching point bisecting the feasible interval for x_2. The resulting variable bounds are shown in Figure 1.7.

Node Selection: We choose the left child of the root node for further exploration. For this node, $x_1 \in [0, 6]$ and $x_2 \in [0, 2]$. We update the relaxation to take advantage of these bounds:

$$\begin{aligned}
(\text{R}_\text{L}) \quad z_{\text{opt}} \geq \min \quad & -x_1 - x_2 \\
\text{s.t.} \quad & (x_1 + x_2)^2 - 6x_1 - 2x_2 \leq 8 \\
& 0 \leq x_1 \leq 6 \\
& 0 \leq x_2 \leq 2
\end{aligned}$$

Solving this relaxation, we obtain a lower bound of -6.74. We partition this node further at $x_2 = 1$.

1.3. ILLUSTRATIVE EXAMPLE

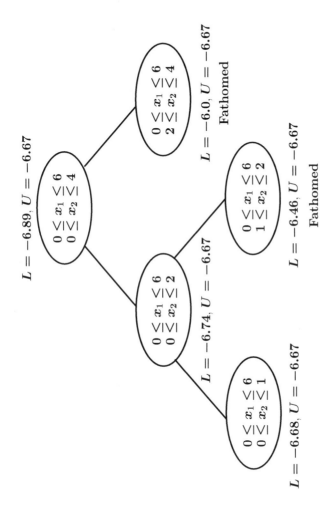

Figure 1.7: Branch-and-bound tree for separable relaxation

Node Selection: We choose the right child of the root node for further exploration. For this node, $x_1 \in [0, 6]$ and $x_2 \in [2, 4]$. We update our relaxation to take advantage of these bounds:

$$
\begin{aligned}
(\text{R}_\text{R}) \qquad z_{\text{opt}} \geq \min \quad & -x_1 - x_2 \\
\text{s.t.} \quad & (x_1 + x_2)^2 - 6x_1 - 6x_2 \leq 0 \\
& 0 \leq x_1 \leq 6 \\
& 2 \leq x_2 \leq 4
\end{aligned}
$$

Solving the relaxation, we obtain a lower bound of -6.00. The best known solution has an objective function value of $-6\frac{2}{3}$. Therefore, the current node does not contain an optimal solution and can safely be fathomed.

Upon further exploration, we observe that the branch-and-bound algorithm proves optimality of $(6, 2/3)$ within an absolute tolerance of 0.01 after conducting five iterations. For a graphical representation of the relaxations, refer to Figure 1.8.

1.3.2 Tighter Relaxation

Here, we demonstrate that a tighter relaxation can be derived using the factorable programming techniques of McCormick (1976). In addition, we show that the tighter relaxation leads to a much superior convergence behavior of the algorithm. If factorable programming techniques (in particular, bilinear envelopes described in Section 1.3.1) are used to relax $x_1 x_2$, then P relaxes to:

$$
\begin{aligned}
(\text{BR}) \qquad \min \quad & -x_1 - x_2 \\
\text{s.t.} \quad & 4x_1 + 6x_2 \leq 28 \\
& 0 \leq x_1 \leq 6 \\
& 0 \leq x_2 \leq 4
\end{aligned}
$$

The optimum for (BR) lies at $(6, 2/3)$, which also provides an upper bound for the problem. Therefore, the problem is solved exactly at the root node. As shown in Figure 1.9, the factorable relaxation provides the convex hull of the nonconvex problem.

1.3.3 Optimality-Based Range Reduction

One of the best-known uses of dual variables in linear programming is in providing sensitivity information with respect to changes in the right-hand-side

1.3. ILLUSTRATIVE EXAMPLE

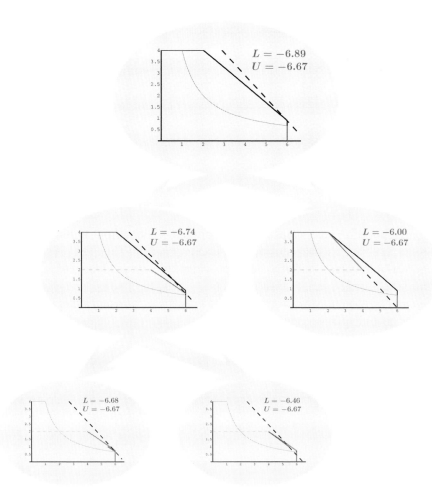

Figure 1.8: Separable relaxations generated for illustrative example

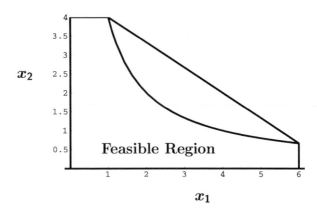

Figure 1.9: Factorable relaxation for illustrative example

of corresponding constraints. For example, if the simple bound constraint $x \leq x^U$ is active in the solution of a linear program, then the corresponding dual variable value (marginal cost) provides the rate of change of the optimal objective function value with respect to changing the upper bound x^U. In the context of nonlinear programming, sensitivity information is provided by Lagrange multipliers. This information can be gainfully employed to improve relaxations of mixed-integer nonlinear programs and expedite convergence of the standard branch-and-bound algorithm as shown by Ryoo & Sahinidis (1996).

These techniques are illustrated, in their simplest form, in Figure 1.10. Once again, consider the simple bound constraint $x \leq x^U$ and assume it is active at the solution of a nonlinear programming relaxation with a corresponding Lagrange multiplier $\lambda > 0$. This multiplier corresponds to the rate of change of the lower bound with respect to changes in x^U. Assuming that L is the optimal objective function value of the relaxed problem and U corresponds to the value of the best known solution of the nonconvex problem, it is simple to argue that a valid lower bound for x is $\kappa = x^U - (U - L)/\lambda$. Indeed, for $x^U < \kappa$, the optimal objective function value of the relaxation is guaranteed to exceed that of the best known solution. As the relaxation provides a lower bound to the nonconvex problem, the region $[x^L, x^U]$ for variable x can be safely reduced to $[\kappa, x^U]$ without loss of solutions better than the one already at hand.

1.3. ILLUSTRATIVE EXAMPLE

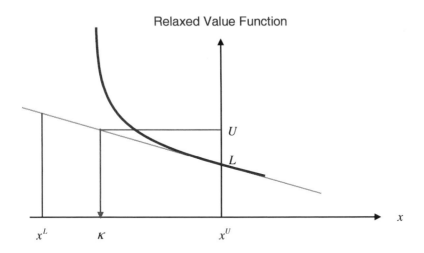

Figure 1.10: Range reduction using marginals

In the context of the illustrative example, the solution of the root node separable relaxation has x_1 at its upper bound with a Lagrange multiplier $\lambda_1 = 0.2$. Applying marginals-based range reduction, it follows that the lower bound for x_1 may be updated to:

$$x_1^L = x_1^U - (U - L)/\lambda_1 \approx 6 - (-6.67 + 6.89)/0.2 \approx 4.86.$$

We note that this reduction strategy is significant and efficient as over 80% of the search space is eliminated for this example at the expense of only three arithmetic operations.

For those variables that are not at bounds at the relaxation solution, a process we refer to as *probing* can be used, whereby these variables are temporarily fixed at bounds and corresponding relaxations are solved to provide linear supports of the value function of the relaxation as illustrated in Figure 1.11.

In the context of the illustrative example, after probing the lower bound of x_2, we get:

$$x_2^L = 0 + 6\frac{2}{3} - 6 = \frac{2}{3}.$$

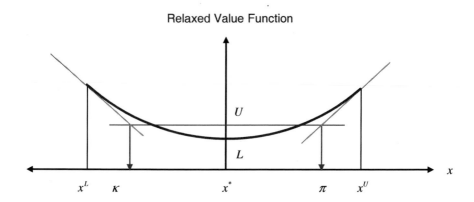

Figure 1.11: Range reduction using probing

Reconstructing the relaxation with the new bounds, we get:

$$\text{(MR)} \quad \min \quad -x_1 - x_2$$
$$\text{s.t.} \quad (x_1 + x_2)^2 - 10.86 x_1 - \frac{14}{3} x_2 + 23.82 <= 0$$
$$4.86 \leq x_1 \leq 6$$
$$0.66 \leq x_2 \leq 4$$

The resulting lower bound is -6.67 and thus global optimality is proven without branching even though the algorithm was based on a weak relaxation.

1.3.4 Drawing Inferences from Constraints

In recent years, the integration of constraint programming and mathematical programming techniques has generated a great deal of excitement in the operations research community. Many problems not otherwise amenable to solution have been solved when constraints are used to draw inferences and reduce the search space (Hooker 2000). In the context of the proposed algorithms for mixed-integer nonlinear programming, this integration was proposed in the early 1990s and governed the development of BARON (Ryoo & Sahinidis 1995, Ryoo & Sahinidis 1996). The main idea is simply to use the nonlinear problem constraints to infer tighter bounds on problem variables.

1.3. ILLUSTRATIVE EXAMPLE

In turn, these tighter bounds give rise to tighter relaxations and expedite convergence of branch-and-bound.

Once again, consider the illustrative example. This time, we will directly work with the nonconvex constraint $x_1 x_2 \leq 4$ and note that, once a valid lower bound of $x_1^L = 4.86$ is established for x_1 through marginals-based range reduction, we can impose an upper bound on x_2 as follows:

$$x_2 \leq \frac{4}{x_1} \leq \frac{4}{x_1^L} \approx 0.82$$

1.3.5 Branching on the Incumbent

The branching technique of Ryoo & Sahinidis (1996) was shown to be finite for concave programs by Shectman & Sahinidis (1998). Even for the current example, it improves the behavior of the branch-and-bound algorithm. Once again, we consider the separable relaxation. This time, instead of branching the root node at $x_2 = 2$, we branch on the incumbent solution $x_2 = 2/3$. After branching, the relaxation at the left node is:

(IR$_L$)
$$\min \; -x_1 - x_2$$
$$\text{s.t.} \; (x_1 + x_2)^2 - 6x_1 - \frac{2}{3}x_2 \leq 8$$
$$0 \leq x_1 \leq 6$$
$$0 \leq x_2 \leq 2/3$$

The lower bound for IR$_L$ is $-6\frac{2}{3}$ and the node is fathomed. For the right child of the root node, the relaxation is:

(IR$_R$)
$$\min \; -x_1 - x_2$$
$$\text{s.t.} \; (x_1 + x_2)^2 - 6x_1 - \frac{14}{3}x_2 \leq \frac{16}{3}$$
$$0 \leq x_1 \leq 6$$
$$2/3 \leq x_2 \leq 4$$

Again, the optimal solution is found at $(6, 2/3)$ and the node is fathomed. The globality of the solution is proven in only three nodes whereas the standard branching scheme required five nodes to terminate with an absolute tolerance of 0.01.

1.4 Outline of this Book

Branch-and-bound is a prototype of a global optimization algorithm and not a formal algorithmic specification since it employs a number of steps that may be tailored to the application at hand. In order to derive an efficient algorithm, it is necessary to study the various techniques for domain reduction, relaxation construction, upper bounding, partitioning, and node selection. Also, a careful choice of data structures and associated algorithms is necessary for the development of a fast implementation.

The algorithms proposed in this book differ from the above simplified prototype in certain ways that reduce memory requirements and/or make them more efficient without compromising the theoretical convergence property. BARON, our implementation, navigates through the branch-and-bound tree using an array of techniques for domain reduction, relaxation construction, partitioning, and tree navigation and expansion. A great deal of emphasis is placed in reducing ranges of variables by drawing inferences from constraints as well as using the marginal values of optimization problems. The former type of inferences may be drawn before the solution of a relaxation at any given node of the search tree. The latter type of inferences may be drawn after the relaxation solution. These pre- and post-processing steps often lead to significant reduction of bounds and computational requirements for solution. For this reason, we refer to the overall strategy as a "branch-and-*reduce*" approach. A flowchart for this algorithm is provided in Figure 1.12.

Readers familiar with the mixed-integer linear programming (MILP) literature will notice that a great deal of research results extend naturally from the linear to the nonlinear case. For example, marginal values from linear programming relaxations are used to fix $0-1$ variables in MILP. The extension of this technique to the MINLP case was outlined above. Similarly, drawing inferences from constraints in MILP is systematically done by several preprocessors that exploit the presence of linear constraints to tighten formulations (Andersen & Andersen 1995). Hence, there is a significant amount of machinery that can be developed for the MINLP case based on ideas that come from the MILP literature.

On the other hand, there are substantial differences between the linear and nonlinear cases. In $0-1$ linear programming, branching on a binary variable creates two subproblems in both of which that variable is fixed. On the contrary, branching on a continuous variable in nonlinear programming may require infinitely many subdivisions. As a result, branch-and-bound

1.4. OUTLINE OF THIS BOOK

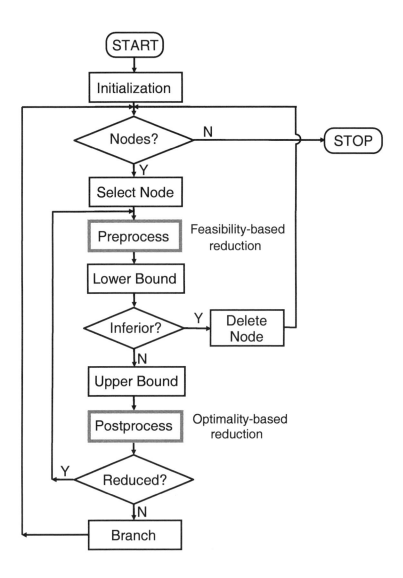

Figure 1.12: Branch-and-reduce algorithm

for 0−1 programs is finite but merely convergent for continuous nonlinear programs.

Another major difference between the linear and nonlinear programming cases lies in the construction of relaxations. While it is straightforward to drop integrality conditions and obtain a linear relaxation of an MILP, dropping integrality conditions may not suffice to obtain a relaxation of an MINLP. Indeed, the resulting NLP may involve nonconvexities that make solution difficult. It is even possible that the resulting NLP may be harder to solve to global optimality than the original MINLP. Furthermore, linear programming technology has reached a level of maturity that provides robust and reliable software for solving linear relaxations of MILP problems. Nonlinear programming solvers, however, often fail even in solving convex problems. At the theoretical level, duality theory provides necessary and sufficient conditions for optimality in linear programming, whereas the KKT optimality conditions in nonlinear programming are not even necessary unless certain constraint qualifications hold (cf. Minoux 1986, Bazaraa, Sherali & Shetty 1993).

The purpose of this book is to develop theory, algorithms, and software for the efficient solution of continuous and mixed-integer nonlinear programs. As such, we consider the development of relaxations, branching schemes, reduction strategies, and implementation issues. Our developments have often been fuelled by complex and important applications that are also presented in this book. In particular, the remainder of the book is organized as follows.

- Chapter 2, *Convex Extensions and Relaxation Strategies*, is based on Tawarmalani & Sahinidis (2002a) and Tawarmalani & Sahinidis (2001) where we provide various advances in relaxation strategies for mixed-integer nonlinear programs. We introduce the concept of convex extensions for lower semicontinuous functions and identify conditions under which convex extensions can be used to constructively determine the convex envelope of the function. Convex extensions yielded, for the first time, the convex envelope of the ratio x/y. This development is presented in this chapter along with tight semidefinite relaxations for fractional programs. These ideas can be easily adapted to build envelopes of a large class of functions.

- Chapter 3, *Product Disaggregation in Global Optimization and Relaxations of Rational Programs*, is based on Tawarmalani, Ahmed &

1.4. OUTLINE OF THIS BOOK

Sahinidis (2002b) and Tawarmalani, Ahmed & Sahinidis (2002a). We first show that distributing the product over the sum in nonlinear programming leads to tighter relaxations, much like variable disaggregation in mixed-integer linear programming. We then use this strategy and apply convex extensions to develop tight linear programming relaxations for 0−1 hyperbolic programs. Computational experience demonstrates that our algorithm performs significantly better than other known techniques for solving p-choice facility location problems that are formulated as 0−1 hyperbolic programs.

- Chapter 4, *Relaxations of Factorable Programs*, is based on Tawarmalani & Sahinidis (1999b) where we synthesize a linear programming based lower bounding scheme for MINLPs by combining factorable programming techniques (McCormick 1976) with the sandwich algorithm (Martelli 1962, Fruhwirth, Burkard & Rote 1989, Burkard, Hamacher & Rote 1992, Rote 1992). In the context of the latter algorithm, we propose a new variant that is based on a projective error rule and prove that this variant exhibits quadratic convergence.

- In Chapter 5, *Domain Reduction*, we develop a theoretical framework for domain reduction applying conjugacy to the dualizing parameterization of an abstract nonlinear program. We propose a new framework for range reduction and show that most of the existing feasibility- and optimality-based range reduction schemes prevalent in the current integer and nonlinear programming literature (Mangasarian & McLinden 1985, Thakur 1990, Hansen, Jaumard & Lu 1991, Hamed & McCormick 1993, Lamar 1993, Savelsbergh 1994, Andersen & Andersen 1995, Ryoo & Sahinidis 1995, Ryoo & Sahinidis 1996, Shectman & Sahinidis 1998, Zamora & Grossmann 1999) are special cases of our approach. Additionally, our range reduction framework naturally leads to a novel learning reduction heuristic and other range reduction techniques that eliminate the search space in the vicinity of fathomed nodes, thereby improving older branching decisions and expediting convergence.

- In Chapter 6, *Node Partitioning Schemes*, we present a new rectangular partitioning procedure for nonlinear programs. A violation transfer scheme is used along with the dual solution of the current relaxation to identify the branching variable. We formulate the problem of finding the branching point as a bilinear program and suggest how variants of this approach can be used to derive exhaustive partitioning schemes for mixed-integer nonlinear programs. Branching schemes for mixed-integer nonlinear programs are not guaranteed to lead to finite algorithms. We argue that, even for two-stage stochastic mixed-integer linear programs, branching schemes that partition the space of the first-stage variables lead to merely convergent and not finite algorithms. We present the finite branching strategy of Ahmed, Tawarmalani & Sahinidis (2000) for the special case of stochastic programs with pure-integer recourse and demonstrate that the research leading to finite branching schemes can significantly impact the computational performance of the branch-and-bound algorithm.

- In Chapter 7, *The Implementation*, we describe the Branch-And-Reduce Optimization Navigator (BARON), which is the implementation of the proposed branch-and-bound global optimization algorithm. BARON is a modular implementation which separates the core branch-and-bound algorithm from problem-specific modules and enhancements. In particular, modules have been developed for solving integer linear programs, separable concave programs, univariate polynomials, fixed-charge programs, multiplicative programs, mixed-integer semidefinite programs, and general factorable integer programs. We describe in this chapter various data structures and algorithms that can be employed to efficiently navigate the branch-and-bound tree. We introduce postponement of nodes in the tree to help reduce the memory requirements of the algorithm and reduce computational requirements when relaxations are solved by expensive iterative procedures.

- In Chapter 8, *Refrigerant Design Problem*, we apply the proposed global optimization algorithms to identify molecules that chemical and physical property estimation techniques classify as candidates worthy of use as automotive refrigerants. The case study, which was solved with the proposed algorithms for the first time by Sahinidis et al. (2002), was initially proposed approximately 15 years ago. We identify not only the global solution, but all feasible solutions to the model using the techniques developed in this book.

- Chapter 9, *The Pooling Problem*, is based on Tawarmalani & Sahinidis (2002b), where we study pooling and blending problems. Using the

1.4. OUTLINE OF THIS BOOK

theory of convex extensions, we present dominance results and structural properties of different formulations of the pooling problem. We demonstrate that, using the algorithms developed here, all benchmark problems in the current literature are solved easily.

- In Chapter 10, *Miscellaneous Problems*, we present extensive computational experience with several problem classes, including separable concave programs, indefinite quadratic programs, multiplicative programs, univariate polynomial programs, and an array of other benchmark factorable continuous, integer, and mixed-integer nonlinear programming problems from a variety of disciplines. In many of these problem classes, we compare our algorithm with specialized implementations reported in the literature demonstrating superior performance of the proposed approach.

- In Chapter 11, *GAMS/BARON: A Tutorial and Empirical Performance Analysis*, we provide modelers, students, and practitioners with easily accessible models and reproducible computational results under the widely adopted GAMS modeling language. By doing so, we demonstrate that the algorithms proposed in this book can be used to solve global optimization models with minimal user intervention. The chapter also serves as a brief introduction to the GAMS/BARON solver that provides convenient access to many of the global optimization algorithms proposed in this book. Its use is illustrated in finding all solutions of systems of nonlinear equations as well as combinatorial optimization problems.

Chapter 2

Convex Extensions and Relaxation Strategies

Synopsis

Central to the efficiency of global optimization methods for nonconvex mathematical programs is the capability to construct tight convex relaxations. In this chapter, we develop the theory of convex extensions of lower semicontinuous (l.s.c.) functions and illustrate its use in building convex envelopes of nonconvex mathematical programs. The techniques developed here amount to a recipe that can be used to construct closed-form expressions of convex and concave envelopes of many classes of functions.

We define a *convex extension* of a lower semicontinuous function to be a convex function that is identical to the given function over a prespecified subset of its domain. Convex extensions are not necessarily constructible or unique. We identify conditions under which a convex extension can be constructed. When multiple convex extensions exist, we characterize the tightest convex extension in a well-defined sense. Using the notion of a generating set, we establish conditions under which the tightest convex extension is the convex envelope. Then, we employ convex extensions to develop a constructive technique for deriving convex envelopes of nonlinear functions. Finally, using the theory of convex extensions we characterize the precise gaps exhibited by various underestimators of x/y over a rectangle and prove that the extensions theory provides convex relaxations that are much tighter than the relaxation provided by the classical outer-linearization of bilinear terms.

Figure 2.1: A nonconvex function and a convex outer-approximation over an interval

2.1 Introduction

As illustrated in the previous chapter, branch-and-bound algorithms can capitalize on convex relaxations. Also illustrated in the previous chapter was the critical importance of the tightness of these relaxations as far as convergence of branch-and-bound is concerned. This chapter addresses the development of tight relaxations for nonconvex mathematical programs.

A nonconvex function is shown in Figure 2.1 in solid line over a certain region of interest. Also in this figure, a convex underestimator and a concave overestimator of the nonconvex function are shown in the bottom dashed line and top dotted lines, respectively, over the same region of interest. Replacing the nonconvex function by its underestimator and overestimator in a mathematical program leads to an outer-approximation of the search space and/or underestimation of the objective function. Thus, an appropriate relaxation is constructed.

For the nonlinear function of Figure 2.1, infinitely many convex underestimators and concave overestimators exist. Figure 2.2 illustrates the tightest possible outer-estimators, namely the convex and concave envelopes of the given function over the region of interest. While any set of outer-estimating functions will lead to a theoretically convergent branch-and-bound algorithm, experience demonstrates the importance of identifying the tightest possible relaxations such as the one shown in this figure.

Figure 2.3 shows a concave function and its convex envelope over an interval. We note that knowledge of the concave function values (circled points) at the two endpoints of the interval suffices in order to construct the envelope. The same observation can be made for the more general nonconvex

2.1. INTRODUCTION

Figure 2.2: A nonconvex function and its convex and concave envelopes over an interval

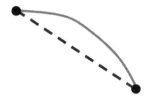

Figure 2.3: A concave function and its convex envelope over an interval

function of Figure 2.4. Once again, knowledge of the function values at the two endpoints of the interval of interest suffices in order to construct the convex envelope of this function over the interval.

Figure 2.5 considers a function in solid line with infinitely many local maximizers. Even in this case, the polyhedral convex envelope is completely characterized once the nonconvex function values at four points are known. These observations indicate that convex envelopes are often characterized by a finite set of points. In the sequel, we will refer to this set of points as the *generating set*. Once this set of points is known, it suffices to extend a convex

Figure 2.4: A nonconvex function and its convex envelope over an interval

Figure 2.5: A nonconvex function and a polyhedral convex envelope

function through these points in order to obtain a convex underestimator of the convex envelope. Provided such extensions are possible, the tightest such extension will provide the convex envelope.

In the sequel, we provide a mathematical formalization of the concepts introduced through the above figures. Our work is closely related to that of Crama (1993) who defined a concave extension of an arbitrary nonlinear function of 0−1 variables as a concave function that extends the given function over the unit hypercube. In Section 2.2, we define the convex extension of an l.s.c. function $\phi(x) : C \mapsto \mathbb{R}$ to be a convex function $\eta : C \mapsto \mathbb{R}$ which is identical to $\phi(x)$ over a prespecified set $X \subseteq C$. We then provide necessary and sufficient conditions under which an extension can be constructed. We also prove that the tightest possible convex/concave extension, whenever constructible, is equivalent to the convex/concave envelope of a restriction of the original l.s.c function.

In Section 2.3, we relate convex extensions and convex underestimators of multilinear functions. For certain nonlinear functions, we develop convexification techniques that make use of convex envelopes of multilinear functions which have been studied extensively by Crama (1993), Rikun (1997), and Sherali (1997).

In Section 2.4, we outline a constructive technique for building convex envelopes that uses convex extensions and disjunctive programming (Rockafellar 1970, Balas 1998, Ceria & Soares 1999). We illustrate the technique by developing the convex and concave envelope of x/y over a rectangular region. We provide a constructive proof for the nonlinear underestimating inequality developed by Zamora & Grossmann (1998) and develop the concave envelope of x/y over the positive orthant using a completely different proof than that in Zamora & Grossmann (1999). We prove that the convex envelope of x/y thus derived over a rectangle is significantly tighter than the underestimator derived by using the outer-linearization of bilinear terms (McCormick 1976,

Al-Khayyal & Falk 1983), which forms the standard technique for relaxing fractional programs. The relaxation that results from including the envelopes for the fractional terms is a semidefinite program which can be solved efficiently using the polynomial-time interior point algorithms (Alizadeh 1995, Vandenberghe & Boyd 1996, Lobo, Vandenberghe, Boyd & Lebret 1998).

Additional applications of the convex extensions theory to developing relaxations for hyperbolic programs and pooling problems will be presented in Chapters 3 and 9, respectively.

2.2 Convex and Concave Extensions of Lower Semicontinuous Functions

In this section, we derive properties of convex envelopes and extensions. Analogous results for concave extensions and concave envelopes of upper semicontinuous functions follow by a simple application of the derived results to the negative of the function. The functions under consideration are assumed to be l.s.c. so that their epigraph is closed. The convex hull of a set S will be denoted by $\text{conv}(S)$ and the affine hull of S by $\text{aff}(S)$. By the projection of X onto $\text{aff}(S)$, we imply the smallest set Y such that $X \subseteq Y + M$, where M is the linear space perpendicular to $\text{aff}(S)$. The set of extreme points of a set S shall be denoted by $\text{vert}(S)$. We denote the relative interior of a set S by $\text{ri}(S)$. The convex envelope of a function $\phi(x) : X \mapsto \mathbb{R}$ is defined to be the tightest convex underestimator of ϕ over X and denoted as $\text{convenv}_X(\phi(x))$ (Horst & Tuy 1996). In the sequel, $\bar{\mathbb{R}}$ denotes $\mathbb{R} \cup \{+\infty\}$.

The following result appears as Theorem 18.3 in Rockafellar (1970).

Theorem 2.1. *Let $C = \text{conv}(S)$, where S is a set of points and directions, and let C' be a nonempty face of C. Then $C' = \text{conv}(S')$, where S' consists of the points in S which belong to C' and the directions in S which are directions of recession of C'.*

We now derive a few properties of the convex envelope as a corollary of the above result.

Corollary 2.2. *Let Φ be the epigraph of $\phi(x)$ and F the epigraph of the convex envelope of ϕ. Let F' be a nonempty face of F. Then $F' = \text{conv}(\Phi')$, where Φ' consists of the points in Φ which belong to F'.*

Proof. Direct application of Theorem 2.1 since $\text{conv}(\Phi) = F$. □

Proposition 2.3. *Let X be a closed convex set and X' be an n-dimensional face of X. Consider a convex function $f : X \mapsto \mathbb{R}$ with epigraph F. There exists a face F' of F such that X' is the projection of F' onto $\text{aff}(X)$ and $\dim(F') = n + 1$.*

Proof. Let X' be expressed as $X \cap M$ where M is an affine space. Consider $F' = F \cap \{(\alpha, x) \mid \alpha \in \mathbb{R}, x \in M\}$, the epigraph of f restricted to X'. We prove by contradiction that F' is a face of F. Assume F' is not a face of F. Then, there exists a closed line segment in F containing a point, say (x^0, f^0), that belongs to $\text{ri}(F')$ and another point, say (x^1, f^1), that does not belong to F'. Since f is real-valued over X, it follows from Theorem 6.8 in Rockafellar (1970) that $x^0 \in \text{ri}(X')$. Clearly, $x^1 \in X$ and $x^1 \notin X'$, which is in contradiction of our assumption that X' is a face of X.

Since X' is n-dimensional, it contains points $x_1, \ldots, x_{n+1}$ that are affinely independent. Let β be a positive number and consider the points $(x_1, f(x_1)), \ldots, (x_{n+1}, f(x_{n+1}))$ and $(x_1, f(x_1) + \beta)$. These are $n + 2$ affinely independent points on F'. Hence, $\dim(\text{aff}(F')) \geq n + 1$.

We prove by contradiction that $\dim(\text{aff}(F')) \leq n + 1$. Assume that the dimension of F' is strictly greater than $n+1$. Then, there exist $n+3$ affinely independent points, say $(x^1, f^1), \ldots, (x^{n+3}, f^{n+3})$, in F'. By definition of affine independence, $(x^1, f^1) - (x^{n+3}, f^{n+3}), \ldots, (x^{n+2}, f^{n+2}) - (x^{n+3}, f^{n+3})$ are linearly independent. Organize the vectors as columns of a matrix and delete the last row of the matrix. Clearly, the rows of the resulting matrix continue to be linearly independent. Therefore, there are $n + 1$ linearly independent columns in $x^1 - x^{n+3}, \ldots, x^{n+2} - x^{n+3}$. Consequently, there are $n + 2$ affinely independent vectors in $x^1, \ldots, x^{n+3}$ contradicting our assumption that $\dim(\text{aff}(X)) = n$. □

Corollary 2.4. *Let X be a convex set and X' be a nonempty face of X. Assume that $f(x)$ is the convex envelope of $\phi(x)$ over X. Then, the restriction of $f(x)$ to X' is the convex envelope of $\phi(x)$ over X'.*

Proof. Let F be the epigraph of $f(x)$. By Proposition 2.3, there exists a face F' of F such that the projection of F' on $\text{aff}(X)$ is X'. Also, from Corollary 2.2, $F' = \text{conv}(\Phi')$, where Φ' is the epigraph of $\phi(x)$ over X'. In other words, $f(x)$ restricted to X' is the convex envelope of ϕ restricted to X'. □

2.2. CONVEX EXTENSIONS OF L.S.C. FUNCTIONS

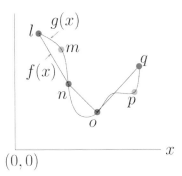

Figure 2.6: Convex extension constructibility

Corollary 2.5. *Let X be a convex set. Assume that $f(x)$ is the convex envelope of $\phi(x)$ over X. Then $f(x) = \phi(x)$ at extreme points of X.*

Proof. Follows directly from Corollary 2.4 as an extreme point of X is a nonempty face of X. □

We now provide a formal definition of convex extensions. This definition is a generalized version of those used by Crama (1993).

Definition 2.6. *Let C be a convex set and $X \subseteq C$. A convex extension of a function $\phi : X \mapsto \mathbb{R}$ over C is any convex function $\eta : C \mapsto \bar{\mathbb{R}}$ such that $\eta(x) = \phi(x)$ for all $x \in X$.*

Note that a convex extension is not always constructible. Consider the univariate function $g(x)$ in Figure 2.6. The function $f(x)$ is a convex extension of $g(x)$ restricted to $\{l, n, o, q\}$. A convex extension of $g(x)$ restricted to $\{l, n, o, p, q\}$ also exists. On the other hand, no convex extension of $g(x)$ restricted to $\{l, m, n, o, q\}$ can be constructed.

In the following result, we provide the conditions under which a convex extension may be constructed.

Theorem 2.7. *A convex extension of $\phi(x) : X \mapsto \mathbb{R}$ over a convex set $C \supseteq X$ may be constructed if and only if*

$$\phi(x) \leq \min \left\{ \sum_i \lambda_i \phi(x_i) \mid \sum_i \lambda_i x_i = x; \sum_i \lambda_i = 1, \lambda_i \in (0,1), x_i \in X \right\}$$

for all $x \in X$

(2.1)

where the above summations consist of finite terms.

Proof. Extend $\phi(x)$ to C as follows:

$$\phi'(x) = \begin{cases} \phi(x), & \text{if } x \in X; \\ +\infty, & \text{if } x \in C \backslash X. \end{cases}$$

Let Φ' be the epigraph of ϕ'.

($\Rightarrow$) Let $\eta(x)$ be the convex envelope of ϕ' and let F be the epigraph of η. By the definition of $\eta(x)$, $\eta(x) \leq \phi'(x) = \phi(x)$ for all $x \in X$. Consider a point $(\eta(x), x) \in F$. Since $F = \text{conv}(\Phi')$, there exist points $(\phi_1, x_1), (\phi_2, x_2) \ldots (\phi_k, x_k)$ in Φ' with $+\infty > \phi_i \geq \phi'(x_i) = \phi(x_i)$ such that $x = \sum_{i=1}^{k} \lambda_i x_i$, $\lambda_i > 0$ and $\sum_{i=1}^{k} \lambda_i = 1$. Hence,

$$\eta(x) = \sum_{i=1}^{k} \lambda_i \phi_i > \sum_{i=1}^{k} \lambda_i \phi(x_i) \geq \phi(x).$$

This shows that $\eta(x) = \phi(x)$ for all $x \in X$. Hence, η is a convex extension of ϕ over C.

($\Leftarrow$) By contradiction. Let us consider a point $(\phi(x), x) \in \Phi'$ with $x \in X$ such that there exist $x_1, x_2, \ldots, x_k$ with $\lambda_1, \lambda_2, \ldots \lambda_k > 0$ such that $\sum_{i=1}^{k} \lambda_i x_i = x$, $\sum_{i=1}^{k} \lambda_i = 1$, $x_i \in X$ and $\phi(x) > \sum_{i=1}^{k} \lambda_i \phi(x_i)$. Let $h(x)$ be any convex function such that $h(x) = \phi(x)$ for all x_i. Then, by convexity of h, $h(x) \leq \sum_{i=1}^{k} \lambda_i \phi(x_i) < \phi(x)$ and hence $h(x) \neq \phi(x)$ for some $x \in X$. □

We shall argue later that the function η constructed in the proof of the above theorem is the tightest possible convex extension of ϕ over C.

If X is a convex set, Theorem 2.7 simply states that a convex extension of $\phi : X \mapsto \mathbb{R}$ over $C \supseteq X$ can only be constructed if ϕ is convex over X. An application of Corollary 2.2 gives us the following result:

Theorem 2.8. *Consider a function ϕ defined over a convex set C and let f be the convex envelope of ϕ. Let Φ and F denote the epigraphs of ϕ and f, respectively. Consider a collection of faces T_I, of F. For any $F' \in T_I$, let $P(F')$ be the projection of F' onto $\text{aff}(C)$. If $X = \bigcup_{F' \in T_I} P(F')$, then a convex extension of ϕ restricted to X may be formed over C if and only if ϕ is convex on $P(F')$ for all $F' \in T_I$.*

Proof. ($\Leftarrow$) Consider a face $F' \in T_I$ such that ϕ is not convex over $P(F')$. It follows then that there does not exist a convex extension of ϕ restricted to

2.2. CONVEX EXTENSIONS OF L.S.C. FUNCTIONS

$P(F')$ over C. Since $X \supseteq P(F')$, there does not exist a convex extension of ϕ restricted to X over C.

($\Rightarrow$) First we show that $f(x) = \phi(x)$ for $x \in X$. Let $F' \in T_I$ be a face of F. From Corollary 2.2, it follows that $F' = \text{conv}(\Phi')$, where Φ' is the set of points in Φ that belong to F. Since Φ' is convex by assumption, it follows that $F' = \Phi'$ for all $F' \in T_I$. Hence, f is a convex extension of ϕ over X. □

Corollary 2.9. *Let C be a convex set and consider an arbitrary collection, F_I, of faces of C. Then, a convex extension of $\phi : C \mapsto \mathbb{R}$ restricted to $\cup_{X \in F_I} X$ can be constructed over C if and only if ϕ is convex over all $X \in F_I$. Further, the convex envelope of ϕ over C is one such convex extension.*

Proof. Follows directly from Proposition 2.3 and Theorem 2.8. □

It should be noted that the convex envelope of ϕ over C is not necessarily the tightest convex extension of ϕ restricted to $\cup_{X \in F_I} X$ over C. This may be illustrated through a simple example. Consider the closed interval $C = [0, 1]$. Let X be the set of its endpoints $\{0, 1\}$. Further, let ϕ be x^2. This function is illustrated in Figure 2.7 and demonstrates that infinitely many convex extensions may exist. The convex envelope of ϕ over C is x^2 while the tightest convex extension of ϕ restricted to X over C is x and $x > x^2$ over $(0, 1)$. Similarly, with $C = [0, 2]$ and $X = \{0, 1, 2\}$, the convex extension of $\phi = x^2$ restricted to X over C is $\max\{x, 3x - 2\}$, which is tighter than x^2 over $C \setminus X$.

It follows that it is possible to construct a convex and concave extension of ϕ restricted to $\cup_{X \in F_I} X$ if and only if ϕ is linear on $X \in F_I$.

Motivated by the above discussion, we investigate the set of all points at which the convex envelope of a function agrees with the function.

Theorem 2.10. *Consider a function $\phi(x) : C \mapsto \mathbb{R}$ with epigraph Φ. Let $f(x)$ be its convex envelope over $\text{conv}(C)$. Assuming that f does not have an infinite two-sided directional derivative at $x_0 \in C$, $f(x_0) = \phi(x_0)$ if and only if there exists ξ such that $\phi(x) \geq \phi(x_0) + \xi^t(x - x_0)$ for all $x \in C$.*

Proof. Since $f(x)$ is a convex function that does not have an infinite two-sided directional derivative, by Theorem 23.3 in Rockafellar (1970), there exists a subgradient ξ to $f(x)$ at x_0 satisfying $f(x) \geq f(x_0) + \xi^t(x - x_0)$.

($\Rightarrow$) Note that $f(x) \leq \phi(x)$. Hence, $\phi(x) \geq f(x_0) + \xi^t(x - x_0)$. If $\phi(x_0) = f(x_0)$, then $\phi(x) \geq \phi(x_0) + \xi^t(x - x_0)$. Therefore, there exists a supporting hyperplane to Φ at x_0.

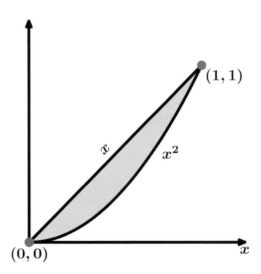

Figure 2.7: Convex extensions of x^2 over $[0,1]$ restricted to $\{0,1\}$

($\Leftarrow$) If there is a support function to Φ at x_0 satisfying $\phi(x) \geq \phi(x_0) + \xi^t(x - x_0)$, then the subgradient is a convex underestimating function of $\phi(x)$. Since $f(x)$ is the pointwise-largest convex underestimating function, $f(x) \geq \phi(x_0) + \xi^t(x - x_0)$. Evaluating the inequality at x_0, $f(x_0) \geq \phi(x_0)$. However, $f(x_0) \leq \phi(x_0)$. Hence, $f(x_0) = \phi(x_0)$. $\square$

Remark 2.11. *Consider a function $\phi(x) : C \mapsto \mathbb{R}$. Let $f(x)$ be its convex envelope over C. It follows from Theorem 2.10 that, for every x_0 such that $\phi(x_0) = f(x_0)$, there exists a vector ξ such that x_0 is a global minimizer of*

$$\tau(x) = \phi(x) - \phi(x_0) - \xi^t(x - x_0)$$

since $\tau(x) \geq 0$ for all $x \in C$ and $\tau(x_0) = 0$.

Let C be a convex set represented as an intersection of convex inequalities $g_i(x) \leq 0$, $i = 1, \ldots, m$ and affine functions $h_i(x) = 0$, $i = 1, \ldots, p$. Consider a point x_0 and an associated index set I defined as $I = \{i \mid g_i(x_0) = 0\}$. Consider a twice-differentiable function $\phi(x) : C \mapsto \mathbb{R}$ with convex envelope $f(x)$ and a point x_0 such that $\phi(x_0) = f(x_0)$. From Remark 2.11, it follows that there exists a vector ξ such that $\tau(x) = \phi(x) - \phi(x_0) - \xi^t(x - x_0)$

2.2. CONVEX EXTENSIONS OF L.S.C. FUNCTIONS

achieves its minimum at x_0. Let u_i and v_i be the vectors of dual multipliers (assuming they exist) corresponding to the inequality and equality constraints respectively obeying the KKT optimality conditions (Bazaraa, Sherali & Shetty 1993, Theorem 4.3.7) for the following mathematical program at x_0:

$$\min\{\tau(x) \mid x \in C\}.$$

Define a Lagrangian function

$$L(x) = \tau(x) + \sum_{i \in I} u_i g_i(x) + \sum_{i=1}^{p} v_i h_i(x).$$

Note that $\nabla^2 \tau(x) = \nabla^2 \phi(x)$. Hence

$$\nabla^2 L(x_0) = \nabla^2 \phi(x_0) + \sum_{i \in I} u_i \nabla^2 g_i(x_0).$$

Then, from second order necessary conditions for local optimality, it follows that $d^t \nabla^2 L(x_0) d \geq 0$ for all $d \in C'$ where

$$C' = \{d \neq 0 \mid \nabla g_i(x_0) d = 0, \forall i \in I, \nabla h_i(x_0) d = 0, i = 1, \ldots, p\}.$$

In particular, if $I = \emptyset$, then the dual multipliers exist (Bazaraa et al. 1993, Theorem 5.1.4) and $\nabla^2 \phi(x_0)$ is positive semidefinite over the affine hull of the equality constraints. This particular subcase affords a much simpler proof. Any function with a Hessian that is not positive semidefinite is locally strictly concave in some direction and hence the point $(x_0, \phi(x_0))$ does not belong to the graph of the convex envelope of ϕ.

The epigraph, hypograph, graph, and level-sets of nonlinear functions are often of interest to us for modeling purposes. The convex extension, if constructible, can replace the nonlinear function in the definition of the above sets thereby yielding natural convex relaxations. For example, the epigraph, $F_{\text{epi}} = \{(f, x) \mid f \geq \phi(x), x \in X\}$, of a nonconvex function $\phi(x) : X \mapsto \mathbb{R}$ is exactly represented by $R_{\text{epi}} = \{(f, x) \mid f \geq \eta(x), x \in X\}$, where η is a convex extension of ϕ restricted to X over $\text{aff}(X)$. A convex relaxation of R_{epi}, for example, $Q_{\text{epi}} = \{(f, x) \mid f \geq \eta(x), x \in \text{conv}(X)\}$, then relaxes F_{epi}. The following simple observations present properties of convex extensions that are quite useful in constructing convex extensions.

The next remark illustrates the preservation of convex extensions under exchange of two functions.

Remark 2.12. Let $\phi(x) = \phi'(x)$ for all $x \in X$. Then, the set of convex extensions of ϕ restricted to X over any set C is identical to the set of convex extensions of ϕ' restricted to X over C.

Remark 2.13. Let f_1 and f_2 be convex extensions of $\phi_1 : X \mapsto \mathbb{R}$ and $\phi_2 : X \mapsto \mathbb{R}$. If the binary operation $\oplus$ preserves convexity, then $f_1 \oplus f_2$ is a convex extension of $\phi_1 \oplus \phi_2$.

The above remark may be easily generalized to operations involving an arbitrary number of functions.

Definition 2.14. A function f is said to be the tightest in a class of convex functions $\mathcal{F}$ if $f \in \mathcal{F}$ and the epigraph of f is a subset of the epigraphs of all other functions in $\mathcal{F}$.

Note that it is not true that a tightest function exists in all classes of convex functions. For example, in the class of functions $\{(x-a)^2\}$, where $a \in R$, none of the members can be classified as the tightest.

The convex extension employed in the proof of Theorem 2.7 is the tightest possible convex extension of $\phi : X \mapsto \mathbb{R}$ over C, where C is a convex superset of X. We are now in a position to prove the following result:

Theorem 2.15. Let C be a convex set. Consider $\phi : C \mapsto \mathbb{R}$ and $X \subseteq C$. Define a function $\phi' : C \mapsto \mathbb{R}$ as follows:

$$\phi'(x) = \begin{cases} \phi(x), & \text{if } x \in X; \\ +\infty, & \text{if } x \notin X. \end{cases}$$

Let f be the convex envelope of ϕ' over C. Then, f is the tightest possible convex underestimator of ϕ restricted to X over C. If $f(x) = \phi(x) \; \forall x \in X$, then $f(x)$ is the tightest convex extension of $\phi(x)$ restricted to X over C, else no convex extension exists.

Proof. In light of Remark 2.12, it is obvious that we may replace function ϕ by ϕ' without affecting the set of convex extensions of ϕ restricted to X over C. Let η be any convex extension of ϕ with epigraph H. Denote the epigraph of ϕ' by Φ'. By the definition of η and the construction of ϕ', it follows that $H \supseteq \Phi'$. Note that H is convex. Then, taking the convex hull on both sides, $H \supseteq \text{conv}(\Phi')$. Let F be the epigraph of f. Then, $F = \text{conv}(\Phi') \subseteq H$. Hence, the tightest convex underestimator is $f(x)$. If $f(x)$ is an extension, then it

2.2. CONVEX EXTENSIONS OF L.S.C. FUNCTIONS

is the tightest since the class of convex extensions is, in this case, a subset of the class of convex underestimators. Further, if $\eta(x)$ is a convex extension when $f(x)$ is not, then there exists a point $x \in X$ such that $\eta(x) \geq f(x)$, a contradiction to the fact that $f(x)$ is the tightest convex underestimator. Hence, if $f(x)$ is not a convex extension then there does not exist any convex extension of ϕ restricted to X. □

We shall now employ the representation theorem of closed convex sets to determine necessary and sufficient conditions under which the tightest convex extension of a function may be developed over a convex set by replacing ϕ' with another function. For any l.s.c. function $\phi(x)$, the epigraph F of its convex envelope $f(x)$ over a closed set is a closed convex set. Let S denote the set of extreme points and extreme directions of F and let L be its lineality space. Then, it follows from the representation theorem of closed convex sets that F can be expressed as the convex sum:

$$F = \text{conv}(S) + L.$$

Let us now restrict the domain of ϕ to a compact set X. Let $f(x)$ be the convex envelope of ϕ over $\text{conv}(X)$ with epigraph F. Then, the convex envelope of ϕ over C is completely specified by the following set:

$$G_C^{\text{epi}}(\phi) = \{x \mid (x, f(x)) \in \text{vert}(F)\}.$$

This set is the generating set of the epigraph of function ϕ. Analogously, we define the $G_C^{\text{hypo}}(\phi)$ as the set of extreme points of the hypograph of the concave envelope of ϕ over C. Whenever we use the term generating set of ϕ without further qualification, it will denote the generating set of the epigraph of ϕ. Note that it follows from Corollary 2.2 that $f(x) = \phi(x)$ for every $x \in G_C^{\text{epi}}(\phi)$. As an example, the generating set of the epigraph of $-x^2$ over $[0, 6]$ consists of the endpoints of the interval (see Figure 2.8).

Theorem 2.16. *Let C be a compact convex set. Consider a function $\phi : C \mapsto \mathbb{R}$, and a set $X \subseteq C$. Further, let f be the convex envelope of ϕ over C. If it is possible to construct a convex extension of ϕ restricted to X over C, then f is the tightest such convex extension possible if and only if $G_C^{\text{epi}}(\phi) \subseteq X$.*

Proof. Let F be the epigraph of f and η be the tightest possible convex extension of ϕ restricted to X over C. Let the epigraph of η be H.

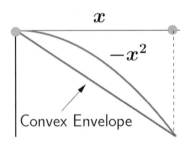

Figure 2.8: Generating set of the epigraph of $-x^2$

($\Rightarrow$) Let us assume that $G_C^{\text{epi}}(\phi) \subseteq X$. It follows that $f(x) = \eta(x)$ for $x \in G_C^{\text{epi}}(\phi)$. From the representation theorem of convex sets it follows that $F \subseteq \text{H}$. Hence
$$\eta(x) \leq f(x) \leq \phi(x) \quad \forall x \in X.$$
As $\eta(x) = \phi(x)$ for $x \in X$, $f(x) = \phi(x)$ for $x \in X$. It follows then that f is a convex extension of ϕ restricted to X over C and, since $F \subseteq \text{H}$, $f(x) \geq \eta(x)$ for any arbitrary convex extension. Thus, f is the tightest convex extension.

($\Leftarrow$) If $G_C^{\text{epi}}(\phi) \not\subseteq X$, then there exists a point $x \in C \setminus X$ such that $(x, f(x))$ cannot be expressed as a convex combination of $(x_i, \phi(x_i))$, $i = 1, \ldots, p$ where each $x_i \in X$. Hence, by Theorem 2.15, $f(x)$ is not the tightest convex extension of $\phi(x)$ restricted to X over C. $\square$

We have shown that the convex envelope of a function ϕ over C restricted to a face of C is the convex envelope of ϕ over that face (Corollary 2.4). Under the assumption of convexity of ϕ over a set of faces of C, a convex extension of ϕ restricted to this set of faces may be constructed over C (Corollary 2.9). Further, a direct application of Theorem 2.16 shows that this is the tightest possible convex extension we can hope to achieve over C if the $G_C^{\text{epi}}(\phi)$ is a subset of the collection of these faces. This condition is not easy to verify in general. We present a result which provides a characterization of $G_C^{\text{epi}}(\phi)$.

Theorem 2.17. *Let $\phi(x)$ be a l.s.c. function on a compact convex set C. Consider a point $x_0 \in C$. Then, $x_0 \notin G_C^{\text{epi}}(\phi)$ if and only if there exists a convex subset X of C such that $x_0 \in X$ and $x_0 \notin G_X^{\text{epi}}(\phi)$. In particular, if for an ϵ-neighbourhood $N_\epsilon \subset C$ of x_0, it can be shown that $x_0 \notin G_{N_\epsilon}^{\text{epi}}(\phi)$, then $x_0 \notin G_C^{\text{epi}}(\phi)$.*

2.2. CONVEX EXTENSIONS OF L.S.C. FUNCTIONS

Proof. ($\Leftarrow$) It follows easily by taking $X = C$.

($\Rightarrow$) Let us denote the epigraph of ϕ over X by Φ_X and that over C by Φ_C. Since $X \subseteq C$, $\Phi_X \subseteq \Phi_C$. Taking the convex hull on both sides, it follows that $\text{conv}(\Phi_X) \subseteq \text{conv}(\Phi_C)$. Further, it follows from Corollary 2.3 that $x_0 \in G_C^{\text{epi}}(\phi)$ only if $(x_0, \phi(x_0)) \in \text{vert}(\text{conv}(\Phi_C))$. Let us assume that $(x_0, \phi(x_0)) \notin \text{vert}(\text{conv}(\Phi_X))$ and $(x_0, \phi(x_0)) \in \text{vert}(\text{conv}(\Phi_C))$. Then, there exist two points (x_1, ϕ_1) and (x_2, ϕ_2) such that $(x_0, \phi(x_0))$ can be expressed as a convex combination of these points and $x_1 \in X$ and $x_2 \in X$. Since $\text{conv}(\Phi_X) \subseteq \text{conv}(\Phi_C)$, (x_1, ϕ_1) and (x_2, ϕ_2) belong to $\text{conv}(\Phi_C)$. In other words, $(x_0, \phi(x_0)) \notin \text{vert}(\text{conv}(\Phi_C))$ which is a contradiction of the assumption. $\square$

Corollary 2.18. *If for any $x \in C$ we can identify a segment $l_x \subseteq C$ such that $x \in \text{ri}(l_x \cap C)$ and $\phi(x)$ is concave over $\text{ri}(l_x \cap C)$, then $x \notin G_C^{\text{epi}}(\phi)$.*

Proof. Direct application of Theorem 2.17 with set $X = \text{ri}(l_x \cap C)$. $\square$

Note that, if the condition in Theorem 2.17 is true for all $x \in \text{ri}(C)$, then $G_C^{\text{epi}}(\phi)$ is a subset of the set of proper faces of C. For example, consider a function $z = -(4 - y^2)^{0.5}$ defined over the x-y plane with x restricted to lie in $[-1, 1]$ and y restricted in $[-2, 2]$. Then, it is easily seen that the generating set is a subset of the faces $x = -1$ and $x = 1$.

As another example, we interpret some results from Rikun (1997). Assume that, for a function $\phi(x) : C \mapsto \mathbb{R}$, it is possible to construct an l_x satisfying the condition in Corollary 2.18 for all $x \notin \text{vert}(C)$. Then, $G_C^{\text{epi}}(\phi)$ is a subset of the set of vertices of C. Further, if x is a vertex of C, then $x \in G_C^{\text{epi}}(\phi)$ since x cannot be represented as a convex combination of points in C, and, by Corollary 2.5, the convex envelope is exact at the extreme points of C. The epigraph of the convex envelope of ϕ over C is in this case finitely-generated and hence polyhedral. It was further shown by Rikun (1997) that general multilinear functions fall under this class of functions and thus have polyhedral envelopes.

We now apply Theorem 2.17 to the Cartesian products of sets. Consider a set C expressible as the Cartesian product of two convex sets $C_1 \times C_2$. Consider a function $\kappa : C \mapsto \mathbb{R}$. At all points (x_1, x_2) such that $x_1 \in C_1$ and $x_2 \in C_2$, consider the corresponding set, $C_x = C_1 + (0, x_2)$. Let $G_{C_x}^{\text{epi}}(\kappa) \subset X + (0, x_2)$ for some $X \subset C_1$ (when $X = \text{vert}(C_1)$, the condition is equivalent to $G_{C_x}^{\text{epi}}(\kappa) \subset \text{vert}(C_x)$). Then, the tightest convex relaxation of κ restricted to $X \times C_2$ is the tightest convex relaxation over $C_1 \times C_2$. This follows directly from Theorem 2.17 since we have demonstrated a set C_x for each point (x_1, x_2), $x_1 \notin X$ such that $(x_1, x_2) \notin G^{\text{epi}}_{C_x}(\kappa)$.

2.3 Envelopes and Extensions of Multilinear functions

Definition 2.19 (Rikun 1997). *A function $L(x_1, \ldots, x_k)$ is said to be a general multilinear function if for each $i = 1, \ldots, k$ the function $L(x_1^0, \ldots, x_i, \ldots, x_k^0)$ linearly depends on vector x_i provided all the other $k - 1$ vector arguments are fixed.*

The polyhedrality of the convex envelope of a general multilinear function over Cartesian product of polytopes was shown by Rikun (1997).

Theorem 2.20. *The general multilinear function $L(x_1, \ldots, x_k)$ has a polyhedral convex and concave envelope over $P = \prod_i P_i$, $i = 1, \ldots, k$, where $x_i \in P_i$, P_i is a polytope. Let F_I be the collection of faces of P over which L is linear. Then, convex and concave extensions of L restricted to F_I may be formed over P. Consider a set X such that $\text{vert}(P) \subseteq X \subseteq F_I$. Then the convex envelope of L over P is the tightest convex extension of L, restricted to X, over P and the concave envelope of L over P is the tightest concave extension of L, restricted to X, over P.*

Proof. Consider a point $\hat{x} \in P$ but $\hat{x} \notin \text{vert}(P)$. Then, there exists an index i such that $\hat{x}_i \notin \text{vert}(P_i)$. Consider $P_s \subseteq P$ containing all x, such that $x_j = \hat{x}_j$, $j \neq i$. P_s is a translate of P_i and L is linear on it. Since $\hat{x}_i \notin \text{vert}(P_i)$, $\hat{x} \notin G^{\text{epi}}_{P_s}(L)$. By Theorem 2.17, $\hat{x} \notin G^{\text{epi}}_P(L)$. Hence, $G^{\text{epi}}_P(L) \subseteq \text{vert}(P)$. As $\text{vert}(P) \subseteq G^{\text{epi}}_P(L)$, it follows $G^{\text{epi}}_P(L) = \text{vert}(P)$. The polyhedrality of the concave envelope follows from the fact that the class of general multilinear functions is closed under negation.

Since $X \subset F_I$, it follows from Corollary 2.9 that the convex and concave envelopes of L over P are the convex and concave extensions of L restricted to X, over P. Also, we showed that $G^{\text{epi}}_P(L) = \text{vert}(P)$ and $G^{\text{epi}}_P(-L) = \text{vert}(P)$ and it was assumed that $\text{vert}(P) \subseteq X$. Then, it follows from Theorem 2.16 that the convex and concave envelopes of L over P are the tightest convex and concave extensions of L restricted to X, over P. □

2.3. MULTILINEAR FUNCTIONS

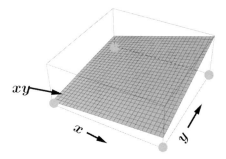

Figure 2.9: Generating set of xy over a rectangle

In the case of the bilinear function defined over a rectangle, Theorem 2.20 shows that the generating set of the epigraph (and hypograph) is the set of extreme points of the rectangle (see Figure 2.9).

We now show that the convex envelope of a multilinear function L is inexact at all points except those belonging to the faces over which L is linear.

Theorem 2.21. *Let $L(x_1, \ldots, x_k)$ be a multilinear function defined over $P = \prod_{i=1}^{k} P_i$ where $x_i \in P_i$ and P_i is a polytope, $i = 1, \ldots, k$. Let $\hat{x} \in \mathrm{ri}(F)$ where F is a face of P. Let $f(x_1, \ldots, x_k)$ be the convex envelope of $L(x_1, \ldots, x_k)$ over P. Then $f(\hat{x}) = L(\hat{x})$ if and only if L is a linear function over F.*

Proof. ($\Leftarrow$) First we show that the general multilinear form reduces to a quadratic form $x_i^t A x_j$ when all x_r, $r \notin \{i, j\}$ are fixed. Let $x_i = [x_{i1}, \ldots, x_{ip}]$ and $x_j = [x_{j1}, \ldots, x_{jm}]$. From the definition of L, it follows that:

$$L(x_1, \ldots, x_k) = \sum_{u=1}^{p} L_{iu} x_{iu}$$

where L_{iu}, $u = 1, \ldots, p$ is a general multilinear function in x_r, where $r \in \{1, \ldots, k\} \setminus \{i\}$. Similarly, each L_{iu} can be expressed as a linear expression in x_j. Collecting terms, it is easy to see that $L(x_1 \ldots, x_k)$ reduces to a quadratic form $x_i^t A x_j$.

The face F can be expressed as $\prod_{u=1}^{k} F_u$ where F_u is a face of P_u. Consider two vector components x_i and x_j such that L is not linear on $F_i \times F_j$. Such components can be found using the following constructive procedure. Take any two components x_v and x_w. If the function is linear on $F_v \times F_w$, combine arguments v and w to form a single vector. Applying the above argument recursively, we detect a quadratic form identifying two components i and j such that L is not linear on $F_i \times F_j$.

Let the dimension of the vector x_i be d_i and the dimension of the face F_i be r_i. Consider a $d_i \times d_i$ matrix T_i such that the first r_i columns form the basis of the unique subspace parallel to F_i augmented with columns of the zero vector. Then, $x_i = T_i y_i + \hat{x}^i$ for all $x_i \in F_i$. Since $\hat{x}_i \in \text{ri}(F_i)$, choosing y_i with $|y_i|$ sufficiently small, produces $x_i \in F_i$. Also,

$$x_i^t A x_j = y_i^t T_i^t A T_j y_j + y_i^t T_i^t A \hat{x}_j + \hat{x}_i^t A T_j y_j + \hat{x}_i^t A \hat{x}_j.$$

Let $B = T_i^t A T_j$. Since $x_i A x_j$ is not linear over $F_i \times F_j$, $B \neq 0$. Define

$$B_f = \begin{pmatrix} 0 & B \\ 0 & 0 \end{pmatrix}$$

such that $y_i^t B y_j = [y_i^t, y_j^t] B_f [y_i, y_j]^t$. B_f is not positive semidefinite by (Bazaraa et al. 1993, Theorem 3.3.12). Hence, from the discussion following Remark 2.11, it is clear that $L(\hat{x}) > f(\hat{x})$.

($\Rightarrow$) From Corollary 2.9, it follows easily that $f(x_0) = L(x_0)$ if L is linear on F. □

If each variable x_i, $i = 1, \ldots, n$ is required to lie in the interval $I_i = [l_i, u_i]$, $i = 1, \ldots, n$, then the feasible region is an n-dimensional hypercube, H^n, expressed as $\prod_{i=1}^{n} I_i$. We shall denote $\text{vert}(H^n)$ as E^n. The unit hypercube $[0,1]^n$ shall be denoted as U^n and the set of its extreme points by B^n.

Theorem 2.22. *Consider a nonlinear function $\phi(x, y) : E^n \times C \mapsto \mathbb{R}$ where $x \in E^n$ and $y \in C$. Assume that $\phi(x, y)$ is convex when x is fixed. Let $f(x, y)$ be the tightest convex extension of ϕ over $H^n \times C$. Then, there exists a function $\phi'(x, y)$ such that:*

1. *$\phi'(x, y) = \phi(x, y)$ for all $(x, y) \in E^n \times C$,*

2. *$\phi'(x, y^0)$ is a uniquely determined multilinear function for every fixed y^0, and*

3. *$\text{convenv}_{H^n \times C}\, \phi'(x, y) = f(x, y)$.*

Proof. It follows directly from Corollary 2.9 that a convex extension of ϕ restricted to $E^n \times C$ may be constructed over $H^n \times C$. We now construct the function ϕ'. Consider a point $x^k = [x_1^k, x_2^k, \ldots, x_n^k] \in E^n$. Define

$$y_{x_i^k}(x) = \begin{cases} (x_i - l_i)/(u_i - l_i), & \text{if } x_i^k = u_i; \\ (u_i - x_i)/(u_i - l_i), & \text{if } x_i^k = l_i. \end{cases}$$

Construct the product term $w_{x^k}(x) = \prod_{i=1}^n y_{x_i^k}(x)$. Consider the function:

$$\phi'(x, y) = \sum_{x^k \in E^n} w_{x^k}(x)\phi(x^k, y).$$

It follows easily that $\phi'(x, y) = \phi(x, y)$ if $(x, y) \in E^n \times C$, since $w_{x^k}(x^k) = 1$ and $w_{x^k}(x) = 0$ if $x \in E^n \setminus x^k$. Further, since each $w_{x^k}(x)$ is a product term, $\phi'(x, y)$ is multilinear when y is fixed. It follows easily from a dimensionality argument that a multilinear expression is uniquely determined by its value at the extreme points of a full dimensional hypercube, Thus, $\phi'(x, y^0)$ is uniquely determined.

From Theorems 2.20 and 2.17, the generating set of $\text{convenv}_{H^n \times C}\, \phi'(x, y)$ is a subset of $E^n \times C$. Hence, from Theorem 2.16, it follows that

$$f(x, y) = \text{convenv}_{H^n \times C}\, \phi'(x, y).$$

$\square$

Corollary 2.23. *Consider a nonlinear function $\phi(x) : E^n \mapsto \mathbb{R}$. Let $f(x)$ be the tightest convex extension of ϕ restricted to E^n over H^n. There exists a unique multilinear function ϕ' such that $\phi'(x) = \phi(x)$ for all $x \in E^n$. The function $f(x)$ is the polyhedral convex envelope of $\phi'(x)$ over H^n.*

Proof. Direct application of Theorem 2.22 establishes the existence of $\phi'(x)$. Polyhedrality of f and its equivalence to the convex envelope of ϕ when it is multilinear, follows from Theorem 2.20. $\square$

2.4 Analysis of Convex Underestimators of x/y

Throughout this section, we consider the function $f(x, y) = x/y$ over a rectangle $[x^L, x^U] \times [y^L, y^U]$ in the positive orthant.

2.4.1 Convex Envelope of x/y

As x/y is linear in x for a fixed y and convex in y for a fixed x, Corollary 2.18 establishes that:
$$G^{\text{epi}}_{[x^L,x^U]\times[y^L,y^U]}(x/y) \subseteq \{(x^L,y) \mid y^L \le y \le y^U\} \cup \{(x^U,y) \mid y^L \le y \le y^U\}.$$

By Theorem 2.15, the convex envelope of x/y is also the convex envelope of
$$f'(x,y) = \begin{cases} x^L/y, & \text{if } x = x^L; \\ +\infty, & \text{if } x^L < x < x^U; \\ x^U/y, & \text{if } x = x^U. \end{cases}$$

Let F' be the epigraph of $f'(x,y)$. F' is thus a union of two convex sets. By applying convex disjunctive programming techniques (Rockafellar 1970), the epigraph of the convex envelope of x/y may be stated as

$$\left.\begin{aligned} z &\ge \frac{x^L}{y_a}(1-\lambda) + \frac{x^U}{y_b}\lambda \\ y^L &\le y_a \le y^U \\ y^L &\le y_b \le y^U \\ y &= (1-\lambda)y_a + \lambda y_b \\ x &= x^L + (x^U - x^L)\lambda \\ 0 &\le \lambda \le 1. \end{aligned}\right\} \quad (2.2)$$

Introduce $y_p = y_a(1-\lambda)$. Then $\lambda y_b = (y - y_p)$. After algebraic manipulations (assuming $0 < \lambda < 1$), (2.2) above may be restated as:

$$\left.\begin{aligned} z &\ge \frac{x^L}{y_p}\left(\frac{x^U - x}{x^U - x^L}\right)^2 + \frac{x^U}{y - y_p}\left(\frac{x - x^L}{x^U - x^L}\right)^2 \\ y^L(x^U - x) &\le y_p(x^U - x^L) \le y^U(x^U - x) \\ y^L(x - x^L) &\le (y - y_p)(x^U - x^L) \le y^U(x - x^L) \end{aligned}\right\} \quad (2.3)$$

Note that (2.3) is valid only when $x^L < x < x^U$, since otherwise a division by zero occurs. As $x \to x^L$, (2.3) converges to the epigraph of x^L/y and,

2.4. ANALYSIS OF CONVEX UNDERESTIMATORS OF X/Y

when $x \to x^U$, (2.3) converges to the epigraph of x^U/y. However, we show next that (2.3) can be easily rewritten to attain the limiting value of x/y as $x \to x^L$ and $x \to x^U$ by introducing z_p such that:

$$z_p \geq \frac{x^L}{y_p}\left(\frac{x^U - x}{x^U - x^L}\right)^2.$$

As a result, we derive the following reformulation of (2.3):

(R)
$$z_p y_p \geq x^L(1-\lambda)^2$$
$$(z_c^e - z_p)(y - y_p) = x^U \lambda^2$$
$$y_p \geq \max\left\{y^L(1-\lambda), y - y^U \lambda\right\}$$
$$y_p \leq \min\left\{y^U(1-\lambda), y - y^L \lambda\right\}$$
$$x = x^L + (x^U - x^L)\lambda$$
$$z_p \geq 0, z_c^e - z_p \geq 0$$
$$0 \leq \lambda \leq 1.$$

Applying Fourier-Motzkin elimination to the above constraint set, we can eliminate y_p from the formulation. Each inequality of the form $zy \geq x^2$ can be rewritten as $\sqrt{zy} \geq x$ to make its convexity of the above set apparent. The convex envelope at a point (x^0, y^0) is then computed by solving:

$$z_c(x^0, y^0) = \min_{y_p, z_p, z_c^e} \{z_c^e \mid x = x^0, y = y^0, \text{R}\}. \tag{2.4}$$

2.4.2 Closed-Form Expression of Convex Envelope

In this section, we analyze the KKT points of (2.4) and derive a closed form expression of the convex envelope of x/y over $[x^L, x^U] \times [y^L, y^U]$.

If we restrict our attention to $x^L < x < x^U$, we can rewrite (2.4) as:

(C) $\quad \min \quad \dfrac{x^L}{y_p}\left(\dfrac{x^U - x}{x^U - x^L}\right)^2 + \dfrac{x^U}{y - y_p}\left(\dfrac{x - x^L}{x^U - x^L}\right)^2$ (2.5)

s.t. $\quad y^L(x^U - x) \leq y_p(x^U - x^L) \leq y^U(x^U - x)$ (2.6)

$\quad y^L(x - x^L) \leq (y - y_p)(x^U - x^L) \leq y^U(x - x^L).$ (2.7)

We first construct a relaxation of C by replacing (2.6) and (2.7) with $0 < y_p < y$ as below:

(Cr) $$\min_{y_p}\{(2.5) \mid 0 < y_p < y\}.$$

Theorem 2.24. *Given that $x \in [x^L, x^U]$ and $y \in (0, \infty)$, the minimum in Cr is:*

$$z_u(x,y) = \frac{1}{y}\left(\frac{x + \sqrt{x^L x^U}}{\sqrt{x^L} + \sqrt{x^U}}\right)^2, \qquad (2.8)$$

and is obtained when:

$$y_p^0 = \frac{\sqrt{x^L}(x - x^U)y}{(\sqrt{x^L} - \sqrt{x^U})(\sqrt{x^L}\sqrt{x^U} + x)}. \qquad (2.9)$$

Proof. The minimum in Cr exists because (2.5) can be easily extended to a lower semicontinuous function over $[0, y]$ by defining it to be $+\infty$ at the end points. When $x = x^L$ or $x = x^U$, (2.5) and (2.8) reduce to x/y and therefore, we restrict our attention to $x \in (x^L, x^U)$. Since Cr is convex, any local minimum of Cr is also a global minimum. It remains to show that (2.5) is locally minimized when $y_p = y_p^0$ is as defined in (2.9). We first identify the candidate solutions by analyzing the KKT conditions of Cr. The derivative of (2.5) with respect to y_p is zero when y_p is a root of the following quadratic equation:

$$y_p^2 x^U (x - x^L)^2 - (y - y_p)^2 x^L (x^U - x)^2 = 0. \qquad (2.10)$$

The two solutions which satisfy (2.10) are given by (2.9) and

$$y_p^1 = \frac{\sqrt{x^L}(x^U - x)y}{(\sqrt{x^L} + \sqrt{x^U})(\sqrt{x^L}\sqrt{x^U} - x)}. \qquad (2.11)$$

Since (2.5) is convex when $y_p \in (0, y)$, the set of local minima form a convex set and can not consist of just the two roots given above. In fact, it can be easily verified that y_p^1 in (2.11) always lies outside $(0, y)$ and y_p^0 in (2.9) always lies in $(0, y)$. Since the feasible set under consideration is an (open) interval, there are no other KKT points. We have thus shown that the minimum of Cr is attained at $y_p = y_p^0$ as given in (2.9) where the objective function value is easily verified to be given by (2.8). □

Corollary 2.25. *(2.8) defines the convex envelope of x/y over $[x^L, x^U] \times (0, \infty)$.*

2.4. ANALYSIS OF CONVEX UNDERESTIMATORS OF X/Y

Proof. When $x^L < x < x^U$, $y^L \to 0$ and $y^U \to +\infty$, C reduces to Cr. The result then follows from Theorem 2.24. □

Corollary 2.26. *(2.8) underestimates x/y over $[x^L, x^U] \times [y^L, y^U]$.*

Proof. Directly from Corollary (2.25). □

This corollary was first proven by Zamora & Grossmann (1998), albeit using a nonconstructive argument.

We now return our attention to developing the closed form expression for the convex envelope of x/y. It is possible that y_p^0 as defined in (2.9) is not feasible to (2.6) and (2.7). In this case, the following result holds:

Theorem 2.27. *The optimum in C is attained at:*

$$y_p^* = \min\{y^U(1-\lambda), y - y^L\lambda, \max\{y^L(1-\lambda), y - y^U\lambda, y_p^0\}\},$$

where $\lambda = (x - x^L)/(x^U - x)$ and y_p^0 is as defined in (2.9). The convex envelope of x/y over $[x^L, x^U] \times [y^L, y^U]$ is given by x^L/y if $y_p^* = y$, x^U/y if $y_p^* = 0$, and

$$\frac{x^L}{y_p^*}\left(\frac{x^U - x}{x^U - x^L}\right)^2 + \frac{x^U}{y - y_p^*}\left(\frac{x - x^L}{x^U - x^L}\right)^2$$

otherwise.

Proof. If y_p^0 is feasible, it minimizes C (Theorem 2.24). Let us denote the feasible region of C as $[y_p^L, y_p^U]$. It follows from the convexity of (2.5) that the optimum is attained at:

$$y_p^* = \min\{y_p^U, \max\{y_p^L, y_p^0\}\}.$$

Since $y_p^U = \min\{y^U(1-\lambda), y - y^L\lambda\}$ and $y_p^L = \max\{y^L(1-\lambda), y - y^U\lambda\}$, the result follows. □

2.4.3 Theoretical Comparison of Underestimators

We study the tightness of various underestimators of x/y over $[x^L, x^U] \times [y^L, y^U]$ using convex extensions. The underestimators we compare are:

1. $z_c(x, y)$ which is defined in (2.4) above;

2. $z_u(x, y)$ defined in (2.8) above;

3. $z_f(x,y)$ which is derived by relaxing $yf(x,y) = x$ using the bilinear envelopes (McCormick 1976, Al-Khayyal & Falk 1983) and is given by:

$$z_f(x,y) = \max\left\{\frac{xy^U - yx^L + x^L y^U}{y^{U^2}}, \frac{xy^L - yx^U + x^U y^L}{y^{L^2}}\right\}; \quad (2.12)$$

4. $z_g(x,y)$ which is defined as:

$$z_g(x,y) = \max\{z_f(x,y), z_u(x,y)\}. \quad (2.13)$$

Theorem 2.28. *The maximum gap between x/y and $z_f(x,y)$ is attained at:*

$$(x_1^*, y_1^*) = \left(\frac{x^L y^{L^{\frac{3}{2}}} + x^U y^{U^{\frac{3}{2}}}}{\sqrt{y^L}\sqrt{y^U}(\sqrt{y^L} + \sqrt{y^U})}, \sqrt{y^L y^U}\right) \quad (2.14)$$

if $x_1^* \leq x^U$. The gap in this case is:

$$g_1 = \frac{(\sqrt{y^L} - \sqrt{y^U})(x^L y^L - x^U y^U)}{y^L y^U(\sqrt{y^L} + \sqrt{y^U})}.$$

Otherwise, the maximum gap is:

$$g_2 = \frac{x^U(y^U - y^L)^2(x^U y^U - x^L y^L)^2}{y^L y^U(2x^U y^U - x^L y^L - x^U y^L)(x^U y^{U^2} - x^L y^{L^2})}$$

and is attained at:

$$(x_2^*, y_2^*) = \left(x^U, y^L + \frac{y^L(y^U - y^L)(x^U y^U - x^L y^L)}{x^U y^{U^2} - x^L y^{L^2}}\right). \quad (2.15)$$

Proof. For any fixed x^0, we denote the intersection of the linear underestimators of z_f in (2.12) by (x^0, y_{x^0}) where y_{x^0} is given by the following relation:

$$y_{x^0} = \frac{-x^0 y^{L^2} y^U - x^L y^{L^2} y^U + x^0 y^L y^{U^2} + x^U y^L y^{U^2}}{x^U y^{U^2} - x^L y^{L^2}}. \quad (2.16)$$

Since $x^0/y - z_f(x^0, y)$ is nonnegative and convex for $y \in [y^L, y_{x^0}]$, it is maximized at either $y = y^L$ or $y = y_{x^0}$. However, $x^0/y^L = z_f(x^0, y^L)$.

2.4. ANALYSIS OF CONVEX UNDERESTIMATORS OF X/Y

Therefore, the maximum is attained at $y = y_{x^0}$. Similarly, the maximum of $x^0/y - z_f(x^0, y)$ for $y \in [y_{x^0}, y^U]$ occurs at $y = y_{x^0}$. In other words:

$$\frac{x^0}{y_{x^0}} - z_f\left(x^0, y_{x^0}\right) \geq \frac{x^0}{y} - z_f\left(x^0, y\right) \quad \forall y \in [y^L, y^U].$$

The gap at (x^0, y_{x^0}) is given by:

$$g(x^0) = \frac{x^0}{y_{x^0}} - z_f\left(x^0, y_{x^0}\right) = \frac{(y^U - y^L)(x^0 y^U - x^L y^L)(x^U y^U - x^0 y^L)(x^U y^U - x^L y^L)}{y^L y^U (x^U y^U - x^0 y^L + x^0 y^U - x^L y^L)(x^U y^{U^2} - x^L y^{L^2})}. \tag{2.17}$$

Analyzing the derivative of $g(x^0)$, it can be shown that $g(x^0)$ increases with x^0 for $x^0 \in [x^L, x_1^*]$ and decreases when $x^0 \geq x_1^*$. Clearly, $x_1^* \geq x^L$. Simplifying (2.16) when $x^0 = x_1^*$, yields y_1^* in (2.14). If $x_1^* \geq x^U$, then the point of maximum gap is obtained by setting $x^0 = x^U$ in (2.16) which yields y_2^*. □

Consider $x^L = 0.1$, $x^U = 4$, $y^L = 0.1$, $y^U = 0.15$. Then, the maximum of $x/y - z_f(x, y)$ is attained at $(x_1^*, y_1^*) = (2.73364, 0.122474)$ where the gap is $g_1 = 3.9735$ (see Figure 2.15). An example illustrating the case when the optimum is attained at (x_2^*, y_2^*) will be provided in Section 2.4.4.

Theorem 2.29. *Let* $X = \{(x^L, y) \mid y^L \leq y \leq y^U\} \cup \{(x^U, y) \mid y^L \leq y \leq y^U\}$ *and* $z_a(x, y)$ *be any convex extension of* $x/y : X \mapsto \mathbb{R}$ *over* $[x^L, x^U] \times [y^L, y^U]$. *Then,* $d(x, y) = z_a(x, y) - z_f(x, y)$ *is maximized at:*

$$(x^*, y^*) = \left(x^U, y^L + \frac{y^L(y^U - y^L)(x^U y^U - x^L y^L)}{x^U y^{U^2} - x^L y^{L^2}}\right) \tag{2.18}$$

and

$$d(x^*, y^*) = \frac{x^U (y^U - y^L)^2 (x^U y^U - x^L y^L)^2}{y^L y^U (2 x^U y^U - x^L y^L - x^U y^L)(x^U y^{U^2} - x^L y^{L^2})}. \tag{2.19}$$

Proof. For any fixed x^0, we denote the intersection of the linear underestimators in the factorable relaxation (2.4) by (x^0, y_{x^0}) where y_{x^0} is given as in (2.16). It can be shown easily that y_{x^0} lies in $[y^L, y^U]$ for all $x^0 \in [x^L, x^U]$. The line segment joining (x^L, y_{x^L}) and (x^U, y_{x^U}) splits the rectangle $[x^L, y^L] \times [y^L, y^U]$ into two quadrilaterals Q_1 and Q_2. The corner points of Q_1

are (x^L, y^L), (x^L, y_{x^L}), (x^U, y_{x^U}), and (x^U, y^L) and the corner points of Q_2 are (x^L, y_{x^L}), (x^L, y^U), (x^U, y^U), and (x^U, y_{x^U}). Since $z_f(x,y)$ is linear over Q_1 and $z_a(x,y)$ is convex, $d(x,y)$ is convex over Q_1. Similarly, $d(x,y)$ is convex over Q_2. The maximum of $d(x,y)$ over $[x^L, x^U] \times [y^L, y^U]$ is therefore attained at one of the corner points of Q_1 or Q_2. Since $z_a(x,y) = z_f(x,y) = x/y$ at the corner points of $[x^L, x^U] \times [y^L, y^U]$, it follows that:

$$\max\{d(x,y) \mid (x,y) \in [x^L, x^U] \times [y^L, y^U]\}$$
$$= \max\{0, d(x^L, y_{x^L}), d(x^U, y_{x^U})\}.$$

Note that $d(x^L, y_{x^L})$ and $d(x^U, y_{x^U})$ are nonnegative because $z_a(x,y) = x/y$ when x is at its bounds and $z_f(x,y) \leq x/y$. Thus, we are left with (x^L, y_{x^L}) and (x^U, y_{x^U}) as the two candidate points for maximizing $d(x,y)$. By direct calculation:

$$d(x^U, y_{x^U}) - d(x^L, y_{x^L})$$
$$= \frac{(x^U - x^L)(y^U - y^L)^2(x^U y^U - x^L y^L)^3}{y^L y^U (2x^U y^U - x^U y^L - x^L y^L)(x^U y^U + x^L y^U - 2x^L y^L)(x^U y^{U^2} - x^L y^{L^2})}$$
$$\geq 0.$$

Thus, the maximum occurs at (x^U, y_{x^U}). Simplifying (2.16), we obtain (2.18). Evaluating $d(x^*, y^*)$, we get (2.19). □

Corollary 2.30. *Define F as the set of proper faces of $[x^L, x^U] \times [y^L, y^U]$. Then, $z_c(x,y)$ and $z_g(x,y)$ are convex extensions of $x/y : F \mapsto \mathbb{R}$ over $[x^L, x^U] \times [y^L, y^U]$. The tightest such convex extension is z_c. The maximum value of $z_c - z_f$, $z_g - z_f$, and $z_u - z_f$ is given by (2.19) and is attained at (2.18).*

Proof. Note that x/y is convex over all proper faces of $[x^L, x^U] \times [y^L, y^U]$. Further, the generating set of x/y is a proper subset of F and $z_c(x,y)$ is the convex envelope of x/y over $[x^L, x^U] \times [y^L, y^U]$. Therefore, it follows from Corollary 2.9 and Theorem 2.15 that $z_c(x,y)$ is the tightest convex extension of $x/y : F \mapsto \mathbb{R}$ over $[x^L, x^U] \times [y^L, y^U]$.

Let $X = \{(x^L, y) \mid y^L \leq y \leq y^U\} \cup \{(x^U, y) \mid y^L \leq y \leq y^U\}$. It can be easily verified that $z_u(x,y)$ is a convex extension of $x/y : X \mapsto \mathbb{R}$ over $[x^L, x^U] \times [y^L, y^U]$. The same result can also be deduced from Corollary 2.9

2.4. ANALYSIS OF CONVEX UNDERESTIMATORS OF X/Y

since $z_u(x,y)$ is the convex envelope of x/y over $[x^L, x^U] \times (0, +\infty)$ (Corollary 2.25). Let $Y = \{(x, y^L) \mid x^L \leq x \leq x^U\} \cup \{(x, y^U) \mid x^L \leq x \leq x^U\}$. It is easy to verify that $z_f(x,y)$ is a convex extension of $x/y : Y \mapsto \mathbb{R}$ over $[x^L, x^U] \times [y^L, y^U]$. The result can also be deduced from Corollary 2.9 since the bilinear term is convex when any of the associated variables is at its bounds. Note that $z_g(x,y) = \max\{z_u(x,y), z_f(x,y)\}$ and $F = X \cup Y$. Therefore, z_g is a convex extension of $x/y : F \mapsto \mathbb{R}$ over $[x^L, x^U] \times [y^L, y^U]$.

The remainder of the result follows directly from Theorem 2.29 since $F \supset X$. □

Corollary 2.31. *Let $z_a(x,y)$ be any convex underestimator of x/y over $[x^L, x^U] \times [y^L, y^U]$ and (x^*, y^*) be given by (2.18). If $z_a(x^*, y^*) = x^*/y^*$, then the maximum value of $z_a(x,y) - z_f(x,y)$ is attained at (x^*, y^*) and is given by (2.19).*

Proof. Since $z_c(x,y) \geq z_a(x,y)$, the result follows directly from Corollary 2.30. □

Theorem 2.32. *The maximum gap between x/y and $z_u(x,y)$ is:*

$$\frac{\left(\sqrt{x^L} - \sqrt{x^U}\right)^2}{4y^L} \quad (2.20)$$

and is attained at:

$$(x_u, y_u) = \left(\frac{x^L + x^U}{2}, y^L\right). \quad (2.21)$$

Further, $z_c(x,y) - z_u(x,y)$, $z_g(x,y) - z_u(x,y)$, and $z_f(x,y) - z_u(x,y)$ are all maximized at (x_u, y_u) with the gap given in (2.20).

Proof. Since x/y^0 is linear for a fixed y^0, $x/y^0 - z_u(x, y^0)$ is a nonnegative concave function which attains its maximum at $((x^L + y^L)/2, y^0)$. By direct calculation:

$$z_a\left(\frac{x^L + y^L}{2}, y\right) - z_u\left(\frac{x^L + y^L}{2}, y\right) = \frac{\left(\sqrt{x^L} - \sqrt{x^U}\right)^2}{4y} \quad (2.22)$$

which is a decreasing function of y and attains its maximum at (x_u, y_u) as in (2.21). Substituting $y = y^L$ in (2.22), we get (2.20). The remainder of the result follows since $z_c(x,y)$, $z_g(x,y)$, and $z_f(x,y)$ underestimate x/y and are exact at (x_u, y_u). □

2.4.4 Numerical Example

Consider the underestimators of x/y described in Section 2.4.3 over the box $[x^L, x^U] \times [y^L, y^U] = [0.1, 4] \times [0.1, 0.5]$.

The maximum values of $x/y - z_f(x,y)$, $z_c(x,y) - z_f(x,y)$, $z_g(x,y) - z_f(x,y)$, $z_u(x,y) - z_f(x,y)$, $x/y - z_u(x,y)$, $z_c(x,y) - z_u(x,y)$, $z_g(x,y) - z_u(x,y)$ and $z_f(x,y) - z_u(x,y)$ can be easily computed using the closed form expressions of Section 2.4.3.

The maximum value of $x/y - z_c(x,y)$ is found by solving:

$$\text{(OC)} \qquad \max \; \frac{x}{y} - z_c^e \qquad \text{s.t.} \; R.$$

The maximum value of $x/y - z_g(x,y)$ is found by solving:

$$\text{(OG)} \qquad \max \; \frac{x}{y} - z_g^e \qquad \text{s.t.} \; z_g^e \geq z_g(x,y).$$

The maximum value of $z_c(x,y) - z_g(x,y)$ is found by solving:

$$\begin{aligned}
\text{(CG)} \quad \max \; & z_c^e - z_g^e \\
\text{s.t.} \; & \lambda^2 x^U/(y - y_p)^2 - x^L(1-\lambda)^2/y_p^2 - r - s + t + u = 0 \\
& ry^L(1-\lambda) - ry_p = 0 \\
& sy - sy^U \lambda - sy_p = 0 \\
& ty^U(1-\lambda) - ty_p = 0 \\
& uy - uy^L \lambda - uy_p = 0 \\
& R \\
& z_g \geq z_g(x,y).
\end{aligned}$$

Note that CG models the KKT conditions of (2.4). Models OC, OG, and CG were solved to global optimality using BARON, the implementation of which is detailed in Chapter 7. Table 2.1 provides the maximum gaps. Figures 2.11, 2.12, 2.13, and 2.14 graph the difference between the various underestimators. It can be readily observed that $z_c(x,y)$ is significantly tighter than the other convex underestimators. The underestimators $z_u(x,y)$ and $z_f(x,y)$ are not only inexact when x and y are at bounds, but exhibit large gaps at such points. Combining them produces a convex extension $z_g(x,y)$ that reduces the gap significantly.

2.4. ANALYSIS OF CONVEX UNDERESTIMATORS OF X/Y

Gap Function	Point (x^*, y^*)	x^*/y^*	Maximum Gap
$x/y - z_c(x,y)$	$(1.6067, 0.1574)$	10.2077	3.3753
$x/y - z_u(x,y)$	$(2.05, 0.1)$	10.2077	7.0877
$x/y - z_f(x,y)$	$(4, 0.17968)$	22.6014	14.1337
$x/y - z_g(x,y)$	$(1.9393, 0.1235)$	15.7028	5.7190
$z_c(x,y) - z_u(x,y)$	$(2.05, 0.1)$	20.5	7.0877
$z_g(x,y) - z_u(x,y)$	$(2.05, 0.1)$	20.5	7.0877
$z_f(x,y) - z_u(x,y)$	$(2.05, 0.1)$	20.5	7.0877
$z_g(x,y) - z_f(x,y)$	$(4, 0.17968)$	22.6014	14.1337
$z_c(x,y) - z_f(x,y)$	$(4, 0.17968)$	22.6014	14.1337
$z_u(x,y) - z_f(x,y)$	$(4, 0.17968)$	22.6014	14.1337
$z_c(x,y) - z_g(x,y)$	$(2.2417, 0.1253)$	17.8906	3.1977

Table 2.1: Underestimators of x/y over $[0.1, 4] \times [0.1, 0.5]$

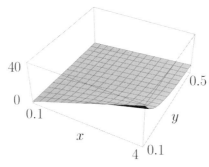

Ratio: x/y

Figure 2.10: Ratio: x/y

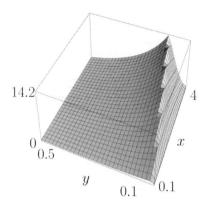

Figure 2.11: $x/y - z_f$

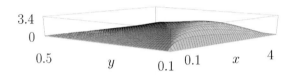

Figure 2.12: $x/y - z_c$

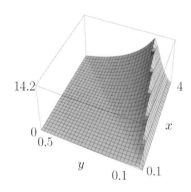

Figure 2.13: $z_c - z_f$

2.4. ANALYSIS OF CONVEX UNDERESTIMATORS OF X/Y

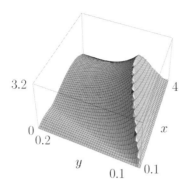

Figure 2.14: $z_c - z_g$ (restricted region for clarity)

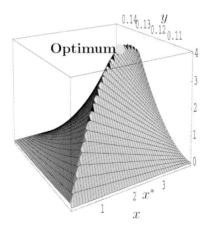

Figure 2.15: $x/y - z_f$ (optimum at x^*, not x^U)

2.4.5 Concave Envelope of x/y

Consider the fractional function x/y over a rectangular subset, $[x^L, x^U] \times [y^L, y^U]$, of the positive quadrant as depicted in Figure 2.10. Some characteristics of this function are:

- at a fixed value of y, the function is linear;
- at a fixed value of x, the function is convex.

An application of Theorem 2.17 shows that the generating set of the concave envelope of x/y consists of the four corners of the hypercube:

$$G^{\text{hypo}}_{[x^L,x^U]\times[y^L,y^U]}(\text{conc}(x/y)) = \{x^L, y^L\} \cup \{x^L, y^U\} \cup \{x^U, y^L\} \cup \{x^U, y^U\}.$$

This follows from the fact that any point, apart from the corner points, can be expressed as a convex combination of neighboring points along either the x axis or the y axis direction. Since the function is convex in both directions, the point in consideration can be eliminated from further consideration and the result follows.

We now develop the concave envelope of x/y. Since the generating set consists of a finite number of points, the concave envelope is polyhedral. A direct application of Theorem 2.22 establishes that the bilinear function which fits the fractional function values at the corner points of the rectangle has the same concave envelope as x/y. Such a bilinear function can be constructed rather easily as:

$$\frac{1}{(x^U - x^L)(y^U - y^L)} \left(\frac{x^L}{y^L}(x^U - x)(y^U - y) + \frac{x^L}{y^U}(x^U - x)(y - y^L) + \frac{x^U}{y^L}(x - x^L)(y^U - y) + \frac{x^U}{y^U}(x - x^L)(y - y^L) \right)$$

and be simplified to:

$$\frac{-xy + xy^L + xy^U}{y^L y^U}.$$

This function is depicted in Figure 2.16. Now, the development of the concave envelope is trivial using the McCormick envelopes (Al-Khayyal & Falk 1983). Algebraically, the concave envelope of x/y over the rectangle

2.4. ANALYSIS OF CONVEX UNDERESTIMATORS OF X/Y

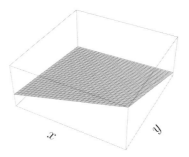

Figure 2.16: $\dfrac{-xy + xy^L + xy^U}{y^L y^U}$

$[x^L, x^U] \times [y^L, y^U]$ is given by:

$$\operatorname{concenv}\left(\frac{x}{y}\right) = \frac{1}{y^L y^U} \min\left\{y^U x - x^L y + x^L y^L, y^L x - x^U y + x^U y^U\right\}. \tag{2.23}$$

The concave envelope of x/y was shown to be given by (2.23) in the work of Zamora & Grossmann (1999). These authors derived the linear inequalities in (2.23) by using the following relations:

$$\left(\frac{x}{y} - \frac{x^L}{y^U}\right)\left(\frac{y}{y^L} - 1\right) \geq 0$$

$$\left(\frac{x^U}{y^L} - \frac{x}{y}\right)\left(1 - \frac{y}{y^U}\right) \geq 0,$$

and then verifying that the above formed the concave envelope of x/y.

2.4.6 Relaxing the Positivity Requirement

In this subsection, we relax the assumption that x and y belong to the positive orthant and develop the convex and concave envelopes of x/y as long as $0 \notin [y^L, y^U]$ (see Figure 2.17). To accomplish this, we must derive the convex and concave envelopes of x/y when $0 \in [x^L, x^U]$ and $y^L > 0$. The development of any one of the convex or concave envelope is adequate, since the other is developed in an identical fashion by substituting $u = -x$. Therefore, without loss of generality, we restrict our attention to the convex envelope characterization.

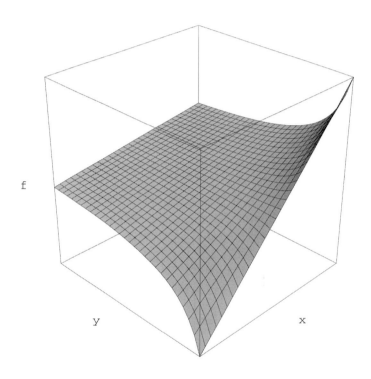

Figure 2.17: x/y when $0 \in [x^L, x^U]$

2.4. ANALYSIS OF CONVEX UNDERESTIMATORS OF X/Y

The function x/y is linear in x for a fixed value of y. Therefore, the generating set of the convex envelope is a subset of the faces $x = x^L$ and $x = x^U$. Further, x/y is concave when $x = x^L$. Therefore, the generating set can be written as the following union:

$$(x^L, y^L) \cup (x^L, y^U) \cup \{(x^U, y) \mid y^L \leq y \leq y^U\}.$$

Note that the generating set is written as a union of three convex sets. We could convexify the function either over all the three sets together or sequentially in two steps. In the present case, we prefer to do this sequentially. We first convexify x/y over $x = x^L$ and use the convexified function to develop the convex envelope of x/y over $[x^L, x^U] \times [y^L, y^U]$. Since the function is concave over $x = x^L$, its convex envelope can be developed easily as:

$$\frac{x^L(y^L + y^U - y)}{y^L y^U}.$$

Now the convex envelope of x/y is given by:

$$z \geq \frac{x^L(y^L + y^U - y_a)}{y^L y^U}(1 - \lambda) + \frac{x^U}{y_b}\lambda$$

$$y^L \leq y_a \leq y^U$$

$$y^L \leq y_b \leq y^U$$

$$y = (1 - \lambda)y_a + \lambda y_b$$

$$x = x^L + (x^U - x^L)\lambda$$

$$0 \leq \lambda \leq 1.$$

Substituting $y_p = y_a(1 - \lambda)$, $y - y_p = \lambda y_b$, and $\lambda = (x - x^L)/(x^U - x^L)$ we get:

$$\left.\begin{array}{l} z_p \geq \dfrac{x^L(x^L y_p - x(y^L + y^U) + x^U(y^L - y_p + y^U))}{(x^U - x^L)y^L y^U} \\ (z - z_p)(y - y_p)(x^U - x^L)^2 \geq x^U(x - x^L)^2 \\ y^L(x^U - x) \leq y_p(x^U - x^L) \leq y^U(x^U - x) \\ y^L(x - x^L) \leq (y - y_p)(x^U - x^L) \leq y^U(x - x^L) \\ z - z_p, z_p \geq 0. \end{array}\right\} \quad (2.24)$$

We have developed the convex and concave envelopes of x/y as long as $y \neq 0$. This assumption is not restrictive. As y approaches zero from above (below) the function takes arbitrarily large (small) values forcing the concave (convex) envelope to infinity ($-$infinity) and the function is not well-defined at $y = 0$.

2.4.7 Semidefinite Relaxation of x/y

We now show that nonlinear convex constraints in R can be represented as linear matrix inequalities using the Schur complement (Vandenberghe & Boyd 1996). We denote the matrix inequality expressing positive semidefiniteness of A, as $A \succeq 0$. Consider the following matrix inequality:

$$A = \begin{pmatrix} y_p(x^U - x^L)^2 & \sqrt{x^L}(x^U - x) \\ \sqrt{x^L}(x^U - x) & z_p \end{pmatrix} \succeq 0. \quad (2.25)$$

This inequality expresses the condition

$$z_p y_p(x^U - x^L)^2 - x^L(x^U - x)^2 \geq 0,$$

since $y_p(x^U - x^L)^2 \geq 0$ and $z_p \geq 0$. Similarly,

$$B = \begin{pmatrix} (y - y_p)(x^U - x^L)^2 & \sqrt{x^U}(x - x^L) \\ \sqrt{x^U}(x - x^L) & z - z_p \end{pmatrix} \succeq 0. \quad (2.26)$$

2.4. ANALYSIS OF CONVEX UNDERESTIMATORS OF X/Y

The above inequality expresses the condition

$$(z - z_p)(y - y_p)(x^U - x^L)^2 - x^U(x - x^L)^2 \geq 0,$$

since $(y - y_p)(x^U - x^L)^2 \geq 0$ and $z \geq z_p$. The following expresses the lower bound on y_p:

$$C = \begin{pmatrix} y_p - y^L \dfrac{x^U - x}{x^U - x^L} & \\ & y_p - y + y^U \dfrac{x - x^L}{x^U - x^L} \end{pmatrix} \succeq 0. \quad (2.27)$$

The constraint below expresses the upper bound on y_p:

$$D = \begin{pmatrix} y^U \dfrac{x^U - x}{x^U - x^L} - y_p & \\ & y - y_p - y^L \dfrac{x - x^L}{x^U - x^L} \end{pmatrix} \succeq 0. \quad (2.28)$$

Cumulatively, the semidefinite programming relaxation of x/y may be expressed as:

$$\begin{pmatrix} A & & & \\ & B & & \\ & & C & \\ & & & D \end{pmatrix} \succeq 0. \quad (2.29)$$

The semidefinite relaxation of x/y described above is second-order cone representable. Using the equivalence (Lobo, Vandenberghe, Boyd & Lebret 1998):

$$yz \geq x^2, y \geq 0, z \geq 0 \iff \left\| \begin{pmatrix} 2x \\ y - z \end{pmatrix} \right\| \leq y + z \quad (2.30)$$

the second-order cone representation can be obtained as:

$$\left.\begin{array}{l} \left\|\begin{pmatrix} 2(1-\lambda)\sqrt{x^L} \\ z_p - y_p \end{pmatrix}\right\| \leq z_p + y_p \\[1em] \left\|\begin{pmatrix} 2\lambda\sqrt{x^U} \\ z - z_p - y + y_p \end{pmatrix}\right\| \leq z - z_p + y - y_p \\[1em] y_p \geq y^L(1-\lambda), y_p \geq y - y^U\lambda \\[0.5em] y_p \leq y^U(1-\lambda), y_p \leq y - y^L\lambda \\[0.5em] x = x^L + (x^U - x^L)\lambda \\[0.5em] z_p, u, v \geq 0, z - z_p \geq 0 \\[0.5em] 0 \leq \lambda \leq 1. \end{array}\right\} \quad (2.31)$$

Using an almost identical procedure, it is easy to show that (2.24) can also be transformed into a semidefinite program using Schur complements or into second-order cone program using (2.30). We have thus shown that fractional programs can be relaxed using semidefinite relaxations. Coupled with similar results for indefinite quadratic programs (Fujie & Kojima 1997, Kojima & Tunçel 1999), our result provides a systematic means for constructing semidefinite relaxations for general factorable programs.

2.5 Generalizations and Applications

The techniques presented in this chapter are fairly general and find applications in developing convex/concave envelopes in a wide variety of situations. We illustrate this by developing convex/concave envelopes of $f(x, y)$ where f is lower semicontinuous concave in x and convex in y. It may be pointed out that we do not assume that y is a scalar. The generating set of the convex envelope of f over a rectangular region is then the set of faces: $x = x^L$ and $x = x^U$. By disjunctive programming techniques, the convex envelope is

2.5. GENERALIZATIONS AND APPLICATIONS

given by:

$$\left.\begin{aligned}
& z \geq f(x^L, y_a)(1-\lambda) + f(x^U, y_b)\lambda \\
& y^L \leq y_a \leq y^U \\
& y^L \leq y_b \leq y^U \\
& y = (1-\lambda)y_a + \lambda y_b \\
& x = x^L + (x^U - x^L)\lambda \\
& 0 \leq \lambda \leq 1.
\end{aligned}\right\} \quad (2.32)$$

In a similar vein to Rockafellar (1970), we define a positively homogenous function g associated with $f(\cdot, y)$ by the following relation:

$$g(\cdot, \lambda, y) = \begin{cases} \lambda f(\cdot, \lambda^{-1}y), & \text{if } \lambda > 0; \\ 0, & \text{if } \lambda = 0; \text{ and } y = 0 \\ +\infty, & \text{if } \lambda = 0, \text{ and } y \neq 0. \end{cases}$$

Since the epigraph of g is convex, g is jointly convex in λ and y. Also, it follows from Theorem 8.2 in Rockafellar (1970) that g is closed if f is bounded in the space under consideration. Introduce the variable $y_p = y_a(1-\lambda)$. After algebraic manipulations, we get the following form of (2.32):

$$\left.\begin{aligned}
& z \geq g\left(x^L, \frac{x^U - x}{x^U - x^L}, y_p\right) + g\left(x^U, \frac{x - x^L}{x^U - x^L}, y - y_p\right) \\
& y^L(x^U - x) \leq y_p(x^U - x^L) \leq y^U(x^U - x) \\
& y^L(x - x^L) \leq (y - y_p)(x^U - x^L) \leq y^U(x - x^L).
\end{aligned}\right\} \quad (2.33)$$

Therefore, whenever there is a way to write the mathematical formulation of g, the convex envelope of $f(x, y)$ can be developed as (2.33). It is possible to generalize the above to the case when x is a vector. However, the generalization is not only unnecessary but restrictive since the same effect is achieved by convexifying the function sequentially using the x variables one at a time. Generalizations to Cartesian products of polytopes (instead of a hypercube) as the feasible space can be easily accomplished. However, such an application does not serve to clarify the proposed concepts any further than already achieved through the previous example.

2.5.1 Envelopes of $(ax + by)/(cx + dy)$

Consider the function $f(x,y) = (ax+by)/(cx+dy)$ over a rectangle $[x^L, x^U] \times [y^L, y^U]$ in the positive orthant. We assume that a, b, c, and d are nonnegative constants and $c + d$ is strictly positive. The function $f(x,y)$ is a slight generalization of x/y as seen by setting $b = c = 0$ and $a = d = 1$. We develop the convex envelope of $f(x,y)$ in this section. The concave envelope can be easily developed by a similar treatment.

The case $ad = bc$ is trivial since the function is either a constant or undefined. Without loss of generality, we assume $ad > bc$. Note that the function $f(x, y^0)$ is concave for a fixed $y^0 > 0$. When $c = 0$, the result is obvious. Otherwise, the concavity of $f(x, y^0)$ follows from the following identity:

$$\frac{ax + by^0}{cx + dy^0} = \frac{a}{c} - \frac{1}{c}\left(\frac{(ad - bc)y^0}{cx + dy^0}\right).$$

For any fixed x^0, $f(x^0, y)$ is convex since:

$$\frac{ax^0 + by}{cx^0 + dy} = \frac{b}{d} + \frac{1}{d}\left(\frac{(ad - bc)x^0}{cx^0 + dy}\right) \qquad (2.34)$$

and $ad > bc$, $d > 0$ and $cx^0 + dy > 0$. The epigraph of the convex envelope of $f(x,y)$ is thus expressible as the convex hull of $A \cup B$ where

$$A = \{(z_a, x^L, y_a) \mid z_a \geq f(x^L, y_a), y^L \leq y_a \leq y^U\}$$

and

$$B = \{(z_b, x^U, y_b) \mid z_b \geq f(x^U, y_b), y^L \leq y_b \leq y^U\}.$$

We denote a point in the epigraph of the convex envelope of $f(x,y)$ by (z, x, y). Introducing $\lambda = (x - x^L)/(x^U - x^L)$, $y_p = y_a(1 - \lambda)$ and $z_p =$

2.5. GENERALIZATIONS AND APPLICATIONS

$z_a(1-\lambda)$, the epigraph of the convex envelope of $f(x,y)$ can be written as:

$$\left.\begin{array}{l} dz_p(cx^L(1-\lambda)+dy_p) = (ad-bc)x^L(1-\lambda)^2 \\ (dz - b - dz_p)(cx^U\lambda + dy - dy_p) = (ad-bc)x^U\lambda^2 \\ y_p \geq \max\left\{y^L(1-\lambda), y - y^U\lambda\right\} \\ y_p \leq \min\left\{y^U(1-\lambda), y - y^L\lambda\right\} \\ x = x^L + (x^U - x^L)\lambda \\ z_p, u, v \geq 0, z - z_p \geq b/d \\ 0 \leq \lambda \leq 1 \end{array}\right\} \qquad (2.35)$$

using a procedure similar to that described in Section 2.4.1. A second-order cone representation of (2.35) is easily derived using (2.30). Using the procedure detailed in Section 2.4.2, the following convex underestimating inequality is derived:

$$f(x,y) \geq \frac{(ad-bc)(x + \sqrt{x^L}\sqrt{x^U})^2}{d(\sqrt{x^L} + \sqrt{x^U})^2(cx+dy)} + \frac{b}{d}, \qquad (2.36)$$

and shown to be the convex envelope of $f(x,y)$ over $[x^L, x^U] \times (0, \infty)$.

2.5.2 Convex Envelope of $f(x)y^2$

We consider the function $f(x)y^2$ over a rectangular region. We assume that $f(x) \geq 0$ over the feasible region. We provide an illustration of $x^{0.8}y^2$ when $0 \leq x \leq 1$, $-1 \leq y \leq 1$ in Figure 2.18. We assume that f is a concave function of x. Then, the function is convex in y and concave in x. Using

(2.33), the convex envelope is expressed as:

$$\left.\begin{array}{l} \min g_1 + g_2 \\ g_1(x^U - x) \geq f(x^L) y_p^2 (x^U - x^L) \\ g_2(x - x^L) \geq f(x^U)(y - y_p)^2 (x^U - x^L) \\ y^L(x^U - x) \leq y_p(x^U - x^L) \leq y^U(x^U - x) \\ y^L(x - x^L) \leq (y - y_p)(x^U - x^L) \leq y^U(x - x^L) \\ g_1, g_2 \geq 0. \end{array}\right\} \quad (2.37)$$

The various candidates for the solution are found by setting the derivative of $g_1 + g_2$ to zero and setting y_p to one of the bounds. The first candidate is found by solving

$$2f(x^L) y_p \frac{x^U - x^L}{x^U - x} - 2f(x^U)(y - y_p) \frac{x^U - x^L}{x - x^L} = 0.$$

The resulting solution is

$$y_p = \frac{f(x^U)(x - x^U) y}{(x - x^L) f(x^L) + (x - x^U) f(x^U)}.$$

Then:

$$z \geq \min \left\{ \vphantom{\frac{a}{b}} \right.$$

$$\frac{(x^L - x^U) y^2 f(x^L) f(x^U)((x^L - x) f(x^L) + (x - x^U) f(x^U))}{((x - x^L) f(x^L) + (x - x^U) f(x^U))^2},$$

$$\frac{(x - x^U)(y^U)^2 f(x^L)}{(x^L - x^U)} - \frac{(x^L y - x y^U - x^U y + x^U y^U)^2 f(x^U)}{(x - x^L)(x^L - x^U)},$$

$$\frac{(x - x^U)(y^L)^2 f(x^L)}{(x^L - x^U)} - \frac{(x^L y - x y^L - x^U y + x^U y^L)^2 f(x^U)}{(x - x^L)(x^L - x^U)},$$

2.5. GENERALIZATIONS AND APPLICATIONS

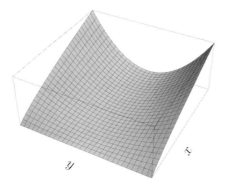

Figure 2.18: $x^{0.8}y^2$

$$\frac{(x^U y - xy^L - x^L y + x^L y^L)^2 f(x^L)}{(x - x^U)(x^L - x^U)} + \frac{(x - x^L)(y^L)^2 f(x^U)}{(x^U - x^L)},$$

$$\left.\frac{(x^U y - xy^U - x^L y + x^L y^U)^2 f(x^L)}{(x - x^U)(x^L - x^U)} + \frac{(x - x^L)(y^U)^2 f(x^U)}{(x^U - x^L)}\right\}. \quad (2.38)$$

Note that, at any given point, out of the five terms only those should be considered which were derived with y_p within the bounds in (2.37). We now investigate the functions $x^a y^2$, $0 < a < 1$ (see Figure 2.18) and $\log_{10}(9x+1)y^2$ (see Figure 2.19) when $x^L = 0$, $x^U = 1$, $y^L = -1$, $y^U = 1$. In this case, (2.38) reduces to:

$$z = \begin{cases} 0, & x = 0; \\ 0, & y + x \leq 1 \text{ and } y \geq 0; \\ (x + y - 1)^2/x, & y + x \geq 1 \text{ and } x \neq 0; \\ 0, & y - x \leq 1 \text{ and } y < 0; \\ (1 - x + y)^2/x & x - y \geq 1 \text{ and } x \neq 0. \end{cases}$$

As in Section 2.4.6, the assumption that $f(x) \geq 0$ can be relaxed. All that is needed is that we convexify the function $f(x)y^2$ over $x = x^L$ if $f(x^L) \leq 0$ and over $x = x^U$ if $f(x^U) \leq 0$. It is easy to see that the relaxation (2.37) can be transformed to linear matrix inequalities and therefore included in a semidefinite relaxation of $f(x)y^2$.

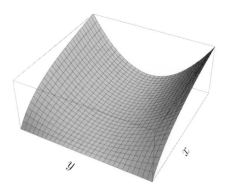

Figure 2.19: $\log_{10}(9x+1)y^2$

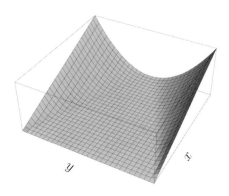

Figure 2.20: Convex envelope of $x^{0.8}y^2$

2.5.3 Convex Envelope of $f(x)/y$

In this section, we consider a slightly more general form of the fractional function x/y. We assume that $y > 0$ and $f(x)$ is a concave function of x. Even though we could follow the same construction as in Section 2.5.2, we shall make use of the convex envelope of the fractional function to develop the convex envelope. This is a simple technique which may be used in more general settings and we use it in this context to illustrate its use. Since the generating set of this function is the same as that of x/y, the following function has the same convex envelope as $f(x)/y$:

$$\frac{1}{y}\left(f(x^L)\frac{x^U - x}{x^U - x^L} + f(x^U)\frac{x - x^L}{x^U - x^L}\right).$$

The convex envelope can then be generating using the convex envelope of x/y.

It is clear that the relaxations developed in this case can also be transformed in a semidefinite program.

2.5.4 Summation of Functions

The following result appears in Rikun (1997) in a slightly different form:

Theorem 2.33 (Rikun 1997). *Consider a hypercube $P = H^{1+\sum_{i=1}^{n} n_i}$. Let $x \in \mathbb{R}$ and $y_i \in \mathbb{R}^{n_i}$. Assume $f_i(x, y_i)$ is a continuous function for each $i \in \{1, \ldots, n\}$. Assume further that each $f_i(x, y_i)$ is concave in x. Then:*

$$\mathrm{convenv}_P\left(\sum_{i=1}^{n} f_i(x, y_i)\right) = \sum_{i=1}^{n} \mathrm{convenv}_P\, f_i(x, y_i).$$

It is easy to prove the above result using convex extensions and we leave that as an exercise to the reader. In Section 2.5, we developed the convex envelope of each of the functions $f_i(x, y_i)$, under the additional assumption that $f_i(x, y_i)$ is convex in y_i. Note that if this assumption is not satisfied we could in principle convexify along the function in the y_i space before proceeding. It follows directly from Theorem 2.33 that using the techniques presented in this chapter, we can develop the convex envelope of functions of the form:

$$\sum_{i=1}^{n} f_i(x, y_i)$$

given that each $f_i(x, y_i)$ is concave in x and convex in y_i. Quite a few functions fall into this category. As an example, consider

$$f(x) \sum_{i=1}^{n} \sum_{j=-p}^{k} a_{ij} y_i^j$$

where f is a concave function, $a_{ij} > 0$ for $i = 1, \ldots, n$; $j = -p, \ldots, k$ and $y_i > 0$.

The techniques in this chapter can be used in a straightforward manner to construct convex envelopes of

$$\frac{l(x)}{\prod_{i=1}^{n} y_i},$$

over a hypercube in $\mathbb{R}^{m+n}$, where $y > 0$ and $l(x)$ is a multilinear function of x. We leave this as an exercise to the reader.

Chapter 3

Product Disaggregation in Global Optimization and Relaxations of Rational Programs

Synopsis

In this chapter, we consider the product of a single continuous variable and the sum of a number of continuous variables. We show that "product disaggregation" (distributing the product over the sum) leads to tighter linear programming relaxations, much like variable disaggregation does in mixed-integer linear programming. We also derive closed-form expressions characterizing the exact region over which these relaxations improve when the bounds of participating variables are reduced.

In a concrete application of product disaggregation, we develop and analyze linear programming relaxations of rational programs. In the process of doing so, we prove that the task of bounding general linear fractional functions of 0−1 variables is $\mathcal{NP}$-hard. Then, we present computational experience to demonstrate that product disaggregation is a useful reformulation technique for global optimization problems. In particular, we apply product disaggregation to develop eight different mixed-integer convex programming reformulations of 0−1 hyperbolic programs. We obtain analytical results on the relative tightness of these formulations and propose a branch-and-bound algorithm for 0−1 hyperbolic programs. The algorithm is used to solve a discrete p-choice facility location problem for locating ten restaurants in the city of Edmonton.

3.1 Introduction

The principal purpose of this chapter is to develop tight relaxations of

$$\phi(f; y_1, \ldots, y_n) = a_0 + \sum_{k=1}^{n} a_k y_k + f b_0 + f \sum_{k=1}^{n} b_k y_k \tag{3.1}$$

where a_k and b_k ($k = 0, \ldots, n$) are given constants, $f \in [f^L, f^U]$, and $y_k \in [y_k^L, y_k^U]$.

It is well-known that, for integer programs, certain formulations are more efficient than others even though they may contain more variables and/or constraints. This has led to the development of variable and constraint disaggregation techniques (cf. Krarup & Bilde 1977, Rardin & Choe 1979, Schrijver 1986, Martin 1987, Nemhauser & Wolsey 1988, Parker & Rardin 1988) that have yielded a variety of tight relaxations in discrete optimization. To bound (3.1), we propose "product disaggregation" (distributing the product over the sum in (3.1)) followed by standard techniques due to McCormick (1976) to bound the resulting n bilinear terms. Compared to defining one single additional variable for the sum in (3.1), and bounding the resulting single bilinear term with the same techniques of McCormick (1976), product disaggregation leads to relaxations with many more variables and constraints. The main contribution of this chapter is to show that this increase in the size of the relaxation comes with an improved tightness as it provides the convex and concave envelopes of (3.1) over the hypercube.

To demonstrate a particular use of product disaggregation, we develop relaxations/reformulations for rational functions. Let $\hat{N} = \{1, \ldots, K\}$ be an arbitrary index set. Consider the rational function below:

$$f(x) = \frac{\prod_{i=1}^{p} \left(\sum_{j \in \mathcal{G}_{ai}} a_{ij} \prod_{k \in \mathcal{M}_{aij}} x_k \right)}{\prod_{i=1}^{m} \left(\sum_{j \in \mathcal{G}_{bi}} b_{ij} \prod_{k \in \mathcal{M}_{bij}} x_k \right)} \tag{3.2}$$

3.1. INTRODUCTION

where $x_k \in [0,1]$, $k \in \hat{N}$, $\mathcal{G}_{ai}$ and $\mathcal{G}_{bi}$ are index sets of product terms and $\mathcal{M}_{aij}$ and $\mathcal{M}_{bij}$ are subsets of $\hat{N}$. We show that, when each x_k is restricted to binary values, product disaggregation leads to a natural mixed-integer linear reformulation of $f(x)$. When x_k is not restricted to be binary, disaggregating the products provides a valid linear relaxation of the rational function over the unit hypercube.

Optimization problems with rational functions of 0−1 variables arise in a variety of applications, including return on investments (Williams 1974), discounts associated with economies of scale (Lawler 1978), sequencing (Sydney 1975), graph density and arboricity (Picard & Queyranne 1982), and optimal clustering (Rao 1971). The special case of ratios of linear functions of 0−1 variables has been used to model problems in cutting stock selection (Gilmore & Gomory 1963), information retrieval systems (Hansen, Poggi de Aragao & Ribeiro 1991), attrition games (Isbelle & Marlow 1956), and discrete facility location (Ghosh, McLafferty & Craig 1995, Tawarmalani et al. 2002a).

Various special instances of 0−1 programs with rational functions have been addressed in the literature. It was shown by Picard & Queyranne (1982) that the unconstrained 0−1 rational function with certain sign restrictions may be maximized over a linear constraint set by a sequence of network-flow problems. Unconstrained optimization of single ratios of linear functions is addressed by Hammer & Rudeanu (1968), Robillard (1971), and Hansen, Poggi de Aragao & Ribeiro (1991). Constrained optimization techniques for the ratio include linearization (Williams 1974), branch-and-bound (Robillard 1971, Aggarwal 1977), cutting plane algorithms (Grunspan & Thomas 1973, Granot & Granot 1977), enumerative methods (Granot & Granot 1976, Arora, Swarup & Puri 1977), and approximation algorithms (Hashizume, Fukushima, Katoh & Ibaraki 1987). See Stancu-Minasian (1997) for detailed descriptions. For certain problems with specially structured constraint sets, strongly polynomial algorithms were developed by Megiddo (1979). Saipe (1975) developed a specialized branch-and-bound algorithm for maximizing a sum of ratios of linear functions subject to a single cardinality constraint. Fractional programming with sums of ratios was recently reviewed by Schaible (1995).

Whereas all the above approaches handle special instances of problems with rational functions of 0−1 variables, a general purpose approach is also possible. Any optimization problem in 0−1 variables can be reduced to a 0−1 polynomial program (Hammer & Rudeanu 1968). Thus, in principle, these problems could be solved using polynomial programming techniques (Glover

& Woolsey 1974, Glover 1975, Balas & Mazzola 1984, Hansen, Jaumard & Mathon 1993). However, these approaches are not directly applicable as the reformulation into a 0−1 polynomial program requires the evaluation of the rational function at all the extreme points of the unit hypercube.

By using variable disaggregation in this context, we develop reformulation schemes for a larger class of rational functions than has been addressed before. In particular, we provide mixed-integer linear reformulations of optimization problems with nonlinearities in the form of sums of rational functions involving ratios of products of linear expressions of 0−1 variables. In the context of rational programming, our work is most closely related to those of Williams (1974), Li (1994), and Wu (1997) who develop reformulations of problems that are special cases of the problem we consider. Williams (1974) developed a scheme for reformulating a single ratio of linear expressions of 0−1 variables into a mixed-integer set. Li (1994) and Wu (1997) considered a 0−1 program involving maximization of a sum of ratios of linear functions over linear constraints and reformulated it into a mixed-integer linear program. We provide mixed-integer linear reformulations of optimization problems involving ratios of products of multilinear functions of 0−1 variables.

The remainder of this chapter is structured as follows. Section 3.2 presents earlier results that we need for our development. In Section 3.3, we derive a reformulation of the rational function into a mixed-integer set and prove that bounding a linear fractional function of 0−1 variables is in general $\mathcal{NP}$-hard. In Section 3.4, we develop the convex and concave envelope of (3.1) over a hypercube by distributing the product over the sum. We also characterize the exact region over which the envelopes strictly improve as a result of tighter bounds on f. In Section 3.5, we show that our reformulation dominates those of Williams (1974), Li (1994), and Wu (1997). In Section 3.6, we present computational experience illustrating the importance of product disaggregation in solving problems with multilinear and rational expressions. Starting with Section 3.7, we apply the developed reformulation scheme to 0−1 hyperbolic programming program. The 0−1 hyperbolic programming program consists of optimizing the sum of ratios of linear functions of binary variables subject to linear constraints. The problem can be stated as follows:

(H) $$\max \sum_{i=1}^{m} \frac{a_{i0} + a_i^T x}{b_{i0} + b_i^T x}$$
$$\text{s.t.} \quad Dx \leq c$$
$$x \in \{0,1\}^n$$

where $D \in \mathbb{R}^{k \times n}, c \in \mathbb{R}^k, a_i \in \mathbb{R}^n, b_i \in \mathbb{R}^n, a_{i0} \in \mathbb{R}$ and $b_{i0} \in \mathbb{R}$. It is assumed that $(b_{i0} + b_i^T x) > 0$ for any i and any feasible x.

The reformulation scheme developed in Section 3.3 is applied in Section 3.7 to derive a number of reformulation schemes for H. The relative tightness of relaxations of these reformulations is characterized in Section 3.8. In Section 3.9, we propose a branch-and-bound algorithm for H. We demonstrate that reformulating the nonconvex problem at every node of the branch-and-bound tree expedites the convergence of the algorithm significantly. We specialize the algorithm to cardinality constrained hyperbolic programs in Section 3.10 and perform computations on cardinality constrained programs in Section 3.11. The computational experiments compare the relaxation bounds of the different reformulations and contrast the solution effort of the proposed branch-and-bound algorithm with the commercial integer programming solver CPLEX 6.0 (ILOG 1997). The p-choice facility location problem is formulated as a cardinality constrained hyperbolic programming program and is used to locate restaurant franchises in the city of Edmonton, Canada.

3.2 Preliminaries

In this section, we present preliminary results that we need for developing the reformulation schemes for rational functions. In the sequel, $\bar{\mathbb{R}}$ denotes $\mathbb{R} \cup \{+\infty\}$. An n-dimensional hypercube will be denoted by H^n. Let $x = [x_1, \ldots, x_n] \in H^n$. The bounds on the variables x_i, $i = 1, \ldots, n$, will be denoted by $I_i = [x_i^L, x_i^U]$. The unit hypercube, $[0, 1]^n$, will be denoted by U^n. The extreme points of a polyhedral set X will be denoted by vert(X). In particular, E^n and B^n will denote vert(H^n) and vert(U^n), respectively. The convex and concave envelopes of $\phi(x)$ over X will be written as convenv$_X(\phi)$ and concenv$_X(\phi)$, respectively. We denote the convex hull of a set X by conv(X). The relative interior of X is denoted ri(X).

Outer-linearization techniques for bilinear expressions were first developed by McCormick (1976). Subsequently, Al-Khayyal & Falk (1983), showed that the outer-linearization forms the convex and concave envelopes of bilinear terms.

Theorem 3.1 (Al-Khayyal & Falk 1983). *Consider a bilinear term $x_1 x_2$ over H^2. Then:*
$$\text{convenv}_{H^2}(x_1 x_2) = \max\{x_1^L x_2 + x_2^L x_1 - x_1^L x_2^L, x_1^U x_2 + x_2^U x_1 - x_1^U x_2^U\}$$

$$\operatorname{concenv}_{H^2}(x_1 x_2) = \min\{x_1^L x_2 + x_2^U x_1 - x_1^L x_2^U, x_1^U x_2 + x_2^L x_1 - x_1^U x_2^L\}.$$

Reformulation techniques for products of 0−1 variables were first proposed by Glover & Woolsey (1974) and shown subsequently by Al-Khayyal & Falk (1983) to constitute their convex and concave envelopes over U^n.

Theorem 3.2 (Crama 1993). *Consider the product term $\varphi = \prod_{k=1}^{n} x_k$. Then:*

$$\operatorname{convenv}_{U^n}(\varphi) = \max\left\{0, \sum_{k=1}^{n} x_k - (n-1)\right\}$$
$$\operatorname{concenv}_{U^n}(\varphi) = \min\{x_k \mid k = 1, \ldots, n\}$$

Nonlinear 0−1 functions can be reformulated as multilinear functions (Hammer & Rudeanu 1968). In Section 2.3, we showed that the convex envelope of the resulting multilinear function is the tightest convex extension of the associated nonlinear function. In particular, Corollary 2.23 is a central result in this regard.

It follows easily from Theorem 3.2 and Corollary 2.23 that:

Corollary 3.3 (Crama 1993). *The epigraph of the tightest convex extension of the product term restricted to B^n is given by the following polyhedron:*

$$y_\varphi \geq \sum_{k=1}^{n} x_k - (n-1)$$
$$y_\varphi \geq 0.$$

The hypograph of the tightest concave extension of the product term restricted to B^n is given by the following polyhedron:

$$y_\varphi \leq x_k \qquad k = 1, \ldots, n.$$

The following result combines results from Theorem 2.20 and Theorem 2.21 in a form more suitable for reformulating rational functions.

Theorem 3.4. *The convex and concave envelopes of a multilinear function $L(x_1, \ldots, x_k)$ over a hypercube, H^k, are polyhedral functions. Let F_I be the collection of faces of H^k over which L is linear. Then, the envelopes are the tightest convex and concave extensions of L restricted to F_I over H^k and the convex and concave envelopes do not support the function at any point not in F_I.*

3.3 Reformulations of a Rational Function

In this section, we reformulate rational functions of 0–1 variables of the form provided in (3.2). We restrict $\mathcal{M}_{aij}$ and $\mathcal{M}_{bij}$ to be subsets of $\hat{N}$ since 0–1 variables are idempotent under multiplication.

Since polynomials are closed under multiplication, the rational function can be written as a ratio of polynomials. Even so, we allow multiple polynomial factors in the numerator and the denominator because the number of terms in a polynomial grows exponentially in the number of factors. Multiplying the polynomials would lead to a prohibitively large reformulation using the schemes we present below.

Definition 3.5. *Consider a function $f(x) : B^n \mapsto \mathbb{R}$. Let L denote an arbitrary set of equality constraints $g_i(x, y, f) = 0$, $i = 1, \ldots, m$ where g_i is a multilinear function of x, y, and f, $x \in B^n$, $y \in \mathbb{R}^m$, $f \in \mathbb{R}$, and g_i is linear in y and f. The set of points, M, feasible to L defines a multilinear encloser of $f(x)$ if the following conditions hold:*

- $\{(x, f) \mid \exists y \text{ s.t. } (x, y, f) \in M\} = \{(x, f) \mid x \in B^n, f = f(x)\}$

- *y is bounded.*

Using the recipe developed in (2.22), multilinear enclosers can be constructed for all nonlinear functions of 0–1 variables even in the absence of y variables. However, the recipe evaluates $f(x)$ at all $x \in B^n$. We make two simple remarks that enable us to construct the multilinear encloser of the rational function (3.2) without resorting to exhaustive enumeration.

Remark 3.6. *Given a function $f(x) : B^n \mapsto \mathbb{R}$, let*

$$\phi(x) = \frac{f(x)}{\prod_{i=1}^{m}\left(\sum_{j \in \mathcal{G}_{bi}} b_{ij} \prod_{k \in \mathcal{M}_{bij}} x_k\right)}, \qquad (3.3)$$

where

$$\prod_{i=1}^{m}\left(\sum_{j \in \mathcal{G}_{bi}} b_{ij} \prod_{k \in \mathcal{M}_{bij}} x_k\right) \neq 0. \qquad (3.4)$$

If F is a multilinear encloser of $f(x)$, then a multilinear encloser of $\phi(x)$ can be constructed as:

$$(x, \phi) \in \left\{ \exists f \text{ s.t. } \phi \prod_{i=1}^{m} \left(\sum_{j \in \mathcal{G}_{bi}} b_{ij} \prod_{k \in \mathcal{M}_{bij}} x_k \right) = f, f \in F \right\}.$$

The functions f and ϕ are defined on all $x \in B^n$ and therefore bounded.

We claimed that $\phi(x)$ in Remark 3.6 is bounded. However, it is not easy to derive bounds on ϕ.

Theorem 3.7. *Deriving finite lower and upper bounds on $1/(b_0 + \sum_{j=1}^{n} b_j x_j)$ where $x \in B^n$ is $\mathcal{NP}$-hard.*

Proof. Consider the function:

$$l(x) = b_0 + \sum_{j=1}^{n} b_j x_j.$$

Finite lower and upper bounds on $1/l(x)$ exist if and only if there does not exist any $x \in B^n$ with $l(x) = 0$. The subset sum problem, a well-known $\mathcal{NP}$-hard problem, is concerned with deciding if there exists a point $x \in B^n$ such that $l(x) = 0$. Any polynomial-time algorithm which derives finite lower and upper bounds on $1/l(x)$ can then be checked for failure to design a polynomial-time algorithm for the subset-sum problem. □

Note that bounding $l(x)$ over $x \in B^n$ is trivial. Therefore, bounds on $1/l(x)$ are easily derived if either $l(x) > 0$ or $l(x) < 0$. Also, if b_j is integral for each $j \in \{1, \ldots, n\}$ and $l(x)$ is never zero on $x \in B^n$, then -1 and 1 are trivial lower and upper bounds on $1/l(x)$. The same argument can be extended to the case when b_j are rational by scaling each b_j appropriately.

Remark 3.8. *Consider a function of the form*

$$\phi(x) = f(x) \prod_{i=1}^{m} \left(\sum_{j \in \mathcal{G}_{ai}} a_{ij} \prod_{k \in \mathcal{M}_{aij}} x_k \right), \tag{3.5}$$

where $x \in B^n$. If F is a multilinear encloser of f, then a multilinear encloser of $\phi(x)$ can be constructed as:

$$(x, \phi) \in \left\{ \exists f \text{ s.t. } \phi = f \prod_{i=1}^{m} \left(\sum_{j \in \mathcal{G}_{ai}} a_{ij} \prod_{k \in \mathcal{M}_{aij}} x_k \right), f \in F \right\}.$$

3.3. REFORMULATIONS OF A RATIONAL FUNCTION

f and ϕ are defined on all $x \in B^n$ and therefore bounded.

The above remarks provide a recipe of reformulating the rational function in terms of multilinear equalities as detailed in our next remark.

Remark 3.9. *The rational function (3.2) can be recursively defined by introducing p functions of the form (3.5) and m functions of the form (3.3) starting with the unit function 1 in $(p+m)!$ different ways. The corresponding applications of Remark (3.6) and Remark (3.8) then provide $(p+m)!$ ways of constructing a multilinear encloser of (3.2).*

A hierarchy of linear relaxations converging to the convex hull of the feasible solutions to a set of multilinear equations was developed in the reformulation-linearization technique (RLT) of Sherali & Adams (1990) and Sherali & Adams (1994) under the additional restrictions that the continuous variables are bounded and the multilinear expressions are linear in the continuous variables. Our definition of the multilinear encloser obeys these restrictions. Therefore, RLT can be extended to include rational functions of $0-1$ variables. RLT produces the tightest convex and concave extensions of the rational function at the nth level. The first level at which RLT produces a mixed-integer reformulation of the rational function is: $d = \max\{\max_{i,j} |\mathcal{M}_{aij}|, \max_{i,j} |\mathcal{M}_{bij}|\}$. Whenever d is large, such a reformulation introduces a large number of constraints. In the sequel, we construct a smaller reformulation by replacing the product terms with their convex and concave extensions described in Corollary 3.3. We then analyze the properties of the resulting reformulation using convex extensions.

First, we define the notion of an exact encloser of a function.

Definition 3.10. *Consider a function $\phi(x) : X \mapsto \mathbb{R}$ where $X \subseteq \mathbb{R}^n$. A closed convex set $\Phi \in \mathbb{R}^{n+1}$ will be called a convex encloser of $\phi(x)$, if the correspondence between the points in X and the points $(x, y) \in \Phi$ where $x \in X$ and $y \in \mathbb{R}$ is one-to-one and is given by $x \leftrightarrow (x, \phi(x))$.*

The motivation behind exact enclosers is that it is possible to replace functions over X by their exact enclosers. Replacing X by some convex superset of itself, we develop a natural convex relaxation of $\phi(x)$. In the sequel, we will call a set the tightest convex encloser of ϕ if it is a subset of all the other convex enclosers of ϕ.

Exact enclosers can also be thought of as an intersection of the epigraph of the convex extension and the hypograph of the concave extension of $\phi(x)$

over some convex superset of X (Rockafellar 1970, Theorem 5.3). The convex and concave extensions of Corollary 3.3 thus characterize the tightest convex encloser of $\phi(x)$ over B^n.

In particular, we can replace the product term by its exact encloser. Such a replacement introduces an additional continuous variable y_ϕ for every product term ϕ. Even though we do not enforce integrality on the y_ϕ variables, they take binary values whenever $x \in B^n$ (Glover & Woolsey 1974). Therefore, we are justified in treating them as binary variables for the purposes of reformulation. Enlarging the set of binary variables, (3.2) is rewritten as:

$$\text{(LF)} \qquad f(x) = \frac{\prod_{i=1}^{p}\left(a_{i0} + \sum_{j \in N} a_{ij}x_j\right)}{\prod_{i=1}^{m}\left(b_{i0} + \sum_{j \in N} b_{ij}x_j\right)},$$

where N is a subset of $2^{\hat{N}}$.

The rational function in (3.2) is reformulated into a set of bilinear equalities by applying the recursive procedure described in Remark 3.9 to LF. As an illustration, p applications of Remark 3.8 followed by m applications of Remark 3.6 yield the following reformulation:

$$\begin{aligned} f_{i-1}a_{i0} + f_{i-1}\sum_{j\in N} a_{ij}x_j - f_i = 0 & \quad i = 1,\ldots,p \\ f_{p+i}b_{i0} + f_{p+i}\sum_{j\in N} b_{ij}x_j - f_{p+i-1} = 0 & \quad i = 1,\ldots,m \end{aligned} \qquad (3.6)$$

where $f_0 = 1$. In the above f_{p+m} models the function $f(x)$. As noted in Remark 3.9, there are $(p+m)!$ ways of carrying out the above reformulation. The generic set of bilinear equalities so generated will be termed as the bilinear encloser of (3.2).

The bilinear envelopes of Theorem 3.1 can be easily verified to be exact whenever any of the variables involved is at its bound. Since $x \in B^n$, the variables x_j are always at their bounds. The bilinear encloser of (3.2) can thus be linearized by replacing each $f_i x_j$ by its convex and concave envelopes. Using the bilinear envelopes to linearize (3.6), we get:

$$\text{(R)} \quad f_{i-1}a_{i0} + \sum_{j\in N} a_{ij}z_{i-1\,j} - f_i = 0 \qquad i = 1,\ldots,p$$

$$f_{p+i}b_{i0} + \sum_{j\in N} b_{ij}z_{p+i\,j} - f_{p+i-1} = 0 \qquad i = 1,\ldots,m$$

3.4. TIGHTNESS OF THE REFORMULATION SCHEME

$$z_{ij} \leq f_i^u x_j \qquad j \in N;\ i = 1, \ldots, p+m-1$$
$$z_{ij} \leq f_i + f_i^l x_j - f_i^l \qquad j \in N;\ i = 1, \ldots, p+m-1$$
$$z_{ij} \geq f_i + f_i^u x_j - f_i^u \qquad j \in N;\ i = 1, \ldots, p+m-1$$
$$z_{ij} \geq f_i^l x_j \qquad j \in N;\ i = 1, \ldots, p+m-1$$
$$x \in B^n.$$

R is clearly an exact encloser of (3.2) and provides a mixed-integer linear reformulation of (3.2) that we sought to develop. Note that valid bounds on $f_1, \ldots, f_{p+m-1}$ are required to construct R. In Section 3.4, we show that the bounds are crucial in deriving a reformulation with a small relaxation gap and provide some recipes for deriving the same.

R is a relaxation of (3.2) when $x \in U^n$ because we have not made use of the fact that $x \in B^n$ other than for showing that the above relaxation is exact when x is restricted to the extreme points of the unit hypercube.

3.4 Tightness of the Reformulation Scheme

In this section, we investigate some properties of the reformulation scheme for rational functions developed in Section 3.3.

The bilinear expressions in the bilinear encloser of (3.2) are of the form:

$$\phi(y, f) = f b_0 + f \sum_{k \in K} b_k y_k - a_0 - \sum_{k \in K} a_j y_j \qquad (3.7)$$

where $f \in [f^L, f^U]$ and $y \in H^{|K|}$. We assume that there exists at least one $k \in K$ such that $b_k \neq 0$. The bilinear expression $f_{p+1} b_{10} + f_{p+1} \sum_{j \in N} b_{1j} x_j - f_p$ appearing in (3.6) is a special case of (3.7) when $f = f_{p+1}$, $K = N \cup \{f_p\}$, $y = (x, f_p)$, $b_k = b_{1j}$ for $k \in N \cup \{0\}$ and $b_{f_p} = 0$, $a_k = 0$ for $k \in N \cup \{0\}$ and $a_{f_p} = 1$. We have chosen a somewhat more general expression in (3.7) than the bilinear expressions in (3.6). The reason for this choice is that the linear fractional function

$$f(x) = \frac{a_0 + \sum_{j \in N} a_j x_j}{b_0 + \sum_{j \in N} b_j x_j} \qquad (3.8)$$

can be rewritten as

$$f(x) b_0 + f(x) \sum_{j \in N} b_j x_j - a_0 - \sum_{j \in N} a_j x_j = 0. \qquad (3.9)$$

Equation (3.9) is a compact representation of the reformulation of (3.8) derived by applying Remark 3.8 followed by Remark 3.6. Notice that (3.7) naturally generalizes the bilinear expression in (3.9) as well as the expressions in (3.6). We have chosen to use y instead of x in definition of $\phi(y,f)$ because our results here are derived for general hypercubes instead of the unit hypercube.

Proposition 3.11. *Consider a function $\phi(y,f) = fb_0 + f\sum_{k \in K} b_k y_k - a_0 - \sum_{k \in K} a_j y_j$ where $y_k \in [y_k^L, y_k^U]$ and $f \in [f^L, f^U]$. Let $H^{|K|+1} = \prod_{k \in K} [y_k^L, y_k^U] \times [f^L, f^U]$. Then:*

$$\text{convenv}_{H^{|K|+1}}\left(fb_0 + f\sum_{k \in K} b_k y_k - a_0 - \sum_{k \in K} a_j y_j\right)$$
$$= fb_0 + \sum_{k \in K} \text{convenv}_{[y_k^L, y_k^U] \times [f^L \times f^U]}(b_k y_k f) - a_0 - \sum_{k \in K} a_j y_j.$$

Proof. For a fixed y, $\phi(y,f)$ is linear in f. Let Φ be the epigraph of the convex envelope of $\phi(y,f)$. Then Φ can be expressed as the convex hull of A and B where $A = \{(\phi^a, y^a) \mid \phi^a \geq \phi(y^a, f^L)\}$ and $B = \{(\phi^b, y^b) \mid \phi^b \geq \phi(y^b, f^U)\}$. In other words,

$$\Phi = \{(z_\phi, y, f) \mid z_\phi \geq (1-\lambda)\phi^a + \lambda\phi^b, \ \phi^a \geq \phi(y^a, f^L), \ \phi^b \geq \phi(y^b, f^U),$$
$$\lambda = (f - f^L)/(f^U - f^L), \ y = y^a(1-\lambda) + y^b\lambda\}\}.$$

Eliminating ϕ^a and ϕ^b, expanding $\phi(y^a, f^L)$ and $\phi(y^b, f^U)$, and collecting terms, we get:

$$\Phi = \left\{(z_\phi, y, f) \mid z_\phi \geq fb_0 + \sum_{k \in K} b_k\left((1-\lambda)f^L y_k^a + \lambda f^U y_k^b\right), -a_0 - \sum_{k \in K} a_j y_j \right.$$
$$\left. \lambda = (f-f^L)/(f^U - f^L), \ y = y^a(1-\lambda) + y^b\lambda\right\}.$$

The minimum value of z_ϕ for a given (y, f) occurs when:

$$\text{convenv}_{[y_k^L, y_k^U] \times [f^L, f^U]}(b_k y_k f) = b_k\left((1-\lambda)f^L y_k^a + \lambda f^U y_k^b\right).$$

□

It may be noted that the proof above uses the techniques for convexification developed in Chapter 2.

3.4. TIGHTNESS OF THE REFORMULATION SCHEME

The standard approach for lower bounding $f\sum_{k\in K} b_k y_k$ is by substituting w for $\sum_{k\in K} b_k y_k$ and then relaxing fw using the convex envelope of fw according to Theorem 3.1 (cf. Quesada & Grossmann 1995a). The required bounds on w are obtained by maximizing/minimizing $\sum_{k\in K} b_k y_k$ over $\prod_{k\in K}[y_k^L, y_k^U]$. It is easy to show that such a construction does not yield the convex envelope of $\phi(f, y)$ developed in Proposition 3.11 above. Consider, for example, $f(2y_1 + 3y_2)$ over $[0,1]^3$. At $f = 0.5$, $y_1 = 1$, and $y_2 = 0$, the convex envelope is exactly equal to $f(2y_1 + 3y_2)$ with a value of one whereas the standard lower bounding procedure gives a value of zero. Just like the standard lower bounding procedure, the convex envelope of Proposition 3.11 also makes use of the bilinear convex envelope of Theorem 3.1. However, the bilinear envelopes are invoked only after the product is distributed over the summation. Distribution of the product over the summation results in disaggregating fw into $\sum_{k\in K} b_k z_k$ where $z_k = fy_k$, much reminiscent of variable disaggregation in mixed-integer linear programming which is also known to provide tight linear relaxations (cf. Schrijver 1986, Nemhauser & Wolsey 1988, Parker & Rardin 1988).

Since (3.7) is closed under negation, Proposition 3.11 also characterizes the concave envelope of (3.7). Combining the results of Proposition 3.11 and Theorem 3.1, it follows that the epigraph of the convex envelope of $fb_0 + f\sum_{k\in K} b_k y_k - a_0 - \sum_{k\in K} a_j y_j$ is given as:

$$\begin{aligned}
z &\geq fb_0 + \sum_{k\in K} b_k z_k - a_0 - \sum_{k\in K} a_k x_k \\
z_k &\geq y_k^U f + f^U y_k - f^U y_k^U & k \in K, \text{ if } b_k > 0 \\
z_k &\geq y_k^L f + f^L y_k - f^L y_k^L & k \in K, \text{ if } b_k > 0 \\
z_k &\leq y_k^L f + f^U y_k - f^U y_k^L & k \in K, \text{ if } b_k < 0 \\
z_k &\leq y_k^U f + f^L y_k - f^L y_k^U & k \in K, \text{ if } b_k < 0
\end{aligned} \quad (3.10)$$

and the hypograph of the concave envelope is:

$$\begin{aligned}
z &\leq fb_0 + \sum_{k\in K} b_k z_k - a_0 - \sum_{k\in K} a_k x_k \\
z_k &\geq y_k^U f + f^U y_k - f^U y_k^U & k \in K, \text{ if } b_k < 0 \\
z_k &\geq y_k^L f + f^L y_k - f^L y_k^L & k \in K, \text{ if } b_k < 0 \\
z_k &\leq y_k^L f + f^U y_k - f^U y_k^L & k \in K, \text{ if } b_k > 0 \\
z_k &\leq y_k^U f + f^L y_k - f^L y_k^U & k \in K, \text{ if } b_k > 0
\end{aligned} \quad (3.11)$$

Comparing R with (3.10) and (3.11), it is easily ascertained that R includes the convex and concave envelopes of all bilinear expressions in (3.6).

Given a nonconvex set X and a hyperplane M, $\operatorname{conv}(X \cap M) \neq \operatorname{conv}(X) \cap M$. Therefore, even though R includes the envelopes of the bilinear expressions in (3.6), it does not follow that R includes the tightest possible representations of the bilinear equalities in (3.6). For example, consider $fy - 1 = 0$, where $y \in [y^L, y^U]$ and $f \in [f^L, f^U]$. The convex hull of the set of feasible solutions of $fy - 1 = 0$ is given by the nonpolyhedral set:

$$\left\{(y, f) \,\Big|\, \frac{1}{y} \leq f \leq \frac{y^U + y^L - y}{y^L y^U}\right\}. \tag{3.12}$$

The convex envelope of $fy - 1$ is polyhedral and yields an outer-approximation of (3.12). More generally, the convex hull of the set of feasible solutions of

$$fb_0 + f \sum_{k \in K} b_k y_k - a_0 - \sum_{k \in K} a_j y_j = 0$$

is given by

$$F_f = \{(y, f) \mid \eta(y) \leq f \leq \chi(y)\}$$

where η and χ are the convex and concave envelopes, respectively, of $(a_0 + \sum_{k \in K} a_k y_k)/(b_0 + \sum_{k \in K} b_k y_k)$ over $\prod_{k \in K} [y_k^L, y_k^U]$. If $x \in B^n$, the convex hull of the set of solutions to (3.9) is a polyhedral set (Theorem 2.23). However, the existence of an intermediate nonpolyhedral convex relaxation of $f(x)$ in (3.8) shows that the convex hull of the set of solutions to (3.9) is a proper subset of the linear relaxation of R.

Note that the bounds on f were not required in our description of F_f. However, the envelopes of the bilinear expression in Proposition 3.11 as well as our reformulation R require bounds on f. In fact, we show next that the quality of the reformulation is very sensitive to the bounds on f. We characterize the precise region over which the convex and concave envelopes of (3.7) improve strictly when improved bounds are available on f.

Lemma 3.12. *Let $(y, f) \in H^{|K|} \times [f^L, f^U]$ where $H^{|K|} = \prod_{k \in K} [y_k^L, y_k^U]$. Define*

$$B = \{(y, f) \mid f \in [f^L, f^U],\ y_k \in \{y_k^L, y_k^U\},\ k \in K\} \cup$$
$$\{(y, f) \mid f \in \{f^L, f^U\},\ y_k \in H^{|K|}\}.$$

Consider the function $\phi(y, f) = fb_0 + f \sum_{k \in K} b_k y_k - a_0 - \sum_{k \in K} a_j y_j$. The convex envelope of $\phi(y, f)$ agrees with $\phi(y, f)$ at (y^0, f^0) if and only if $(y^0, f^0) \in B$.

3.4. TIGHTNESS OF THE REFORMULATION SCHEME

Proof. Consider an arbitrary point $(y^0, f^0) \in H^{|K|} \times [f^L, f^U]$ and let F be the face of $H^{|K|} \times [f^L, f^U]$ such that $(y, f) \in \text{ri}(F)$. Since $\phi(y, f)$ is linear over F if and only if $f \in B$, the result follows directly from Theorem 2.20. □

Let $\alpha(y, f)$ be the convex envelope of $\phi(y, f)$ over $H^{|K|} \times [f^L, f^U]$ given as in Proposition 3.11. Also, let $\alpha_n(y, f)$ be the convex envelope of $\phi(y, f)$ over $H^{|K|} \times [f^L, f_n^U]$ where $f_n^U < f^U$. Consider a point (y^0, f_n^U). It is clear from Lemma 3.12 that $\phi(y^0, f_n^U) = \alpha_n(y^0, f_n^U) > \alpha(y^0, f_n^U)$. Thus, we have shown that an improved bound strictly tightens the convex envelope of $\phi(y, f)$. We now present the exact characterization of the region over which the convex envelope of $\phi(y, f)$ improves when a better bound is derived on f.

Lemma 3.13. *Consider a bilinear function $\beta(y, f) = yf$ where $y \in [y^L, y^U]$. Let $\alpha(y, f)$ be the convex envelope of $\beta(y, f)$ when $f \in [f^L, f^U]$ and $\alpha_n(y, f)$ be the convex envelope of $\beta(y, f)$ when $f \in [f^L, f_n^U]$, where $f_n^U < f^U$. Then, $\alpha_n(y, f) > \alpha(y, f)$ in the hypercube $[y^L, y^U] \times [f^L, f_n^U]$ if and only if*

(PU) $$(y^U - y^L)f + (f_n^U - f^L)y - f_n^U y^U + f^L y^L > 0$$
$$y < y^U.$$

Proof. The convex envelopes of yf (Theorem 3.1) are given by:

$$\alpha(y, f) = \max\{y^U f + f^U y - f^U y^U, y^L f + f^L y - f^L y^L\}$$
$$\alpha_n(y, f) = \max\{y^U f + f_n^U y - f_n^U y^U, y^L f + f^L y - f^L y^L\}.$$

$\alpha_n(y, f) > \alpha(y, f)$ if and only if

$$y^U f + f_n^U y - f_n^U y^U > y^U f + f^U y - f^U y^U \tag{3.13}$$
$$y^U f + f_n^U y - f_n^U y^U > y^L f + f^L y - f^L y^L. \tag{3.14}$$

Rearranging the terms in equation (3.13),

$$(f^U - f_n^U)(y - y^U) < 0.$$

Since $f^U - f_n^U > 0$, $y < y^U$. Also, from equation (3.14),

$$(y^U - y^L)f + (f_n^U - f^L)y - f_n^U y^U + f^L y^L > 0.$$

□

The region PU in Lemma 3.13 is the triangular region with (y^L, f_n^U), (y^U, f^L) and (y^U, f_n^U) as its corner points. The edge joining (y^U, f^L) and (y^U, f_n^U) and the edge joining (y^U, f^L) and (y^L, f_n^U) are however excluded.

Lemma 3.14. *Consider a bilinear term $\beta(y,f) = -yf$ where $y \in [y^L, y^U]$. Let $\alpha(y,f)$ be the convex envelope of $\beta(y,f)$ when $f \in [f^L, f^U]$ and $\alpha_n(y,f)$ be the convex envelope of $\beta(y,f)$ when $f \in [f^L, f_n^U]$, where $f_n^U < f^U$. Then, $\alpha_n(y,f) > \alpha(y,f)$ in the hypercube $[y^L, y^U] \times [f^L, f_n^U]$ if and only if*

(MU) $\qquad (y^U - y^L)f + (f^L - f_n^U)y + f_n^U y^L - f^L y^U > 0$
$\qquad\qquad y > y^L.$

Proof. Similar to the proof of Lemma 3.13. □

Lemma 3.15. *Consider a bilinear function $\beta(y,f) = yf$ where $y \in [y^L, y^U]$. Let $\alpha(y,f)$ be the convex envelope of $\beta(y,f)$ when $f \in [f^L, f^U]$ and $\alpha_n(y,f)$ be the convex envelope of $\beta(y,f)$ when $f \in [f_n^L, f^U]$, where $f_n^L > f^L$. Then, $\alpha_n(y,f) > \alpha(y,f)$ in the hypercube $[y^L, y^U] \times [f_n^L, f^U]$ if and only if*

(PL) $\qquad (y^U - y^L)f + (f^U - f_n^L)y + f^U y^U - f_n^L y^L > 0$
$\qquad\qquad y > y^L.$

Proof. Similar to the proof of Lemma 3.13. □

Lemma 3.16. *Consider a bilinear function $\beta(y,f) = -yf$ where $y \in [y^L, y^U]$. Let $\alpha(y,f)$ be the convex envelope of $\beta(y,f)$ when $f \in [f^L, f^U]$ and $\alpha_n(y,f)$ be the convex envelope of $\beta(y,f)$ when $f \in [f_n^L, f^U]$, where $f_n^L > f^L$. Then, $\alpha_n(y,f) > \alpha(y,f)$ in the hypercube $[y^L, y^U] \times [f_n^L, f^U]$ if and only if*

(ML) $\qquad (y^U - y^L)f + (f_n^L - f^U)y - f_n^L y^U + f^U y^L < 0$
$\qquad\qquad y < y^U.$

Proof. Similar to the proof of Lemma 3.14. □

Theorem 3.17. *Consider the function $\phi(y,f) = fb_0 + f\sum_{k \in K} b_k y_k - a_0 - \sum_{k \in K} a_j y_j$ where $y \in H^{|K|} = \prod_{k \in K}[y_k^L, y_k^U]$. Let $\alpha(y,f)$ be the convex envelope of $\phi(y,f)$ when $f \in [f^L, f^U]$ and $\alpha_n(y,f)$ be the convex envelope of $\phi(y,f)$ when $f \in [f^L, f_n^U]$, where $f_n^U < f^U$. Let $K^+ = \{k \mid b_k > 0\}$ and $K^- = \{k \mid b_k < 0\}$. Then, $\alpha_n(y^0, f^0) > \alpha(y^0, f^0)$ if and only if $(y^0, f^0) \in S$*

3.4. TIGHTNESS OF THE REFORMULATION SCHEME

where S is given by:

$$S = \bigcup_{k \in K^+} \{(y,f) \mid (y_k^U - y_k^L)f + (f_n^U - f^L)y_k - f_n^U y_k^U + f^L y_k^L > 0, y_k < y_k^U\} \cup$$
$$\bigcup_{k \in K^-} \{(y,f) \mid (y_k^U - y_k^L)f + (f^L - f_n^U)y_k + f_n^U y_k^L - f^L y_k^U > 0, y_k > y_k^L\}.$$

Proof. From Theorem 3.11,

$$\text{convenv}_{H^{|K|} \times [f^L, f^U]} \left(fb_0 + f \sum_{k \in K} b_k y_k - a_0 - \sum_{k \in K} a_j y_j \right)$$

$$= fb_0 + \sum_{k \in K} \text{convenv}_{[y_k^L, y_k^U] \times [f^L \times f^U]}(b_k y_k f) - a_0 - \sum_{k \in K} a_j y_j.$$

Hence, $\alpha_n(y^0, f^0) > \alpha(y^0, f^0)$ if and only if for some $k \in K$, $\text{convenv}(b_k y_k f)$ strictly improves. If $k \in K^+$, then by Lemma 3.13 the convex envelope of $b_k y_k f$ strictly improves over the region PU. Similarly, for $k \in K^-$, the convex envelope of $b_k y_k f$ strictly improves over MU. Taking the union over $k \in K$, if follows that $\alpha_n(y^0, f^0) > \alpha(y^0, f^0)$ if and only if $(y^0, f^0) \in S$. Hence, the assertion is proven. □

Corollary 3.18. *Consider the function* $\phi(y, f) = fb_0 + f \sum_{k \in K} b_k y_k - a_0 - \sum_{k \in K} a_j y_j$ *where* $y \in H^{|K|} = \prod_{k \in K} [y_k^L, y_k^U]$. *Let* $\alpha(y, f)$ *be the convex envelope of* $\phi(y, f)$ *when* $f \in [f^L, f^U]$ *and* $\alpha_n(y, f)$ *be the convex envelope of* $\phi(y, f)$ *when* $f \in [f_n^L, f_n^U]$, *where* $f_n^L > f^L$ *and* $f_n^U < f^U$. *Let* $K^+ = \{k \mid b_k > 0\}$ *and* $K^- = \{k \mid b_k < 0\}$. *Then,* $\alpha_n(y^0, f^0) > \alpha(x^0, f^0)$ *if and only if there exists a* $k \in K^+ \cup K^-$ *such that* $y_k^0 \notin \{y_k^L, y_k^U\}$.

Proof. It follows from Lemma 3.13 and Lemma 3.15 that for $k \in K^+$, the convex envelope of $b_k y_k f$ strictly improves at (y_k^0, f^0) if and only if $(y_k^0, f^0) \in D \cap [y^L, y^U] \times [f_n^L, f_n^U]$ where

$$D = \{(y_k, f) \mid (y_k^U - y_k^L)f + (f_n^U - f^L)y_k - f_n^U y_k^U + f^L y_k^L > 0, y_k < y_k^U\} \cup \{(y_k, f) \mid (y_k^U - y_k^L)f + (f^U - f_n^L)y_k + f^U y_k^L - f_n^L y_k^U > 0, y_k > y_k^L\}.$$

Simplifying $D \cap [y^L, y^U] \times [f_n^L, f_n^U]$, we get $\{(y_k, f) \mid y_k \in (y_k^L, y_k^U), f \in [f_n^L, f_n^U]\}$. Similarly, using Lemma 3.14 and Lemma 3.16, it can be shown that the convex envelope of $b_k y_k f$, $k \in K^-$ improves at (y_k^0, f^0) if and only if $(y_k^0, f^0) \in \{(y_k, f) \mid y_k \in (y_k^L, y_k^U), f \in [f_n^L, f_n^U]\}$. □

A mathematical program with rational functions of 0−1 variables can be reformulated into a mixed-integer linear program using the reformulation scheme developed in Section 3.3 and then solved using a branch-and-bound algorithm. Theorem 3.17 and Corollary 3.18 highlight the importance of bounds on f in deriving a convex relaxation of the solutions to $\phi(y, f) = 0$. A tightened relaxation, in turn, reduces the relaxation gap for the mixed-integer linear program. Therefore, it is critical that the reformulation scheme of Section 3.3 be applied at every node of the branch-and-bound tree using the tightest derivable bounds on each fractional term. The validity and finiteness of such a procedure follows from the validity of the reformulation and finiteness of branch-and-bound for mixed-integer programs respectively.

We showed in our discussion following Remark 3.6 that it is in general $\mathcal{NP}$-hard to derive bounds on the fractional functions of 0−1 variables. Thus, it may not be computationally tractable to reformulate the rational function at every node of the branch-and-bound tree. However, as we mentioned earlier, there exist special cases when the fractional function can be bounded efficiently. We will rely on an important result due to Megiddo (1979) in this context.

Theorem 3.19 (Megiddo 1979). *Consider the following two problems:*

(A)
$$\max \quad \sum_{j=1}^{n} c_j x_j$$
$$\text{s.t.} \quad (x_1, \ldots, x_n) \in D$$

(B)
$$\max \quad \frac{\sum_{j=1}^{n} a_j x_j}{\sum_{j=1}^{n} b_j x_j}$$
$$\text{s.t.} \quad (x_1, \ldots, x_n) \in D$$

(assuming the denominator is always positive).

If problem A is solvable within $O(p(n))$ comparisons and $O(q(n))$ additions, then problem B is solvable in time $O(p(n)(q(n) + p(n)))$.

The proof of the above theorem is constructive in the sense that, given an algorithm for problem A, Megiddo provides a recipe for developing an algorithm for problem B. The recipe exploits the following well-known result:

3.4. TIGHTNESS OF THE REFORMULATION SCHEME

Lemma 3.20 (Stancu-Minasian 1997). *Let x^* be an optimal solution to B and let*

$$t^* := \frac{a_0 + \sum_{j=1}^n a_j x_j^*}{b_0 + \sum_{j=1}^n b_j x_j^*}.$$

Define

$$F(t) = \max_{(x_1,\ldots,x_n) \in D} \left(a_0 + \sum_{j=1}^n a_j x_j - t \left(b_0 + \sum_{j=1}^n b_j x_j \right) \right).$$

Then

(a) $F(t) > 0$ if and only if $t < t^*$,

(b) $F(t) = 0$ if and only if $t = t^*$,

(c) $F(t) < 0$ if and only if $t > t^*$.

The algorithm for solving B developed in the proof of Theorem 3.19 determines t^* by parametrically solving $F(t) = 0$ using the algorithm for A. A comparison is performed by algorithm A to choose one out of two possible computation paths. Since $F(t)$ defines a parametric family of problems of type A, the comparison must decide if a linear function of the form $\alpha + \beta t$ is less than or greater than zero. It follows from Lemma 3.20 that $t^* > \alpha/\beta$ whenever $F(\alpha/\beta) > 0$ and $t^* < \alpha/\beta$ whenever $F(\alpha/\beta) < 0$. Thus, we disregard all values of t greater than α/β if $F(\alpha/\beta) < 0$ and disregard all values of t less that α/β if $F(\alpha/\beta) > 0$. The choice of the appropriate computation path is now unique and can be easily performed.

As mentioned by Megiddo (1979), the above algorithm can be accelerated when the algorithm A only performs comparisons of input elements. In this case, the break-points of $F(t)$ are searched using the median finding algorithm to locate the linear segment of $F(t)$ containing the optimal t^*. As an example, we apply this accelerated scheme to the unconstrained single ratio 0–1 hyperbolic programming program to develop an algorithm with $\Theta(n)$ complexity. The single ratio unconstrained 0–1 hyperbolic programming program is formulated as:

(P) max $\dfrac{a_0 + \sum_{j=1}^n a_j x_j}{b_0 + \sum_{j=1}^n b_j x_j}$

s.t. $x_j \in \{0,1\}$ $\qquad j = 1,\ldots,n$

where the denominator is assumed to be positive. The function $F(t)$ in Lemma 3.20 reduces to:

$$F(t) = \max_{x_j \in \{0,1\}} \left(a_0 + \sum_{j=1}^n a_j x_j - t \left(b_0 + \sum_{j=1}^n b_j x_j \right) \right).$$

Megiddo's scheme exploits the solution algorithm for the following problem:

(L) $$\max a_0 + \sum_{j=1}^n a_j x_j - t b_0 - t \sum_{j=1}^n b_j x_j$$
$$\text{s.t. } x_j \in \{0, 1\}.$$

Problem L can be easily solved by setting x_j equal to 1 if $(a_j - tb_j) > 0$ and 0 otherwise. Thus, a_j/b_j for all j are the critical values for t which need to be tested and form the break-points in Megiddo's scheme. Since $F(t)$ is piecewise linear between any two values of a_j/b_j, we search the n breakpoints to locate t^*. We now state a slightly modified version of Megiddo's algorithm for solving P that records certain calculations while solving problems of the type L to improve the efficiency of the algorithm:

Step 0. $J \leftarrow \{0, \ldots, n\}$.
Step 1. Let $T = \{a_j/b_j \mid j \in J\}$. If $|J| = 2$, then let x_j^* to zero for all $j \in J \setminus \{0\}$ such that $a_j \leq \min\{T\} b_j$, $x_j^* = 1$ for all other $j \in J \setminus \{0\}$ and terminate.
Step 2. Calculate $\hat{t} = \text{median}\{T\}$.
Step 3. Let $J_g = \{j \in J \mid a_j > \hat{t} b_j\} \cup \{0\}$ and $J_1 = \{j \in J \mid a_j \geq \hat{t} b_j, j \neq 0\}$. Calculate $F(\hat{t}) = \sum_{j \in J_g} a_j - \hat{t} \sum_{j \in J_g} b_j$.
Step 4. If $F(\hat{t}) = 0$, let $x_j^* = 1$ for all $j \in J_g$, $x_j^* = 0$ for all $j \in J \setminus J_g$ and terminate. If $F(\hat{t}) > 0$, let $x_j^* = 0$ for all $j \in J \setminus J_g$, $J \leftarrow J_g$ and return to Step 1. If $F(\hat{t}) < 0$, let $x_j^* = 1$ for all $j \in J_1$, $J \leftarrow J \setminus J_1$, $a_0 \leftarrow a_0 + \sum_{j \in J_1} a_j$, $b_0 \leftarrow b_0 + \sum_{j \in J_1} b_j$, and return to Step 1.

Proposition 3.21. *Megiddo's algorithm as applied to the unconstrained single ratio 0–1 hyperbolic programming program with positive denominator has $\Theta(n)$ complexity.*

Proof. The effort in each iteration of the above algorithm consists of determining the median of the current set T of critical t values and solving problem $F(t)$. Note that the cardinality of T and J is halved in each iteration. Thus, the total time spent by the algorithm in finding medians (Blum, Floyd, Pratt, Rivest & Tarjan 1973) and evaluating $F(t)$ is $n + n/2 + n/4 + \cdots$ or $O(n)$. The total effort required in evaluating $F(t)$ is then $O(n)$. Clearly, the lower bound on the total time required is $\Omega(n)$, and so the result follows. $\square$

The above algorithm developed by applying the constructive scheme of Megiddo is equivalent to the $\Theta(n)$ algorithm of Hansen, Jaumard & Mathon (1993) for the unconstrained single ratio 0–1 hyperbolic programming program with a positive denominator.

Saipe (1975) developed an $O(np)$ algorithm for single ratio 0–1 hyperbolic programs with a single cardinality constraint, i.e.

$$D := \left\{ x_j \in \{0,1\}, j = 1, \ldots, n \,\Big|\, \sum_{j=1}^{n} x_j = p \right\}.$$

Problem A with D defined as above can be solved by first determining the order $-p$ statistic, $\hat{c}$, of the coefficients, c_j, by using the linear median finding algorithm of Blum et al. (1973) in $O(n)$ comparisons. Let $J := \{j | c_j > \hat{c}\}$, the optimal solution value of problem A is then given by $\sum_{j \in J} c_j + (p - |J|)\hat{c}$, which requires $O(p)$ additions. Thus, by Theorem 3.19, Megiddo's scheme would give rise to an algorithm for problem B of $O(np + n^2)$. Saipe (1975) takes advantage of the result that the variable x_j corresponding to the index $j = \operatorname{argmax}\{a_i/b_i \mid i = 1, \ldots, m\}$ is 1 in an optimal solution to P.

For problems with general constraint sets, one can derive the bounds in polynomial time by relaxing the integrality requirements and solving the continuous hyperbolic programming program by using the Charnes-Cooper linear programming reformulation (Charnes & Cooper 1962).

3.5 Special Instances of the Reformulation

In this section, we provide conditions under which including only the convex or concave extension provides a valid reformulation scheme for the fractional term. We then show that the inclusion of certain nonlinear inequalities may tighten the reformulation of Section 3.3. Finally, we relate some of the previous work done on special instances of the 0−1 rational functions with our work.

Remark 3.22. Let

$$z \leq \frac{a_0 + \sum_{j \in N} a_j x_j}{b_0 + \sum_{j \in N} b_j x_j} \qquad (3.15)$$

be an inequality constraint where $x \in B^n$. Assume $b_0 + \sum_{j \in N} b_j x_j > 0$ over the feasible region. Then, the above inequality may be written as

$$z b_0 + z \sum_{j \in N} b_j x_j - a_0 - \sum_{j \in N} a_j x_j \leq 0.$$

Inequality (3.15) can then be replaced by forming only the convex extension of $z b_0 + z \sum_{j \in N} b_j x_j + a_0 + \sum_{j \in N} a_j x_j$.

Remark 3.23. Let

$$z \geq \frac{a_0 + \sum_{j \in N} a_j x_j}{b_0 + \sum_{j \in N} b_j x_j} \qquad (3.16)$$

be an inequality constraint where $x \in B^n$. Assume $b_0 + \sum_{j \in N} b_j x_j > 0$ over the feasible region. Then, the above inequality may be written as

$$z b_0 + z \sum_{j \in N} b_j x_j - a_0 - \sum_{j \in N} a_j x_j \geq 0.$$

Inequality (3.16) can then be replaced by forming only the concave extension of $z b_0 + z \sum_{j \in N} b_j x_j + a_0 + \sum_{j \in N} a_j x_j$.

Remark 3.24. Let

$$z \geq \frac{1}{b_0 + \sum_{j \in N} b_j x_j} \qquad (3.17)$$

be an inequality constraint where $x \in B^n$. Assume $b_0 + \sum_{j \in N} b_j x_j > 0$ over the feasible region. Then, the above inequality is itself convex and hence the exact encloser described in Section 3.3 is weaker than the nonlinear inequality itself.

We now present a simpler, though weaker, technique of constructing exact enclosers for a fractional term. As a consequence, we show that our reformulation scheme reduces to that of Li (1994) and Wu (1997) when the lower bounds on the intermediate linear fractional functions are ignored. Assume that $f_i \geq 0$ in the following discussion. Note that a convex encloser can be derived from (3.6) if we can model the following disjunction:

$$z_{ij} = \begin{cases} 0, & \text{if } x_j = 0; \\ f_i, & \text{if } x_j = 1. \end{cases}$$

Using standard techniques in modeling integer programs (cf. Schrijver 1986, Balas 1988, Nemhauser & Wolsey 1988, Parker & Rardin 1988), it follows that the above disjunctive constraints may be written as:

(RM) $\quad z_{ij} - f_i \geq -M(1 - x_j)$
$\quad\quad\quad z_{ij} - f_i \leq 0$
$\quad\quad\quad z_{ij} \leq M x_j$
$\quad\quad\quad z_{ij} \geq 0$

The relaxation RM is weaker than convex encloser of $f_i x_j$ from Theorem 3.1. In fact, the bilinear envelopes in Theorem 3.1 reduce to RM when $f_i^L = 0$, $f_i^U = M$ and $x_j^L = 0$ and $x_j^U = 1$. The relaxation of fractional function of 0–1 variables in Williams (1974) results if we reformulate the function by applying Remark 3.8 followed by Remark 3.6 and then use RM to replace $f_i x_j$ with z_{ij}. In light of Remark 3.22, the size of the relaxation could have been reduced considerably requiring only two constraints for every z_{ij} instead of four as used by Williams (1974). The reformulation of Li (1994) and Wu (1997) is derived by reformulating the rational function using Remark 3.6 followed by Remark 3.8. Since Li (1994) and Wu (1997) dealt with minimizing a summation of fractional functions of 0–1 variables over linear constraint sets, Remark 3.23 can be used to reduce the size of their reformulation. In fact, in this case Remark 3.24 yields tighter nonlinear relaxations.

As was shown in Theorem 3.17 and Corollary 3.18, the bounds f_i^l and f_i^u are critical for the tightness of the relaxation. Hence, using RM instead of the bilinear envelopes is highly detrimental to the quality of the relaxation.

3.6 Examples of the Reformulation Scheme

We present three examples to illustrate the impact of product disaggregation in solving nonlinear global optimization problems with multilinear and rational expressions. The problems are solved using BARON's NLP module, which can handle integrality restrictions and is detailed in Chapter 7.

We use CPLEX 7.0 (ILOG 2000) to solve linear programming subproblems and MINOS 5.5 (Murtagh & Saunders 1995) for nonlinear programming subproblems. All experiments were done on an IBM/RS 6000 43P workstation with 128MB RAM and a LINPACK rating of 59.9. The branch-and-bound algorithm used terminates when the lower and upper bounds provided are within an absolute tolerance of 10^{-6}.

3.6.1 Example 1: Hock & Schittkowski (1981)

Our first example is Problem 62 from Hock & Schittkowski (1981):

$$\min \quad -32.174 \left(255 \ln \left(\frac{x_1 + x_2 + x_3 + 0.03}{0.09 x_1 + x_2 + x_3 + 0.03} \right) \right.$$
$$+ 280 \ln \left(\frac{x_2 + x_3 + 0.03}{0.07 x_2 + x_3 + 0.03} \right)$$
$$\left. + 290 \ln \left(\frac{x_3 + 0.03}{0.13 x_3 + 0.03} \right) \right)$$

$$\text{s.t.} \quad x_1 + x_2 + x_3 - 1 = 0$$

$$(x_1, x_2, x_3) \in [0, 1]^3.$$

The problem was solved in 3.47 CPU seconds after 1,220 branch-and-bound nodes. The maximum number of nodes in the memory was 117. The lower bound derived on the root node was $-69,577.2$ whereas the optimal objective function value is $-26,272.6$ and is achieved at $(0.619066, 0.327145, 0.0537887)$.

After disaggregating the products, the problem can be rewritten as:

$$\min \quad -32.174 \Big(255 \ln(x_1 l_1 + x_2 l_1 + x_3 l_1 + 0.03 l_1)$$
$$+ 280 \ln(x_2 l_2 + x_3 l_2 + 0.03 l_2) + 290 \ln(x_3 l_3 + 0.03 l_3) \Big)$$

$$\text{s.t.} \quad x_1 + x_2 + x_3 - 1 = 0$$

$$0.09 x_1 l_1 + x_2 l_1 + x_3 l_1 + 0.03 l_1 - 1 = 0$$

$$0.07 x_2 l_2 + x_3 l_2 + 0.03 l_2 - 1 = 0$$

$$0.13 x_3 l_3 + 0.03 l_3 - 1 = 0$$

3.6. EXAMPLES OF THE REFORMULATION SCHEME

$$(x_1, x_2, x_3) \in [0, 1]^3.$$

The above reformulated problem was solved in 2.97 seconds on the same computer with the same tolerance as before. The number of nodes required this time was only 667 while the memory requirements reduced to 52 nodes. The lower bound derived at the root node was $-51,150.5$, corresponding to a 43% reduction in the relaxation gap.

3.6.2 Example 2: Nuclear Reactor Reload Pattern Design

A nuclear reactor contains nuclear fuel elements which are reshuffled at the end of every core operation cycle and the empty positions in the core grid are filled with new unburnt elements. Mixed-integer nonlinear programming models have been developed by Quist (2000a) for carrying out the nuclear reactor reload pattern design. We refer the reader to Quist (2000a) for additional details including prior attempts to solve this problem to global optimality. Quist (2000b) reports on a number of unsuccessful attempts by global optimization groups to solve this problem. To the best of our knowledge, the results of this section represent the first successful application of global optimization algorithms to this problem.

The model assumes that the reload pattern is the same for each cycle and each bundle of nuclear core follows a certain trajectory before it is discharged. Let x_{ilm} be defined as:

$$x_{ilm} = \begin{cases} 1, & \text{if node } i \text{ contains a bundle of age } l \text{ in trajectory } m; \\ 0, & \text{otherwise}. \end{cases}$$

The basic model is then (Quist 2000a):

$$\max \quad k_T^{\text{eff}}$$

$$\text{s.t.} \quad \sum_{i=1}^{I} V_i x_{ilm} = 1 \qquad l = 1, \ldots, L, \; m = 1, \ldots, M$$

$$\sum_{l=1}^{L} \sum_{m=1}^{M} x_{ilm} = 1 \qquad i = 1, \ldots, I$$

$$k_{il}^{\infty} = \sum_{m=1}^{M} x_{ilm} k^{\text{fresh}}$$

$$+ \sum_{l=2}^{L} \sum_{m=1}^{M} x_{ilm} \sum_{j=1}^{l} V_j x_{jl-1m} k_{jT}^{\infty} \qquad i = 1, \ldots, I \qquad (3.18)$$

$$k_t^{\text{eff}} R_{it} = \sum_{j=1}^{l} G_{ij} k_{jt}^{\infty} R_{jt} \qquad i = 1, \ldots, I; \, t = 1, \ldots, T$$

$$\sum_{i=1}^{I} V_i k_{it}^{\infty} R_{it} = 1 \qquad t = 1, \ldots, T$$

$$k_{j,t+1}^{\infty} = k_{it}^{\infty} - \alpha \Delta_t k_{it}^{\infty} R_{it} \qquad i = 1, \ldots, I; \, t = 1, \ldots, T-1$$

$$k_{it}^{\infty} R_{it} \leq \frac{f^{\lim}}{\sum_{j=1}^{I} V_j} \qquad i = 1, \ldots, I; \, t = 1, \ldots, T$$

$$k_t^{\text{eff}} \geq 0 \qquad t = 1, \ldots T$$

$$k_{it}^{\infty}, R_{it} \geq 0$$

$$x_{ilm} = x_{jlm} \qquad l = 1, \ldots, L; \, m = 1, \ldots, M$$

$$(i,j) \text{ is a diagonal pair}$$

$$x_{ilm} \in \{0,1\} \qquad i = 1, \ldots, I; \, l = 1, \ldots, L;$$

$$m = 1, \ldots, M.$$

The variables in this formulation are as follows: k_t^{eff} is the effective multiplication factor at time t, k_{it}^{∞} is the average infinite multiplication factor at node i at time t, and R_{it} is the normalized power at node i at time t.

The example we solve here is the *tiny* model from Quist (2000b) with four nodes in the core. The core stays in the reactor for two cycles and the number of trajectories is two. When the above model is solved with BARON, the global optimum is verified after evaluating 3,728 nodes in 1,110 seconds of CPU time. The maximum size of the branch-and-bound tree at any time was 244 nodes. The optimal solution to the problem was verified to be 1.00594.

If we disaggregate constraints (3.18) as follows:

$$k_{il}^{\infty} = \sum_{m=1}^{M} x_{ilm} k^{\text{fresh}}$$

$$+\sum_{l=2}^{L}\sum_{m=1}^{M}\sum_{j=1}^{I} V_j x_{ilm} x_{jl-1m} k_{jT}^{\infty}, \quad i = 1, \ldots, I$$

then the problem is solved after 1,485 nodes in 649 seconds. In addition to significant CPU time reductions, disaggregation leads to reduced memory requirements. The maximum number of nodes in memory for this run reduced from 244 to 116.

3.6.3 Example 3: Catalyst Mixing for Packed Bed Reactor

In this example, we consider a classical chemical reaction problem originating from the early sixties (Gunn & Thomas 1965). The system involves three components, A, B, and C. The reactions A⇔B→C take place in a packed bed catalytic reactor. Catalysts I and II apply to A⇔B and B→C, respectively. A feed of pure A enters the reactor bed. The problem is to find the optimal mix of catalysts I and II along the reactor length so as to maximize production of C at the end of the reactor.

Let t denote the distance along the reactor. By $u(t)$, $z^A(t)$, and $z^B(t)$ we denote the fraction of catalyst I, the mole fraction of A, and the mole fraction of B, respectively, at position t. Finally, let k_{AB}, k_{BA}, and k_{BC} denote the rate constants of the reactions. If the reactor is of length t_f, the optimization problem can then be stated as follows:

$$\max \quad 1 - z^A(t_f) - z^B(t_f)$$

$$\text{s.t.} \quad \frac{dz^A}{dt} = u(k_{BA} z^B - k_{AB} z^A)$$

$$\frac{dz^B}{dt} = u(k_{AB} z^A - k_{BA} z^B) - (1-u) z^B$$

$$z^A(0) = 1$$

$$z^B(0) = 0$$

$$0 \leq u(t) \leq 1 \qquad\qquad 0 \leq t \leq t_f$$

Various means of discretization may be employed to convert the above differential problem into an algebraic nonlinear program. Here, we follow

Logsdon & Biegler (1989) who used orthogonal collocation on finite elements. In particular, consider a discretization consisting of $NCOL$ collocation points on each of NFE finite elements. Let $C = \{1, \ldots, NCOL\}$ and $F = \{1, \ldots, NFE\}$ denote the set of collocation points and finite elements, respectively. Define $C^+ = C \bigcup \{0\}$ and $F^- = F \setminus \{1\}$. Also, let α_i denote the length of finite element i ($i \in F$), and define the polynomial basis functions, $\varphi(\tau)$ as:

$$\varphi_j(\tau) = \prod_{k \in C^+, k \neq j} \frac{\tau - \tau_k}{\tau_j - \tau_k}$$

$$\bar{\varphi}_j(\tau) = \prod_{k \in C, k \neq j} \frac{\tau - \tau_k}{\tau_j - \tau_k}$$

where τ_j, for $j \in C^+$, denote the roots of $NCOL$-order Legendre polynomials shifted to $[0, 1]$. Then, the discretized problem is as follows:

$$\max \quad 1 - z_f^A - z_f^B$$

$$\text{s.t.} \quad \sum_{k \in C^+} z_{ik}^A \varphi_k(\tau_j) - \alpha_i u_{ij}(k_{BA} z_{ij}^B - k_{AB} z_{ij}^A) = 0 \qquad i \in F, j \in C \quad (3.19)$$

$$\sum_{k \in C^+} z_{ik}^B \varphi_k(\tau_j) - \alpha_i \left[u_{ij}(k_{AB} z_{ij}^A - k_{BA} z_{ij}^B) \right.$$

$$\left. - (1 - u_{ij}) z_{ij}^B \right] = 0 \qquad i \in F, j \in C \quad (3.20)$$

$$z_{i0}^A = \sum_{k \in C^+} z_{i-1,k}^A \varphi_k(1) \qquad i \in F^- \quad (3.21)$$

$$z_{i0}^B = \sum_{k \in C^+} z_{i-1,k}^B \varphi_k(1) \qquad i \in F^- \quad (3.22)$$

$$z_f^A = \sum_{k \in C^+} z_{NFE,k}^A \varphi_k(1) \qquad i \in F^- \quad (3.23)$$

$$z_f^B = \sum_{k \in C^+} z_{NFE,k}^B \varphi_k(1) \qquad i \in F^- \quad (3.24)$$

$$\sum_{i \in F} \alpha_i = t_f \quad (3.25)$$

$$-10^{-4} \leq \sum_{k \in C^+} z_{ik}^A \dot{\varphi}_k(1) - \alpha_i \left(\sum_{k \in C^+} u_{ik} \bar{\varphi}(1) \right)$$

3.6. EXAMPLES OF THE REFORMULATION SCHEME

$$(k_{BA} z^B_{i+1,0} - k_{AB} z^A_{i+1,0}) \leq 10^{-4} \qquad i \in F \qquad (3.26)$$

$$-10^{-4} \leq \sum_{k \in C^+} z^B_{ik} \dot{\varphi}_k(1) - \alpha_i \left[\left(\sum_{k \in C^+} u_{ik} \bar{\varphi}(1) \right) \right.$$

$$(k_{AB} z^A_{i+1,0} - k_{BA} z^B_{i+1,0})$$

$$\left. - \left(1 - \sum_{k \in C^+} u_{ik} \right) \bar{\varphi}(1) z^B_{i+1,0} \right] \leq 10^{-4} \quad i \in F \qquad (3.27)$$

$$z^A_{1,0} = 1 \qquad (3.28)$$

$$z^B_{1,0} = 0 \qquad (3.29)$$

$$0 \leq z^A_{ij} \leq 1 \qquad i \in F, j \in C \quad (3.30)$$

$$0 \leq z^B_{ij} \leq 1 \qquad i \in F, j \in C \quad (3.31)$$

$$0 \leq u_{ij} \leq 1 \qquad i \in F, j \in C \quad (3.32)$$

$$\alpha_i \geq 0 \qquad i \in F \qquad (3.33)$$

The variables in this formulation are the values of the concentrations z^A_{ij}, z^B_{ij}, and catalyst I fraction u_{ij} at the collocation points ($i \in F, j \in C$), as well as the lengths of the finite element intervals α_i ($i \in F$). Equations (3.19) and (3.20) are the discretized versions of the two governing differential equations. Equations (3.21) and (3.22) enforce continuity of the concentration profiles between different finite elements. The concentrations of A and B at the end of the reactor are defined in (3.23) and (3.24) in terms of the model parameters. Equation (3.25) equates the sum of the lengths of all finite elements to the length of the reactor. The inequality constraints (3.26) and (3.27) restrict the discretization approximation error. Finally, constraints (3.28)–(3.33) impose obvious bounds on the variables. We refer the reader to Logsdon & Biegler (1989) and Biegler & Tjoa (1991) for more details.

While the differential equations are linear in z^A and z^B, the algebraic program is nonlinear for two reasons. First, there are products between the catalyst fraction and concentrations in the right-hand-side of the differential equations. Second, dt is replaced by α_i (the finite element lengths) that multiply the right-hand-side of the equations (or divide the left-hand-side).

The particular instance of the problem we consider has $k_{AB} = 10$, $k_{BA} =$

$k_{BC} = 1$, $t_f = 0.4$, and $NFE = NCOL = 2$. A GAMS code for the problem we solve is provided in Biegler & Tjoa (1991). The problem has 32 variables and 33 constraints. Without product disaggregation, BARON obtains a lower bound of -0.4 at the root node. Then, after exploring over 13 million nodes in 100 hours of CPU time, a lower bound of -0.104861 and an upper bound of -0.01637 are obtained. In other words, the problem was not solvable with this formulation. The main obstacle was memory, as the computer's memory was reached and the solver switched from a breadth-first search to a depth search that apparently is not very efficient for this problem.

On the other hand, after disaggregating all variable products in the above formulation, the lower bound obtained at the root node was -0.125073. This represents a 68% reduction of the relaxation gap of the original formulation. Using the same termination tolerance on the same computer, BARON on the disaggregated formulation terminated with an upper and lower bound of -0.01637 after 259,548 nodes and 112 minutes of CPU time. This is clearly an instance of an engineering problem for which disaggregation is necessary to prove global optimality in a reasonable amount of time.

3.7 Reformulations of Hyperbolic Programs

The remainder of this chapter will concentrate on the application of product disaggregation in the context of 0–1 hyperbolic programs. In Section 3.3, we provided two different ways of reformulating the fractional function:

(LF) $$g = \frac{a_0 + \sum_{j \in N} a_j x_j}{b_0 + \sum_{j \in N} b_j x_j}$$

When Remark 3.8 is applied first, LF is rewritten as:

(M1) $$f a_0 + f \sum_{j \in N} a_j x_j = g$$
$$f b_0 + f \sum_{j \in N} b_j x_j = 1,$$

otherwise it is rewritten as:

(M2) $$g b_0 + g \sum_{j \in N} b_j x_j - a_0 - \sum_{j \in N} a_j x_j = 0.$$

3.7. REFORMULATIONS OF HYPERBOLIC PROGRAMS

In either case, the functions generated are of the form of $\phi(x, f)$ in (3.7) and can be replaced by their convex and concave extensions given in (3.10) or (3.11).

Note that problem H consists of a sum of several fractional terms, each in the form of LF. Using the form M1 along with the convex and concave envelopes from (3.10) and (3.11), we get the following reformulation of H:

$$
\begin{aligned}
\text{(R1)} \quad \max \quad & \sum_{i=1}^{m} f_i a_{i0} + \sum_{i=1}^{m} \sum_{j=1}^{n} a_{ij} u_{ij} \\
\text{s.t.} \quad & Dx \leq c \\
& b_{i0} f_i + \sum_{j=1}^{n} b_{ij} u_{ij} = 1 \quad i = 1, \ldots, m
\end{aligned}
$$

$$
\begin{aligned}
u_{ij} &\leq f_i^u x_j & i &= 1, \ldots, m; j = 1, \ldots, n & (3.34) \\
u_{ij} &\leq f_i + f_i^l x_j - f_i^l & i &= 1, \ldots, m; j = 1, \ldots, n & (3.35) \\
u_{ij} &\geq f_i + f_i^u x_j - f_i^u & i &= 1, \ldots, m; j = 1, \ldots, n & (3.36) \\
u_{ij} &\geq f_i^l x_j & i &= 1, \ldots, m; j = 1, \ldots, n & (3.37)
\end{aligned}
$$
$$x \in \{0, 1\}^n$$

where variable f_i is used for the term $1/(b_{i0} + b_i^T x)$ and u_{ij} is used for the term $f_i x_j$, and f_i^l and f_i^u are valid lower and upper bounds for f_i. Note that, along with the integrality requirement, constraints (3.34) and (3.35) form the concave extension and constraints (3.36) and (3.37) form the convex extension. Li (1994) and Wu (1997) have suggested special cases of this reformulation where the lower bounds on the fractional terms are assumed to be zero.

Remark 3.25. *If for each $i = 1, 2, \ldots, m$, $b_{i0} + b_i^T x > 0$ on the feasible region of H, then the constraints*

$$f_i \geq \frac{1}{b_{i0} + \sum_{j=1}^{n} b_{ij} x_j} \quad i = 1, \ldots, m \quad (3.38)$$

are convex. In such a case, the concave extension of $f_i b_{i0} + f_i \sum_{j=1}^{n} b_{ij} x_j$ is implied and may be replaced by the above convex inequalities in R1.

Following the above remark, a reformulation R2 is constructed replacing constraints (3.36) and (3.37) by the convex nonlinear inequalities (3.38).

R1 is a maximization problem, thus, if $a_{i0} + a_i^T x > 0$ for all i, we only need the concave extensions of the fractional terms as far as the optimal solution is concerned. The concave extension of the fractional term is provided by the convex extension of $f_i b_{i0} + f_i \sum_{j=1}^n b_{ij} x_j$. Hence, the following claim is true:

Remark 3.26. *If $a_{i0} + a_i^T x > 0$ over the feasible region of the mathematical program, then the concave extension of $f_i b_{i0} + f_i \sum_{j=1}^n b_{ij} x_j$ can be dropped from R1.*

Next, we employ the form M2 to derive another reformulation of H. Since H is a maximization problem, it may be reformulated as:

$$
\text{(H)} \quad \max \sum_{i=1}^m g_i
$$
$$
\text{s.t.} \quad g_i \leq \frac{a_{i0} + \sum_{j=1}^n a_{ij} x_j}{b_{i0} + \sum_{j=1}^n b_{ij} x_j} \quad (3.39)
$$
$$
Dx \leq c
$$
$$
x \in \{0,1\}^n
$$

Since $b_{i0} + \sum_{j=1}^n b_{ij} x_j > 0$, inequality (3.39) may be rewritten as

$$g_i b_{i0} + g_i \sum_{j \in N} b_{ij} x_j - a_0 - \sum_{j \in N} a_{ij} x_j \leq 0.$$

Hence, replacing $g_i b_{i0} + g_i \sum_{j \in N} b_{ij} x_j$ by the epigraph of its convex extension is adequate for the reformulation and the concave extension is redundant as far as the optimal solution is concerned. We can thus claim:

Remark 3.27. *When M2 is employed to reformulate H, the concave extension may be dropped in the reformulation.*

As a consequence of Remark 3.27, it follows that H can be reformulated as:

$$
\text{(R3)} \quad \max \sum_{i=1}^m g_i
$$
$$
\text{s.t.} \quad Dx \leq c
$$
$$
b_{i0} g_i + \sum_{j=1}^n b_{ij} v_{ij} = a_{i0} + \sum_{j=1}^n a_{ij} x_{ij} \quad i = 1, \ldots, m
$$

$$
\begin{align}
v_{ij} &\leq g_i^u x_j & i &= 1,\ldots,m; j = 1,\ldots,n; b_{ij} < 0 & (3.40)\\
v_{ij} &\leq g_i + g_i^l x_j - g_i^l & i &= 1,\ldots,m; j = 1,\ldots,n; b_{ij} < 0 & (3.41)\\
v_{ij} &\geq g_i + g_i^u x_j - g_i^u & i &= 1,\ldots,m; j = 1,\ldots,n; b_{ij} > 0 & (3.42)\\
v_{ij} &\geq g_i^l x_j & i &= 1,\ldots,m; j = 1,\ldots,n; b_{ij} > 0 & (3.43)
\end{align}
$$
$$x \in \{0,1\}^n$$

3.7. REFORMULATIONS OF HYPERBOLIC PROGRAMS

where variable g_i is used for the term $(a_{i0} + a_i^T x)/(b_{i0} + b_i^T x)$ and v_{ij} is used for the term $g_i x_j$, and g_i^l and g_i^u are valid lower and upper bounds for g_i. Note that, depending upon the sign of b_{ij}, the constraints (3.40) and (3.41), and (3.42) and (3.43) form the concave extensions.

The reformulations R1 and R3 can also be combined to construct the following:

(R4) $\max \quad \sum_{i=1}^{n} g_i$

s.t. $Dx \leq c$

$g_i = f_i a_{i0} + \sum_{j=1}^{n} a_{ij} u_{ij} \qquad i = 1, \ldots, m$

$b_{i0} f_i + \sum_{j=1}^{n} b_{ij} u_{ij} = 1 \qquad i = 1, \ldots, m$

$u_{ij} \leq f_i^u x_j \qquad i = 1, \ldots, m; j = 1, \ldots, n$
$u_{ij} \leq f_i + f_i^l x_j - f_i^l \qquad i = 1, \ldots, m; j = 1, \ldots, n$
$u_{ij} \geq f_i + f_i^u x_j - f_i^u \qquad i = 1, \ldots, m; j = 1, \ldots, n$
$u_{ij} \geq f_i^l x_j \qquad i = 1, \ldots, m; j = 1, \ldots, n$

$b_{i0} g_i + \sum_{j=1}^{n} b_{ij} v_{ij} = a_{i0} + \sum_{j=1}^{n} a_{ij} x_{ij} \qquad i = 1, \ldots, m$

$v_{ij} \leq g_i^u x_j \qquad i = 1, \ldots, m; j = 1, \ldots, n; b_{ij} < 0$
$v_{ij} \leq g_i + g_i^l x_j - g_i^l \qquad i = 1, \ldots, m; j = 1, \ldots, n; b_{ij} < 0$
$v_{ij} \geq g_i + g_i^u x_j - g_i^u \qquad i = 1, \ldots, m; j = 1, \ldots, n; b_{ij} > 0$
$v_{ij} \geq g_i^l x_j \qquad i = 1, \ldots, m; j = 1, \ldots, n; b_{ij} > 0$

$x \in \{0, 1\}^n$

To obtain tighter relaxations of the reformulated problem, we can introduce additional valid inequalities obtained by multiplying the original constraint set of H by the variables corresponding to the fractional terms. Specifically, we can introduce the following constraints to R1:

$$\sum_{j=1}^{n} d_{rj} u_{ij} - f_i^l \sum_{j=1}^{n} d_{rj} x_j \leq c_r f_i - c_r f_i^l \qquad r = 1, \ldots, k; \ i = 1, \ldots, m \quad (3.44)$$

$$f_i^u \sum_{j=1}^{n} d_{rj} x_j - \sum_{j=1}^{n} d_{rj} u_{ij} \leq c_r f_i^u - c_r f_i \qquad r = 1, \ldots, k; \ i = 1, \ldots, m \quad (3.45)$$

and the following constraints to R3:

$$\sum_{j=1}^{n} d_{rj}v_{ij} - g_i^l \sum_{j=1}^{n} d_{rj}x_j \leq c_r g_i - c_r g_i^l \quad r=1,\ldots,k;\ i=1,\ldots,m \quad (3.46)$$

$$g_i^u \sum_{j=1}^{n} d_{rj}x_j - \sum_{j=1}^{n} d_{rj}v_{ij} \leq c_r g_i^u - c_r g_i \quad r=1,\ldots,k;\ i=1,\ldots,m \quad (3.47)$$

where d_{rj} are the elements of the constraint matrix D and c_r are the elements of the right-hand-side vector c. Let R5 be obtained by adding constraints (3.44) and (3.45) to R1 and R7 be obtained by adding constraints (3.46) and (3.47) to R3. It is obvious that R5 and R7 are valid reformulations of H. In R5 and R7, we retain the concave as well as the convex extensions since in the presence of constraints (3.44), (3.45), (3.46) or (3.47) the concave extensions are no longer redundant when the integrality requirements are dropped.

Another reformulation, R6, is constructed from R5 by adding the convex nonlinear constraints (3.38). Our final reformulation R8 is constructed by combining R5 and R7 in a similar manner as R4 is constructed from R1 and R3.

Note that no additional binary variables are introduced in the reformulations. However, a large number of continuous variables are introduced. Table 3.1 compares the size of the various reformulations R1-R8 in terms of m, n and k. The relative importance of these reformulations depends on the strength the bounds offered by their relaxations.

Reformulation	Number of Continuous variables	Number of constraints	
		Linear	Nonlinear
R1	$m + mn$	$k + m + 4mn$	0
R2	$m + mn$	$k + m + 2mn$	m
R3	$m + mn$	$k + m + 2mn$	0
R4	$2(m + mn)$	$k + 3m + 6mn$	0
R5	$m + mn$	$k + 2km + m + 4mn$	0
R6	$m + mn$	$k + 2km + m + 2mn$	m
R7	$m + mn$	$k + 2km + m + 4mn$	0
R8	$2(m + mn)$	$k + 4km + 3m + 8mn$	0

Table 3.1: Size of the reformulations

3.8. UPPER BOUNDING OF 0−1 HYPERBOLIC PROGRAMS

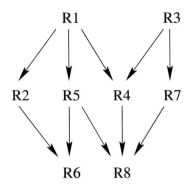

Figure 3.1: Comparison of reformulations: Ri → Rj ⇔ $V_{\text{LRi}} \geq V_{\text{LRj}}$.

3.8 Upper Bounding of 0−1 Hyperbolic Programs

Relaxing the integrality requirement in the reformulations R1-R8 results in convex upper bounding problems suitable for use within a branch-and-bound framework. We denote these relaxations by LR1-LR8 respectively. The optimal value of a mathematical program R will be denoted by V_R. The upper bounds obtained from the reformulations are compared in Figure 3.1. These relationships are clear from the construction of the reformulations.

As mentioned earlier, the reformulations proposed in Li (1994) and Wu (1997) are special cases of R1. Thus, it is clear from the relationships in Figure 3.1, that the bounds obtained from the reformulations R2, R4, R5, R6, and R8 dominate those obtained from the reformulations of Li (1994) and Wu (1997). As an example of the relative tightness of the bounds from the different reformulations, we consider the following numerical example from Li (1994):

$$
\begin{aligned}
\max \quad & -\frac{1+x_1}{0.5x_1 + x_2} - \frac{2 - 3x_2}{2x_2 + x_3 + x_4} \\
\text{s.t.} \quad & x_1 + x_2 + x_3 \geq 2 \\
& 2x_1 + 3x_2 - 4x_4 \leq 2 \\
& 0.5x_1 + x_2 > 0
\end{aligned}
$$

$$2x_2 + x_3 + x_4 > 0$$
$$x_1, x_2, x_3, x_4 \in \{0, 1\}$$

The reformulation proposed by Li (1994) provides a linear programming relaxation bound of -0.5. Reformulations R5 and R7 provide relaxation bounds of -0.6923 and -0.6563, respectively. In all three cases, the reformulations were constructed using bounds on the fractional terms that were determined by solving unconstrained single ratio 0–1 hyperbolic programs. The optimal solution to the problem is -0.75. Thus, R5 and R7 have relaxation gaps of 7.69% and 12.49% respectively, as compared to the reformulation of Li (1994), which has a gap of 33.33%. Further computational results on the comparison of the bounds from the various reformulations are presented in Section 3.11.

Note that the objective function of H may be expressed as $\sum_{i=1}^{m} g_i(x)$ where

$$g_i(x) = \frac{a_{i0} + \sum_{j=1}^{n} a_{ij} x_j}{b_{i0} + \sum_{j=1}^{n} b_{ij} x_j}.$$

Let g_i^u $i = 1, \ldots, m$ be the tightest upper bounds available on the following maximization problem:

$$\max\{g_i(x) \mid Dx \leq c, x \in \{0, 1\}^n\}.$$

Such bounds may be obtained by solving single ratio hyperbolic programs and will be discussed in the next section. Upper bounds on $g_i(x)$ may also be obtained over the linear relaxation of the constraint set of H using the Charnes-Cooper transformation for single ratio fractional programs (Charnes & Cooper 1962). Let us denote these bounds by q_i^u, $i = 1, \ldots, m$. We then define $V_{\text{UB}} = \sum_{i=1}^{m} g_i^u$ and $V_{\text{CC}} = \sum_{i=1}^{m} q_i^u$. It is easy to see that V_{UB} and V_{CC} provide valid upper bounds for V_{H}. We shall now compare the reformulation bounds to these simple bounding strategies.

Theorem 3.28. $V_{\text{H}} \leq V_{\text{LR3}} \leq V_{\text{UB}}$. *If the extreme points of the feasible region of H are binary-valued and $V_{\text{H}} < V_{\text{UB}}$, then $V_{\text{LR3}} < V_{\text{UB}}$.*

Proof. Since LR3 is a valid upper bounding scheme for H, $V_{\text{H}} \leq V_{\text{LR3}}$. Since $g_i \leq g_i^u$, it follows that $V_{\text{LR3}} \leq V_{\text{UB}}$.

If there is an optimal solution to LR3 which is integral, then $V_{\text{H}} = V_{\text{LR3}}$ and hence $V_{\text{H}} < V_{\text{UB}}$ implies $V_{\text{LR3}} < V_{\text{UB}}$.

3.8. UPPER BOUNDING OF 0−1 HYPERBOLIC PROGRAMS

Now assume that there does not exist an integral solution to LR3. Let x^0 be the optimal solution of LR3. If $V_{\text{LR3}} = V_{\text{UB}}$, then each g_i is at its upper bound. It follows from Theorem 2.7, that R3 exactly represents the fractional program when each g_i is at its bound. Let F be the smallest face of U^n such that $x^0 \in F$. Then $x^0 \in \text{ri}(F)$. Since, for all i, $g_i(x)$ is an explicit quasi-convex function which achieves its global maximum in $\text{ri}(F)$, each $g_i(x)$ is constant over F (Stancu-Minasian 1997, Theorem 2.3.7). Hence, $g_i(x)$, $i = 1, \ldots, m$ attain their optimal value at every $x \in F$. By assumption, the extreme points of F are binary vectors. Hence, there exists an integral optimum of LR3 thereby contradicting the assumption. □

The above result implies that, if $V_{\text{LR3}} = V_{\text{UB}}$, then $V_{\text{H}} = V_{\text{LR3}} = V_{\text{UB}}$. Moreover, from Figure 3.1, it also follows that the bounds from R4, R7, and R8 dominate V_{UB}.

Proposition 3.29. $V_{\text{H}} \leq V_{\text{LR5}} \leq V_{\text{CC}}$.

Proof. Consider the inequalities:

$$\sum_{j=1}^{n} \hat{a}_{ij} u_{ij} - \sum_{j=1}^{n} \hat{a}_{ij} x_j f_i^l \leq \hat{b}_i (f_i - f_i^l) \quad i = 1, \ldots, k \quad (3.48)$$

$$\sum_{j=1}^{n} \hat{a}_{ij} x_j \leq \hat{b}_i \quad i = 1, \ldots, k. \quad (3.49)$$

Since $f_i^l \geq 0$, from (3.49)

$$-\sum_{j=1}^{n} \hat{a}_{ij} x_j f_i^l \geq -\hat{b}_i f_i^l \quad i = 1, \ldots, k. \quad (3.50)$$

Then, by (3.50) and (3.48), it follows that

$$\sum_{j=1}^{n} \hat{a}_{ij} u_{ij} \leq \hat{b}_i f_i \quad i = 1, \ldots, k.$$

Therefore, the Charnes-Cooper transformation is embedded in R5 implying that $V_{\text{H}} \leq V_{\text{LR5}} \leq V_{\text{CC}}$. □

From Figure 3.1, it follows that the bounds from R6 and R8 dominate V_{CC}. The above results will be used in Section 3.11.3 to show that the proposed reformulations provide better bounds than those in Saipe (1975) for the cardinality constrained 0−1 hyperbolic programming program.

3.9 A Branch-and-Bound Algorithm

In this section, we develop a branch-and-bound algorithm for the 0−1 hyperbolic programming program. A naive approach to solving H would be to apply standard branch-and-bound techniques to its MILP reformulation. As has been pointed out in Section 3.4, the gap between the reformulation and its continuous relaxation is extremely sensitive to the quality of the bounds on the fractional terms. Also, since in a branch-and-bound framework, the feasible space of the subproblems at each node is a subset of the feasible space of the problem at the root node, it is possible to derive tighter bounds on the fractional terms over these smaller feasible spaces. Therefore, in the branch-and-bound algorithm for H, we propose to reformulate the problem at every node of the branch-and-bound tree using the best available bounds. Henceforth, we shall refer to this technique as *node tightening*.

We now present a formal statement of the algorithm. The following notation is used:

k	Iteration number
i	Current node of the branch-and-bound tree
$\mathcal{L}$	List of active nodes
R_i	Reformulated problem at the ith node
$U(R_i)$	Upper bound on the solution of R_i
U^k	Upper bound on the solution of H at the kth iteration
L^k	Lower bound on the solution of H at the kth iteration

Initialization:

Set $k = 0$, $L^k = -\infty$.

Construct R_0, the reformulation at the root node, and set $U(R_0) = +\infty$.

Include the root node in the list: $\mathcal{L} = \{0\}$.

Main Step(Iteration k)
Step 1. Termination:

Set the upper bound $U^k := \max_{i \in \mathcal{L}}\{U(R_i)\}$.

Set $\mathcal{L} = \mathcal{L}\setminus\{i\}$ for all i with $U(R_i) \leq L^k$.

3.9. A BRANCH-AND-BOUND ALGORITHM

If $\mathcal{L} = \phi$ stop, the current best solution is optimal. Else, set $k = k+1$, $U^k = U^{k-1}$ and $L^k = L^{k-1}$.

Step 2. Node Selection:

Select node i from $\mathcal{L}$ according to some node selection rule.

Set $\mathcal{L} = \mathcal{L}\setminus\{i\}$.

Step 3. Node Tightening:

Get tight bounds on the fractional terms.

Reconstruct R_i using these bounds.

Step 4. Upper Bounding:

If the relaxation of R_i is desired to be solved, solve it and obtain $U(R_i)$ and go to Step 5. Else, go to Step 6.

Step 5. Lower Bounding:

If $U(R_i) \leq L^k$, go to Step 1.

Use a heuristic method to construct a feasible solution and update L^k. Record the best known solution.

Step 6. Branching:

Use a branching strategy to obtain a set of new problems $R_{i_1}, \ldots, R_{i_q}$ from R_i.

Update node list $\mathcal{L} = \mathcal{L} \cup \{i_1, i_2, \ldots, i_q\}$.

Go to Step 1.

As mentioned earlier, the key difference between the above algorithm and standard branch-and-bound schemes for global optimization is in the node tightening step. The next two sections present specific implementations of the algorithm.

3.10 Hyperbolic Programs with Cardinality Constraints

Cardinality constrained hyperbolic programs are of the form:

$$\text{(CCH)} \quad \max \quad \sum_{i=1}^{m} \frac{a_{i0} + \sum_{j=1}^{n} a_{ij} x_j}{b_{i0} + \sum_{j=1}^{n} b_{ij} x_j}$$

$$\text{s.t.} \quad \sum_{j=1}^{n} x_j = p$$

$$x_j \in \{0, 1\} \quad j = 1, \ldots, n.$$

We shall demonstrate the efficacy of the proposed branch-and-bound algorithm for 0−1 hyperbolic programs through a specific implementation for CCH. The applications of cardinality constrained hyperbolic programming program include scheduling common carriers (Saipe 1975) and p-choice facility location which we address in Section 3.11.3.

Saipe (1975) developed a specialized branch-and-bound algorithm for CCH using the upper bound V_{UB} described in Section 3.8, where the upper bound for each fractional term is obtained employing Saipe's algorithm for single ratio fractionals as discussed in Section 3.4. We show next that LR3-LR8 provide better bounds than that used by Saipe (1975).

Corollary 3.30. $V_{\text{CCH}} \leq V_{\text{LR3}} \leq V_{\text{UB}}$. If $V_{\text{CCH}} < V_{\text{UB}}$, then $V_{\text{LR3}} < V_{\text{UB}}$.

Proof. Direct application of Theorem 3.28. □

Corollary 3.31. For CCH, $V_{\text{UB}} = V_{\text{CC}}$ and $V_{\text{CCH}} \leq V_{\text{LR5}} \leq V_{\text{UB}}$.

Proof. The set of feasible solutions given by

$$D = \left\{ x_j \in [0, 1], j = 1, \ldots, n \mid \sum_{j=1}^{n} x_j = p \right\}$$

is a polytope with integral vertices. Hence, $V_{\text{CC}} = V_{\text{UB}}$. Then, it follows from Proposition 3.29 that $V_{\text{CCH}} \leq V_{\text{LR5}} \leq V_{\text{UB}}$. □

The above results establish the dominance of bounds obtained from the reformulations R3 and R5 over V_{UB}. Recall, from Figure 3.1, that R4 and R7 provide tighter bounds than R3, and R6, and R8 provide tighter bounds than R5. Thus, R3-R8 provide tighter bounds than those used by Saipe (1975).

Next, we apply the proposed reformulation technique to a numerical instance of CCH from the literature. We consider Example 3 of Saipe (1975) for which $m = 3, n = 6$ and $p = 3$. The values of the parameters a_{ij} and b_{ij} are shown in Table 3.2. The bounding scheme employed in Saipe (1975) at the root node provides an upper bound of 5.119. The problem was eventually solved to optimality in 7 nodes. The optimal objective value is 4.729. Thus, there is a gap of 8.25% at the root node. Upon applying reformulation R7 to this example with the bounds on the fractional terms determined by Saipe's algorithm for single ratio problems, we find that the linear programming relaxation of the reformulation provides the optimal solution.

	1	2	3	4	5	6
a_{1j}	9	2	3	5	8	7
b_{1j}	6	2	8	2	9	1
a_{2j}	9	8	5	2	1	3
b_{2j}	8	8	7	3	6	1
a_{3j}	9	2	2	4	3	5
b_{3j}	7	2	7	6	4	5

Table 3.2: Data for Example 3 in Saipe (1975)

3.11 Computational Results for Cardinality Constrained Hyperbolic Programs

In this section, we provide detailed computational results for the cardinality constrained hyperbolic programming program CCH. We first compare the bounding strengths of the various reformulation schemes proposed in Section 3.4. Next, we describe an implementation of the proposed branch-and-bound strategy to solve this problem to global optimality. Finally, the proposed algorithm is used to solve a facility location problem.

3.11.1 Comparison of Bounds

To compare the bounds obtained from the various reformulations, we solve relaxations for randomly generated instances of CCH. The coefficients a_{ij} and b_{ij} for problems with different values of m, n and p were generated from a uniform distribution with range $[1, 1000]$. The reformulations were modeled using GAMS (Brook, Kendrick & Meeraus 1988) while the linear programming relaxations were solved using OSL (IBM 1995) and the nonlinear programming relaxations were solved using MINOS (Murtagh & Saunders 1995). The optimal solution of each problem was obtained by solving the MILP reformulation of type R8 using (IBM 1995). The minimum, average and maximum percentage relaxations gaps and the average CPU times, in seconds, for solving the relaxations are compared in Tables 3.3, 3.4, 3.5, and 3.6. The CPU times reported do not include the time spent in deriving bounds on the fractional terms. The averages are computed based upon five instances for each problem type. All computations were carried out on an IBM RS/6000 43P workstation with 64MB RAM.

The theoretical dominance of the various relaxation bounds as discussed in Section 3.8 can be clearly observed in Tables 3.3 and 3.4. For some cases the nonlinear relaxation of R2 could not be solved to optimality owing to numerical difficulties with MINOS. This fact results in the discrepancy between the bounds obtained from R1 and R2. For these cases, the numbers in parenthesis indicate the number of instances solved. For problem $20-20-12$, the nonlinear relaxation of LR6 could not be solved for any of the five instances. The CPU times reflect the relative sizes of the various reformulations as compared in Table 3.1. The relaxations LR2 and LR6 presumably take more time to optimize due to the presence of nonlinearities in these relaxations. From Tables 3.3, 3.4, 3.5 and 3.6, it is clear that even though the time taken to solve LR1-LR8 is more than the linear programming relaxation of Li's reformulation (Li 1994), the significant improvement in the bounds derived using LR1-LR8 justify their use in a branch-and-bound algorithm.

3.11.2 Performance of the Proposed Algorithm

The proposed branch-and-bound algorithm was implemented for CCH using BARON (Sahinidis 1996), which we describe in detail in Chapter 7. In addition to providing several ready to use modules for certain classes of global optimization problems, BARON is capable of solving any global optimization

3.11. COMPUTATIONAL RESULTS FOR CCH PROGRAMS

Problem	Li (1994)				R1				R2			
m-n-p	min	ave	max	CPU s	min	ave	max	CPU s	min	ave	max	CPU s
5-10-3	25.02	61.57	91.34	0.22	0.00	5.03	11.85	0.34	0.00	5.03	11.85	0.28
5-10-5	61.45	92.25	138.82	0.24	2.30	4.34	6.55	0.33	2.08	4.30	6.55	0.40
5-10-7	69.70	118.20	177.85	0.25	0.03	1.28	3.67	0.33	0.03	1.28	3.67	0.36
5-20-8	60.01	155.06	298.81	0.46	3.53	6.66	13.33	0.80	3.51	6.65	13.33	1.19
5-20-10	59.77	165.75	302.13	0.46	2.29	4.81	7.61	0.81	2.29	4.81	7.61	1.16
5-20-12	63.46	172.50	299.45	0.49	1.53	2.97	4.39	0.84	1.53	2.96	4.39	1.11
Problem	R3				R4				R5			
m-n-p	min	ave	max	CPU s	min	ave	max	CPU s	min	ave	max	CPU s
5-10-3	0.43	14.01	24.48	0.32	0.00	4.75	10.62	0.87	0.00	2.83	5.24	0.37
5-10-5	2.83	9.68	14.08	0.37	1.09	3.47	5.76	0.98	0.00	2.65	6.55	0.38
5-10-7	2.70	3.66	5.76	0.38	0.02	1.12	3.35	0.97	0.00	0.82	2.37	0.38
5-20-8	11.20	20.46	34.67	0.92	3.22	6.60	13.33	2.67	1.43	5.25	11.67	0.96
5-20-10	8.77	13.96	22.77	0.91	2.29	4.77	7.61	2.98	1.45	3.63	7.03	0.93
5-20-12	5.85	8.31	12.67	0.94	1.53	2.92	4.39	2.78	1.17	2.19	3.11	0.94
Problem	R6				R7				R8			
m-n-p	min	ave	max	CPU s	min	ave	max	CPU s	min	ave	max	CPU s
5-10-3	0.00	2.83	5.24	0.35	0.00	2.69	4.96	0.36	0.00	1.87	3.21	0.97
5-10-5	0.00	2.65	6.55	0.43	0.00	2.31	4.25	0.39	0.00	1.67	3.08	1.00
5-10-7	0.00	0.81	2.37	0.45	0.04	0.67	1.69	0.37	0.00	0.34	0.92	1.01
5-20-8	1.43	5.25	11.67	1.97	1.30	4.96	11.08	0.99	0.82	4.23	10.30	3.04
5-20-10	1.45	3.63	7.03	2.02	1.30	3.47	7.01	0.97	0.90	2.96	6.18	3.15
5-20-12	1.17	2.19	3.11	1.92	0.89	1.70	2.52	0.94	0.67	1.43	2.27	3.11

Table 3.3: Comparison of various bounds ($m = 5$)

problem as long as the problem specific upper and lower bounding subroutines are supplied by the user. It is the latter feature of this code we build upon here.

In the implementation of the proposed algorithm for 0−1 hyperbolic programs, we have augmented the upper bounding step with a node tightening step. For CCH, tight bounds on the fractional terms were obtained by solving single ratio 0−1 hyperbolic programs using Saipe's algorithm. These bounds were used in turn to construct tight reformulations of the problem at every node of the branch-and-bound tree. The continuous relaxation of the reformulation was then solved to obtain the upper bound. The lower bounding was carried out by heuristically solving CCH using the Genetic Algorithm (GA) described in Haque & Ahmed (1998). In our implementation, the GA was applied to the original formulation H which permits easy representation of the solutions in the form of binary vectors. In addition to fitness-based updating of the solution population, at each iteration of the branch-and-bound algorithm, the best solution to the relaxation was rounded to construct a feasible solution to CCH and appended to the solution population maintained by the GA. Introduction of these solutions considerably improved the performance of the GA heuristic. After a prespecified number (100) of children generation, the GA was terminated and the best solution from the population was used to update the lower bound in the branch-and-bound algorithm. The GA served to provide good solutions early on in the branch-and-bound tree. However, no substantial deterioration of the algorithm was observed without this heuristic lower bounding scheme. In this implementation, the conventional integer programming scheme of branching on the variables taking fractional values was used.

Comparing the average gap and the solution times of the various reformulations for CCH in Tables 3.3 and 3.4, it appears that for these problems, reformulations R5 and R7 are the most effective. We chose to use the reformulation R7 within our branch-and-bound strategy. Consequently, the upper bounding was carried out by solving the linear programming relaxation of R7 by CPLEX 6.0 (ILOG 1997).

To demonstrate the advantages of node tightening, the proposed algorithm has been compared to solving the MIP reformulation R7 derived at the root node using the commercial MIP solver of CPLEX 6.0 (ILOG 1997). The comparison was carried out on a set of random test problems generated similarly as in Section 3.11.1. All computations were carried out on an IBM RS/6000 43P workstation with 64MB RAM.

3.11. COMPUTATIONAL RESULTS FOR CCH PROGRAMS

The results of the comparison are presented in Tables 3.7, 3.8 3.9 and 3.10. For each of the five instances of the same problem size, the tables show the total number of nodes explored, the node at which the optimal solution was found, the maximum number of nodes held in memory during the search and the total CPU seconds required for both algorithms. In all cases, the proposed algorithm required fewer nodes than CPLEX. The differences were much greater for problems with a larger number of binary variables. In several cases, CPLEX was unable to solve the problems to optimality within 1,000,000 iterations. These results demonstrate the importance of node tightening in obtaining tighter bounds for the problems and hence faster convergence.

3.11.3 p-Choice Facility Location

We describe, as an application of CCH, the p-choice facility location problem. In facility location, the attractiveness of a particular site is often measured in terms of the market share associated with it. In p-choice location models (Ghosh & McLafferty 1987, Ghosh, McLafferty & Craig 1995), the market share of a location is measured in terms of the ratio of the utility of that location to the sum of the utilities of all available locations to the consumers. Consider the problem of locating p facilities in n possible locations to service m customer locations with the objective of maximizing market share. Let U_{ij} denote the utility of location j to the customers at i, d_i be the demand at customer location i and w_j be a preferential weight for a particular location j. Then, the problem of determining the set of facility locations S to maximize the weighted market share can be formulated as follows (Haque & Ahmed 1998):

$$\text{(P)} \quad \max \quad \sum_{j \in S} w_j \sum_{i=1}^{m} \frac{U_{ij}}{\sum_{k \in S} U_{ik}} d_i$$
$$\text{s.t.} \quad |S| = p$$
$$S \subseteq \{1, 2, \ldots, n\}$$

Introducing binary variables x_j equal to 1 if location j is chosen and 0 otherwise, the formulation becomes:

$$\text{(P)} \quad \max \quad \sum_{i=1}^{m} \frac{\sum_{j=1}^{n} U_{ij} w_j d_i x_j}{\sum_{j=1}^{n} U_{ij} x_j}$$

$$\text{s.t.} \quad \sum_{j=1}^{n} x_j = p$$
$$x_j \in \{0,1\} \quad j = 1, 2, \ldots, n$$

The proposed algorithm was applied to the problem of locating a set of ten restaurant franchises in the city of Edmonton, Canada, using the formulation P. The city of Edmonton is divided into 886 enumeration areas (EA), or zones for Federal Census purposes. Using geographic information system (GIS) data, the centroids of each of these zones and their inter-spatial distances were calculated. 100 of the most populated EA centroids were chosen as demand points and 58 of these as candidate locations for restaurants. The criterion for choosing these sites was the existence of at least three other restaurants at the location. For each of the demand points, the total expenditure on existing restaurants was assumed to represent the demand for that location. The utility of a prospective location to a customer at a particular demand point was estimated using a multiplicative competition interaction criteria (Nakanishi & Cooper 1974) as a function of distance, accessibility, and a number of attractiveness factors. The preferential weight of a location was estimated using the index of retail saturation (Ghosh & McLafferty 1987) which is a function of population and per capita expenditure. Detailed description of the data for this problem is presented in Haque & Ahmed (1998).

Using the above data, the resulting p-choice model consists of a sum of 100 fractional terms, 58 binary variables and a cardinality requirement of ten. Using reformulation R7 the resulting MIP has 5,900 continuous variables, 58 binary variables, and 23,501 constraints. The problem was solved to global optimality using the proposed algorithm on a single processor of an IBM RS/6000 SP2 with 512MB RAM in 5.5 hours. 97% of the CPU time was consumed by CPLEX 6.0 in solving the linear programming relaxations. In particular, 10% of the total CPU time was spent in solving the linear programming relaxation at the root node. The linear programming relaxation gap at the root node was 4.34%. Global optimality was proven in a total of 79 nodes with at most nine nodes stored in memory at any point during the search. The optimal solution was located at the tenth node. To the best of our knowledge this algorithm was the first to solve this problem to global optimality.

3.11. COMPUTATIONAL RESULTS FOR CCH PROGRAMS   117

Problem	Li (1994)				R1				R2			
m-n-p	min	ave	max	CPU s	min	ave	max	CPU s	min	ave	max	CPU s
10-10-3	62.13	90.27	151.22	0.44	0.64	7.70	14.69	0.77	0.59	7.67	14.60	1.30
10-10-5	68.47	109.22	170.35	0.45	2.96	4.01	5.59	0.82	2.94	4.00	5.59	1.19
10-10-7	88.64	134.67	185.09	0.45	0.73	1.27	1.57	0.78	0.68	1.22	1.55	1.02
10-20-8	117.95	137.90	169.09	0.93	8.08	14.76	22.60	2.69	8.08	14.76	22.60	7.99
10-20-10	116.83	142.34	164.42	0.99	5.38	10.01	16.29	2.73	5.38	10.01	16.29	9.10
10-20-12	114.88	146.84	174.72	0.96	4.01	5.38	8.95	2.62	4.00	5.55	8.95	5.52

Problem	R3				R4				R5			
m-n-p	min	ave	max	CPU s	min	ave	max	CPU s	min	ave	max	CPU s
10-10-3	12.42	24.28	37.02	0.90	0.30	7.39	14.21	2.46	0.00	5.53	11.60	0.91
10-10-5	12.02	14.56	16.24	1.01	2.95	4.00	5.58	2.45	0.44	1.91	4.12	0.95
10-10-7	3.01	3.89	5.30	0.90	0.59	1.22	1.48	2.72	0.37	0.64	1.06	0.85
10-20-8	24.75	33.60	44.20	2.87	8.07	14.74	22.57	10.11	7.04	13.57	20.48	2.78
10-20-10	15.78	21.46	28.94	2.95	5.19	9.93	16.29	11.23	4.06	8.95	14.86	3.02
10-20-12	9.80	12.08	15.86	2.94	3.74	5.24	8.95	11.30	2.55	4.51	7.54	2.78

Problem	R6				R7				R8			
m-n-p	min	ave	max	CPU s	min	ave	max	CPU s	min	ave	max	CPU s
10-10-3	0.00	5.53	11.59	1.52	0.00	5.24	11.72	1.00	0.00	4.05	9.15	2.61
10-10-5	0.44	1.91	4.12	1.48	0.03	1.35	3.33	1.24	0.00	1.20	3.01	2.97
10-10-7	0.33	0.60	1.05	1.46	0.08	0.63	1.16	0.84	0.07	0.38	0.90	2.88
10-20-8	7.04	13.57	20.48	8.69	6.80	13.12	20.11	3.22	6.08	12.22	18.96	11.27
10-20-10	4.06	8.95	14.86	10.82	4.08	8.46	14.50	3.19	3.57	7.76	13.56	12.10
10-20-12	2.55	4.51	7.54	8.70	2.41	4.27	7.56	3.10	2.13	3.85	6.85	11.27

Table 3.4: Comparison of various bounds ($m = 10$)

Problem	Li (1994)			R1				R2			
m-n-p	min	ave	max	CPU s	min	ave	max	CPU s	min	ave	max	CPU s
15-10-3	96.23	191.06	214.77	1.74	8.58	8.83	9.82	4.93	8.58	8.83	9.81	19.59
15-10-5	109.39	109.95	110.08	0.74	6.77	6.94	6.98	1.50	6.77	6.94	6.98	3.43
15-10-7	93.85	153.04	324.63	0.73	0.55	2.31	6.77	1.40	0.52	2.30	6.77	2.55
15-20-8	140.33	167.02	205.17	1.69	16.27	23.00	31.28	5.59	22.52	26.90	31.28	34.77[2]
15-20-10	132.53	166.85	211.20	1.80	10.18	13.83	18.87	6.70	10.18	10.34	10.49	33.19[2]
15-20-12	130.00	167.63	214.77	2.05	3.38	6.47	8.58	5.66	3.38	6.08	8.58	18.21[4]

Problem	R3				R4				R5			
m-n-p	min	ave	max	CPU s	min	ave	max	CPU s	min	ave	max	CPU s
15-10-3	19.18	22.73	36.93	4.72	8.58	8.80	9.68	21.49	6.84	7.57	7.75	4.63
15-10-5	16.23	16.40	16.44	1.70	6.76	6.93	6.98	5.74	4.98	5.27	5.34	1.52
15-10-7	2.19	6.25	16.23	1.77	0.21	2.22	6.76	6.15	0.00	1.30	4.98	1.61
15-20-8	37.92	44.19	55.47	5.82	16.26	22.98	31.28	22.50	13.78	19.89	27.19	5.33
15-20-10	21.76	27.25	34.34	6.12	10.18	13.83	18.87	27.75	8.50	12.21	16.74	5.85
15-20-12	10.49	14.87	19.18	5.90	3.38	6.47	8.58	26.41	2.84	5.67	7.75	5.48

Problem	R6				R7				R8			
m-n-p	min	ave	max	CPU s	min	ave	max	CPU s	min	ave	max	CPU s
15-10-3	6.84	7.57	7.75	23.22	7.75	8.42	8.58	6.19	6.14	7.43	7.75	26.01
15-10-5	4.98	5.27	5.34	3.88	4.71	4.96	5.02	1.98	4.52	4.59	4.60	6.54
15-10-7	0.00	1.30	4.98	3.58	0.00	1.30	4.71	1.66	0.00	1.09	4.52	6.67
15-20-8	13.78	19.89	27.19	20.67	15.26	20.04	26.40	6.82	13.29	18.87	25.68	25.75
15-20-10	8.50	12.20	16.74	25.08	8.53	12.25	16.58	6.99	8.24	11.69	15.90	27.50
15-20-12	2.84	5.66	7.75	24.84	2.67	5.87	8.58	6.25	2.22	5.21	7.29	26.88

Table 3.5: Comparison of various bounds ($m = 15$)

3.11. COMPUTATIONAL RESULTS FOR CCH PROGRAMS 119

Problem	Li (1994)				R1				R2			
m-n-p	min	ave	max	CPU s	min	ave	max	CPU s	min	ave	max	CPU s
20-10-3	79.81	85.89	91.19	0.97	0.30	7.37	13.47	2.08	0.25	7.33	13.40	4.77
20-10-5	92.94	125.77	203.86	0.91	7.56	8.59	9.51	2.56	7.56	8.58	9.50	6.77
20-10-7	95.48	140.33	223.93	1.07	1.82	2.15	2.81	2.27	1.76	2.11	2.73	4.20
20-20-8	135.65	204.11	320.16	2.67	14.67	23.19	27.86	10.36	14.67	19.94	25.21	58.74[2]
20-20-10	142.09	206.42	327.37	2.57	10.57	16.21	20.63	11.98	10.57	10.57	10.57	48.86[1]
20-20-12	147.12	207.51	331.69	2.96	6.14	8.51	9.90	10.86	9.49	9.49	9.49	32.72[1]

Problem	R3				R4				R5			
m-n-p	min	ave	max	CPU s	min	ave	max	CPU s	min	ave	max	CPU s
20-10-3	19.13	27.52	36.75	2.30	0.30	7.26	12.97	8.01	0.00	5.58	12.14	2.46
20-10-5	16.06	17.18	19.72	2.68	7.56	8.50	9.50	10.17	5.70	6.73	7.67	2.63
20-10-7	4.15	4.95	5.79	2.55	1.71	2.07	2.74	10.24	0.97	1.41	1.86	2.48
20-20-8	35.80	46.34	51.98	10.00	14.67	23.17	27.86	44.06	12.20	19.82	24.88	9.03
20-20-10	23.46	29.41	32.97	9.80	10.57	16.21	20.63	47.78	9.03	14.07	18.45	10.22
20-20-12	13.66	16.70	18.45	10.15	6.14	8.51	9.90	48.96	5.14	7.47	8.84	11.11

Problem	R6				R7				R8			
m-n-p	min	ave	max	CPU s	min	ave	max	CPU s	min	ave	max	CPU s
20-10-3	0.00	5.57	12.08	6.05	0.00	5.80	11.50	2.72	0.00	4.72	10.20	9.97
20-10-5	5.70	6.73	7.67	8.29	4.65	6.51	7.69	2.94	4.12	5.82	6.63	11.75
20-10-7	0.96	1.41	1.86	6.27	1.08	1.41	1.93	2.52	0.62	1.04	1.43	12.72
20-20-8	12.20	19.82	24.88	34.19	12.17	20.32	25.17	11.73	11.36	19.09	23.58	50.56
20-20-10	9.03	14.07	18.45	42.85	8.86	14.31	18.44	12.40	8.34	13.85	18.21	54.64
20-20-12	—	—	—	—	5.05	7.65	9.27	12.81	4.71	7.06	8.55	54.15

Table 3.6: Comparison of various bounds ($m = 20$)

Problem		CPLEX 6.0				BARON			
m-n-p	No.	N_{tot}	N_{opt}	N_{mem}	CPU s	N_{tot}	N_{opt}	N_{mem}	CPU s
5-10-3	1	1	1	1	0.12	1	1	1	0.20
5-10-3	2	16	7	13	0.28	7	1	3	0.70
5-10-3	3	18	5	17	0.23	3	1	2	0.40
5-10-3	4	12	8	11	0.20	3	1	2	0.50
5-10-3	5	1	1	1	0.14	1	1	1	0.20
5-10-5	1	1	1	1	0.16	1	1	1	0.20
5-10-5	2	10	6	5	0.23	5	1	2	0.60
5-10-5	3	8	6	7	0.20	5	1	2	0.50
5-10-5	4	17	9	17	0.23	7	1	3	0.60
5-10-5	5	1	1	1	0.16	1	1	1	0.20
5-10-7	1	3	3	3	0.17	1	1	1	0.30
5-10-7	2	9	7	9	0.22	5	1	2	0.50
5-10-7	3	7	3	2	0.20	3	2	2	0.50
5-10-7	4	5	5	5	0.14	1	1	1	0.30
5-10-7	5	9	6	8	0.22	3	1	2	0.50

Table 3.7: Computational results for CCH ($m = 5$)

3.11. COMPUTATIONAL RESULTS FOR CCH PROGRAMS

Problem		CPLEX 6.0				BARON			
m-n-p	No.	N_{tot}	N_{opt}	N_{mem}	CPU s	N_{tot}	N_{opt}	N_{mem}	CPU s
5-20-8	1	127	83	125	2.89	15	1	5	2.40
5-20-8	2	24	20	22	0.66	5	4	2	1.10
5-20-8	3	29	22	25	1.05	5	1	2	1.20
5-20-8	4	15	12	13	0.53	3	2	2	1.00
5-20-8	5	73	34	60	1.43	21	1	4	2.90
5-20-10	1	103	73	83	2.91	21	1	6	3.10
5-20-10	2	29	17	28	0.77	3	1	2	1.10
5-20-10	3	58	45	46	1.40	15	12	4	2.60
5-20-10	4	23	12	22	0.67	5	4	2	1.20
5-20-10	5	41	36	41	1.13	7	1	2	1.50
5-20-12	1	42	37	40	1.21	9	1	2	1.60
5-20-12	2	17	11	15	0.75	5	4	2	1.00
5-20-12	3	36	29	36	1.18	7	6	3	1.50
5-20-12	4	17	13	17	0.79	5	1	2	1.20
5-20-12	5	40	19	39	1.07	11	1	4	1.90
5-50-23	1	45949	30293	30595	2185	574	145	108	222
5-50-23	2	702881	187494	344341	28247	7715	10	1401	2988
5-50-23	3	40042	32275	37739	2536	667	10	130	241
5-50-23	4	—	—	—	—	6938	5164	1457	2488
5-50-23	5	615930	179449	179471	31977	2790	622	448	940
5-50-25	1	69187	33848	54399	3020	975	10	169	374
5-50-25	2	506225	125633	184451	19372	6959	6728	1230	2534
5-50-25	3	32846	24686	32001	1905	877	19	150	329
5-50-25	4	—	—	—	—	4349	28	714	1609
5-50-25	5	—	—	—	—	5261	1788	946	1943
5-50-27	1	36609	18897	36609	1622	641	19	112	233.10
5-50-27	2	203792	44385	192151	7328	4813	37	816	1702
5-50-27	3	23602	15465	20149	1343	637	1	115	251
5-50-27	4	—	—	—	—	5079	3420	893	1803
5-50-27	5	—	—	—	—	11811	19	1909	3704

Table 3.8: Computational results for CCH ($m = 5$)

Problem		CPLEX 6.0				BARON			
m-n-p	No.	N_{tot}	N_{opt}	N_{mem}	CPU s	N_{tot}	N_{opt}	N_{mem}	CPU s
10-10-3	1	9	6	8	0.55	5	1	2	1.20
10-10-3	2	1	1	1	0.34	1	1	1	0.50
10-10-3	3	19	10	18	0.94	7	1	2	1.60
10-10-3	4	15	8	15	0.88	7	1	3	1.50
10-10-3	5	2	2	2	0.37	1	1	1	0.60
10-10-5	1	8	5	6	0.60	5	1	2	1.40
10-10-5	2	16	12	15	0.95	3	1	2	0.80
10-10-5	3	35	21	35	1.42	7	1	2	1.50
10-10-5	4	11	7	10	0.72	5	1	2	1.30
10-10-5	5	9	7	8	0.62	3	1	2	0.90
10-10-7	1	24	18	21	0.73	7	1	3	1.40
10-10-7	2	9	7	8	0.58	1	1	1	0.60
10-10-7	3	14	5	4	0.57	3	1	2	1.10
10-10-7	4	6	5	6	0.38	1	1	1	0.70
10-10-7	5	8	6	7	0.57	5	1	2	1.20
10-20-8	1	1521	534	1518	104.91	234	231	54	70.50
10-20-8	2	616	310	597	58.28	85	54	17	31.30
10-20-8	3	60	44	55	7.38	13	1	4	5.90
10-20-8	4	663	389	636	54.06	99	1	19	35.00
10-20-8	5	145	133	144	14.66	17	1	4	6.90
10-20-10	1	1804	558	735	121.68	247	1	49	75.80
10-20-10	2	185	83	84	16.51	39	1	11	14.80
10-20-10	3	87	74	86	10.04	13	1	4	5.50
10-20-10	4	452	219	451	39.09	61	1	13	24.30
10-20-10	5	216	168	204	20.99	21	14	6	8.90
10-20-12	1	611	200	611	45.29	79	24	15	31.30
10-20-12	2	59	34	35	6.82	17	1	6	7.00
10-20-12	3	79	45	79	6.86	19	1	4	7.10
10-20-12	4	104	65	101	10.67	27	1	6	10.80
10-20-12	5	96	42	95	7.42	31	31	7	11.20

Table 3.9: Computational results for CCH ($m = 10$)

3.11. COMPUTATIONAL RESULTS FOR CCH PROGRAMS

Problem m-n-p	No.	CPLEX 6.0				BARON			
		N_{tot}	N_{opt}	N_{mem}	CPU s	N_{tot}	N_{opt}	N_{mem}	CPU s
20-20-8	2	2373	770	1057	536	267	228	55	246
20-20-8	3	2430	783	739	524	229	1	50	242
20-20-8	4	598	160	150	167	103	32	26	98
20-20-8	5	1072	441	1004	304	117	116	28	120
20-20-10	1	3080	1177	1575	782	299	248	60	297
20-20-10	2	1668	813	1658	489	177	76	36	170
20-20-10	3	6988	2332	3431	1415	565	1	111	533
20-20-10	4	670	201	181	176	93	1	22	85
20-20-10	5	1293	671	1192	400	125	1	31	128
20-20-12	1	1456	476	508	383	189	1	35	197
20-20-12	2	518	266	285	164	99	94	16	95
20-20-12	3	2381	780	2326	552	251	30	46	246
20-20-12	4	349	151	330	113	45	34	10	49
20-20-12	5	666	277	632	207	85	1	18	83
20-30-12	1	178111	67041	86196	55702	3689	1	715	5548
20-30-12	2	56521	19253	33484	20289	1241	1	229	2073
20-30-12	3	68910	20357	13104	24720	1727	1	357	2344
20-30-12	4	248326	75427	117556	76551	4643	1	905	7593
20-30-12	5	41233	13527	41233	14523	1302	995	280	2178
20-30-14	1	173519	59322	91570	56777	3841	3252	765	5510
20-30-14	2	59084	21858	57927	23099	1636	1613	317	2361
20-30-14	3	47991	20725	47076	22763	1135	1	216	1811
20-30-14	4	141853	62496	72827	53159	2489	2278	473	4127
20-30-14	5	92518	36266	44205	33678	2351	1	455	3476
20-30-16	1	104255	44747	99248	41081	3187	271	611	4547
20-30-16	2	27097	8781	10772	11057	1033	1	193	1493
20-30-16	3	22456	9307	18964	10175	585	520	102	1087
20-30-16	4	55943	22708	29132	22709	1501	1	265	2521
20-30-16	5	38832	16381	36904	17047	1353	1	256	1912

Table 3.10: Computational results for CCH ($m = 20$)

Chapter 4

Relaxations of Factorable Programs

Synopsis

In Chapter 2, we developed a theory of convex extensions and formalized a technique for building tight relaxations. However, the size of the resulting relaxations can be exponential in the number of variables. In this chapter, we present a slightly modified version of the factorable programming technique due to McCormick (1976) that, when used in conjunction with our relaxation techniques, constructs relaxations that are tight as well as manageable in size and can be generated in an automated fashion. To enable the use of efficient LP software, in Section 4.2 we build linear programming relaxations of the nonlinear convex relaxations using the sandwich algorithm (Rote 1992).

4.1 Recursive Decomposition and Relaxation Construction

Factorable functions are defined as functions that can be formed by taking recursive sums and products of univariate functions. The factorable programming problem is formally stated as:

(FP)
$$\min \ f(x,y)$$
$$\text{s.t.} \ g(x,y) \leq 0$$
$$x \in X \subseteq \mathbb{Z}^p$$

$$y \in Y \subseteq \mathbb{R}^n$$

where $f : (X, Y) \mapsto \mathbb{R}$ and $g : (X, Y) \mapsto \mathbb{R}^m$ and $f(x, y)$ and $g_i(x, y)$, $i = 1, \ldots, m$ are factorable functions. The technique we use for relaxing FP is, like most other prevalent ones, a variant of the relaxation strategies developed in the seminal work of McCormick on separable and factorable programming (McCormick 1972, McCormick 1976, McCormick 1983). We shall, in the sequel, describe our relaxation technique and derive properties that aid in the construction of the relaxation, prove its tightness relative to other schemes or point out generalizations and extensions to the proposed scheme. Let us start with the description of Relax, a recursive algorithm that performs the initial decomposition of factorable functions in terms of sums and products of univariate functions.

Algorithm Relax f(x)

If $f(x)$ is a function of single variable x_i then
    Construct underestimators and overestimators for $f(x)$ over $[x_i^l, x_i^u]$,
else if $f(x) = g(x)/h(x)$, then
    Fractional_Relax (f, g, h)
    end of if
else if $f(x) = \prod_{i=1}^{l} f_i(x)$, then
    for $i := 1$ to l do
        Introduce variable y_{f_i}, such that $y_{f_i} = $ Relax $f_i(x)$
    end of for
    Introduce variable y_f, such that $y_f = $ Multilinear_Relax $\prod_{i=1}^{l} y_{f_i}$
else if $f(x) = \sum_{i=1}^{l} f_i(x)$, then
    for $i := 1$ to l do
        Introduce variable y_{f_i}, such that $y_{f_i} = $ Relax $f_i(x)$
    end of for
    Introduce variable y_f, such that $y_f = \sum_{i=1}^{l} y_{f_i}$
else if $f(x) = g(h(x))$, then
    Introduce variable $y_h = $ Relax $h(x)$
    Introduce variable $y_f = $ Relax $g(y_h)$
end of if

Note that the decomposition of $f(x)$ as a product/sum of functions, may be chosen so that there is a unique univariate function of each variable x_i. Such

4.1. NONLINEAR RELAXATION CONSTRUCTION

decompositions provide tighter relaxations. The introduction of a variable in Algorithm Relax is done only if the same relation was not introduced earlier. Otherwise, a unification operation identifies the earlier variable with all subsequent uses of the new variable.

Before we detail Fractional_Relax and Multilinear_Relax, let us take a moment to analyze the tightness of relaxation of $f(x)$ expressible as a composition of functions, $g(h(x))$. The following result appears in Chapter 18 (page 389) of McCormick (1983).

Theorem 4.1 (McCormick 1983). *Consider a function $f(x) = g(h(x))$ where $h : \mathbb{R}^n \mapsto \mathbb{R}$ and $g : \mathbb{R} \mapsto \mathbb{R}$. Let $S \subseteq \mathbb{R}^n$ be a convex domain of h and let $a \leq h(x) \leq b$ over S. Assume $c(x)$ and $C(x)$ are, respectively, a convex underestimator and a concave overestimator of $h(x)$ over S. Also, let $e(\cdot)$ and $E(\cdot)$ be the convex and concave envelopes of $g(h)$ for $h \in [a, b]$. Let $h_{\min}$ and $h_{\max}$ be such that $g(h_{\min}) = \inf_{h \in [a,b]}(g(h))$ and $g(h_{\max}) = \sup_{h \in [a,b]}(g(h))$. Then,*
$$e\left(\mathrm{mid}\left[c(x), C(x), h_{\min}\right]\right)$$
is a convex underestimating function and
$$E\left(\mathrm{mid}\left[c(x), C(x), h_{\max}\right]\right)$$
is a concave overestimating function for $f(x)$.

We argue that the simple scheme included in Algorithm Relax implies the above underestimators and overestimators as long as $e(\cdot)$ and $E(\cdot)$ are used as the underestimators and overestimators for $g(y_h)$. Consider the $n+2$ dimensional space of points of the form (x, h, f). Construct the surface $M = \{(x, h, f) \mid f = g(h)\}$, the set $F = \{(x, h, f) \mid e(h) \leq f \leq E(h)\}$, and the set $H = \{(x, h, f) \mid c(x) \leq h \leq C(x), x \in S\}$. Then, the feasible region as constructed by Algorithm Relax is the projection of $F \cap H$ on the space of x and f variables assuming $h(x)$ does not occur elsewhere in the factorable program. The resulting set is convex since intersection and projection operations preserve convexity. For a given x, let us characterize the set of feasible points, (x, f). Note that $f \leq e(h)$ for all $h \in [c(x), C(x)]$. If $h_{\min} \leq c(x)$, the minimum value of $e(h)$ is attained at $c(x)$, by quasi-convexity of e. Similarly, if $c(x) \leq h_{\min} \leq C(x)$, then the minimum is attained at $h_{\min}$ and, if $C(x) \leq h_{\min}$, then the minimum is attained at $C(x)$. The resulting function is exactly the same as described in Theorem 4.1. The overestimating function follows by a similar argument. We have thus proven the following result:

Theorem 4.2. *The underestimator and overestimator described in Theorem 4.1 are implied in the relaxation derived through the application of Algorithm Relax as long as $e(\cdot)$ and $E(\cdot)$ are used as the underestimators and overestimators for $g(y_h)$.*

We chose to relax the composition of functions in the way described in Algorithm Relax instead of constructing underestimators based on Theorem 4.1, because our procedure is simpler to implement, the resulting relaxations are differentiable and can be tighter if $h(x)$ occurs elsewhere in the mathematical program FP. However, the scheme implied in Theorem 4.1 has an advantage in that it introduces one variable lesser than that in Algorithm Relax for every composite function relaxed by the algorithm.

We shall now concentrate on the relaxation of the multilinear term $y = \prod_{i=1}^{l} y_{f_i}$. Various relaxation strategies for multilinear terms over a hypercube have been proposed by Crama (1993), Rikun (1997), Sherali (1997), and Ryoo & Sahinidis (2001). We employ a recursive interval arithmetic scheme for this purpose:

Algorithm Multilinear_Relax $\prod_{i=1}^{l} \mathbf{y_{r_i}}$ (Recursive Arithmetic)
for $i := 2$ to l do
    Introduce variable $y_{r_1,\ldots,r_i} = $ Bilinear_Relax $y_{r_1,\ldots,r_{i-1}} y_{r_i}$
end of for

The fractional term is relaxed using Algorithm Fractional_Relax. Two different ways to relax the fractional term were developed by Tawarmalani et al. (2002b), Tawarmalani et al. (2002a), and Tawarmalani & Sahinidis (2001) and were presented in Chapters 2 and 3. The technique of Chapter 3 is detailed below:

Algorithm Fractional_Relax $(\mathbf{f, g, h})$ (Convex Extensions)
Introduce variable y_f
Introduce variable $y_g = $ Relax $g(x)$
If $h(x) = \sum_{i=1}^{l} h_i(x)$, then
    for $i := 1$ to l do
        Introduce variable $y_{f,h_i} = $ Relax $(y_f h_i(x))$
    end of for
    Introduce relation $y_g = \sum_{i=1,\ldots,l} y_{f,h_i}$

4.1. NONLINEAR RELAXATION CONSTRUCTION

    else
        Introduce variable y_h = Relax $h(x)$
        Introduce relation y_g = Bilinear_Relax $y_f y_h$
    end of if

The bilinear term can be relaxed by first performing a separable reformulation (McCormick 1972) or by directly using the known convex/concave envelopes (McCormick 1983, Al-Khayyal & Falk 1983). Clearly, the semidefinite relaxation resulting from the techniques of Section 2.5 can be used instead of Algorithm Fractional_Relax to relax $h(x)/g(x)$. The reader may remember that we proved in Chapter 2 that the semidefinite relaxation so constructed is guaranteed to be tighter than the one generated by Algorithm Fractional_Relax when h is a linear function.

Algorithm Bilinear_Relax $y_i y_j$ (Convex/Concave Envelope)

Bilinear_Relax $y_i y_j$ $\geq$ $y_i^u y_j + y_j^u y_i - y_i^u y_j^u$
Bilinear_Relax $y_i y_j$ $\geq$ $y_i^l y_j + y_j^l y_i - y_i^l y_j^l$
Bilinear_Relax $y_i y_j$ $\leq$ $y_i^u y_j + y_j^l y_i - y_i^u y_j^l$
Bilinear_Relax $y_i y_j$ $\leq$ $y_i^l y_j + y_j^u y_i - y_i^l y_j^u$

Algorithm Bilinear_Relax $y_i y_j$ (Separable)
Bilinear_Relax $y_i y_j$ = Relax $\left(\frac{1}{4}(y_i + y_j)^2 - \frac{1}{4}(y_i - y_j)^2\right)$

We are now left with the task of underestimating and overestimating the univariate functions. At present, we do not attempt to detect individual functional properties, but decompose the univariate function as a recursive sum and product of monomial, logarithmic, and power terms. This keeps the implementation simple while at the same time includes most of the functions typically encountered in global optimization problems. More functions may be added to those listed above without much difficulty.

Algorithm Univariate_Relax $f(x_j)$ (Recursive Sums and Products)
if $f(x_j) = cx_j^p$ then
    Introduce variable y_f = Monomial_Relax cx_j^p
else if $f(x_j) = cp^{x_j}$ then
    Introduce variable y_f = Power_Relax cp^{x_j}
else if $f(x_j) = c\log(x_j)$ then

Introduce variable y_f = Logarithmic_Relax $c\log(x_j)$
else if $f(x_j) = \prod_{i=1}^{l} f_i(x_j)$, then
    for $i := 1$ to l do
        Introduce variable y_{f_i}, such that y_{f_i} = Relax $f_i(x)$
    end of for
    Introduce variable y_f, such that y_f = Multilinear_Relax $\prod_{i=1}^{l} y_{f_i}$
else if $f(x_j) = \sum_{i=1}^{l} f_i(x_j)$, then
    for $i := 1$ to l do
        Introduce variable y_{f_i}, such that y_{f_i} = Relax $f_i(x_j)$
    end of for
    Introduce variable y_f, such that $y_f = \sum_{i=1}^{l} y_{f_i}$
end of if

We do not discuss Monomial_Relax, Power_Relax and Logarithmic_Relax in detail, as they may be easily derived by constructing the convex/concave envelopes over $[x_j^l, x_j^u]$. Monomial_Relax needs a slightly more careful analysis than the other procedures since the function is not necessarily convex or concave and can be concavoconvex/convexoconcave under certain conditions. We detail next a generic procedure for relaxing concavoconvex functions which with slight modifications also works for convexoconcave functions.

4.1.1 Concavoconvex Functions

A univariate function $f(x_j)$ is defined to be concavoconvex over an interval $[x_j^l, x_j^u]$ if, for some x_j^m, $f(x_j)$ is concave over $[x_j^l, x_j^m]$ and is convex over $[x_j^m, x_j^u]$. As an example, consider x^3 over $[-1, 1]$. We define:

$$\check{f}(x_j) = \begin{cases} f(x_j), & x_j \in [x_j^m, x_j^u]; \\ +\infty, & \text{otherwise.} \end{cases}$$

We are interested in locating a point x_j^ξ with a subdifferential x^* to $f(x_j)$ at x_j^ξ such that the corresponding tangent hyperplane passes through $(x_j^l, f(x_j^l))$ (see Figure 4.1). Let us consider the somewhat more general problem of identifying the set X of all x_j^ξ such that there exists a tangent to $\check{f}(x_j)$ at $x_j^\xi \in X$ that passes through the point $(\overline{x}, \overline{y})$ where $\overline{x} < x_j^m$. X is the set of points x_j^ξ for which there exists an $x^* \in \partial \check{f}(x_j^\xi)$ such that $g^*(x^*) = \overline{y}$, where

4.1. NONLINEAR RELAXATION CONSTRUCTION

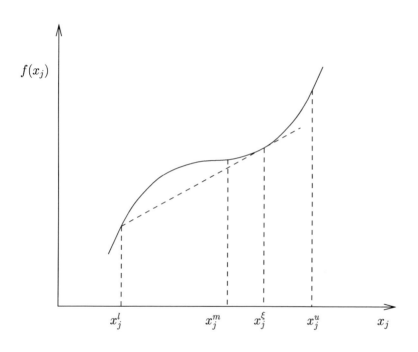

Figure 4.1: Convex envelope for concavoconvex functions

$g(x) = \check{f}(x + \overline{x})$. Inverting, we get $x^* = g^{*-1}(\overline{y})$. Since $x_j - \overline{x} > 0$ for all $x_j \in [x_j^m, x_j^u]$, $x \in \partial g^*(x^*) > 0$. It follows that $g^*(x^*)$ is a strictly increasing function of x^*. Therefore, the inverse g^{*-1} is unique. Since $\partial g^* = (\partial g)^{-1}$,

$$X = (\partial \check{f}^*)(g^{*-1}(\overline{y})) = (\partial \check{f}^*)\Big((\check{f}^* - \langle \cdot, \overline{x} \rangle)^{-1}(\overline{y})\Big).$$

When $f(x)$ is differentiable, it may be simpler to use the gradient expression:

$$\overline{y} - f(x_j) + \langle \nabla f(x_j), \overline{x} - x_j \rangle = 0.$$

Note that the left-hand side of the above equation is an increasing function of x_j and, hence, any numerical technique like the secant method or bisection method may be used for its solution. Consider the function

$$\eta(t) = \int_{x_j^m}^t \overline{y} dx_j - \int_{x_j^m}^t (f(x_j) + \nabla f(x_j)(\overline{x} - x_j)) dx_j.$$

Then, $x_j^\xi = t^*$, where t^* is any optimal solution of

$$\min\{\eta(t) \mid t \in [x_j^m, x_j^u]\}.$$

Similar techniques may be used to derive the concave envelope—negative of the convex envelope of $-f(x)$—and the convex and concave envelopes of a convexoconcave function—convex in $x_j \in [x_j^l, x_j^m]$ and concave in $x_j \in [x_j^m, x_j^u]$.

4.2 Polyhedral Outer-Approximation

The relaxations to factorable programs derived in Section 4.1 are typically nonlinear and include inequalities of the form $g(x) \leq 0$ where $g(x)$ is a convex function. However, nonlinear relaxations are harder to solve and are associated with more numerical issues than their linear counterparts. To develop a linear relaxation, a convex nonlinear programming relaxation of the factorable program is first constructed using the techniques of Section 4.1. The nonlinear relaxation is then further relaxed to a linear one. In order to linearize convex nonlinear relations, an outer-approximation method is required. We investigate the different possibilities in this regard.

4.2. POLYHEDRAL OUTER-APPROXIMATION

Consider a convex function $g(x)$ and an associated convex inequality $g(x) \leq 0$. The subgradient inequality defines a hyperplane at every point $\bar{x}, s \in \partial g(\bar{x})$ such that

$$0 \geq g(x) \geq g(\bar{x}) + \langle s, x - \bar{x} \rangle. \tag{4.1}$$

The primary concern is the identification of points where the subgradients may be constructed.

We are mainly interested in outer-approximating univariate convex functions which lead to the following inequality:

$$\phi(x_j) - y \leq 0$$

(see Figure 4.2). Then, (4.1) reduces to:

$$0 \geq \phi(x_j) - y \geq \phi(\bar{x}_j) + s(x_j - \bar{x}_j) - y.$$

We would like to locate points x_j in the interval $[x_j^l, x_j^u]$ such that the subgradient inequalities approximate the function $\phi(x_j)$ closely.

Typical outer-approximation methods accumulate cuts at the relaxation solution value. The most straightforward way to maintain a constant number of outer-approximators would be to construct subgradients at n equally spaced points between x_j^l and x_j^u. We argue that this is not necessarily a good scheme. Consider, for example, the function:

$$\phi(x_j) = \begin{cases} e^x & x_j \in [0, 5] \\ xe^5 - 4e^5 & x_j \in (5, 1000]. \end{cases}$$

Outer-approximating $\phi(x_j)$ at any point in $[5, 1000]$ generates the same subgradient whereas unique subgradients exist for all points in $[0, 5]$. Since only 0.5% of the equally spaced points lie within $[0, 5]$, outer-approximation with equi-spaced points does not perform well on the above function. The example illustrates that an outer-approximation based on equi-spaced points is deficient in that it does not take into account the nature of the convex function.

The problem of outer-approximating convex functions is closely related to that of polyhedral approximations to convex sets (Gruber & Kenderov 1982, Gruber 1993). It is known that the distance between a planar convex figure and its best approximating $n-$gon is $O(1/n^2)$ under various error measures including the Hausdorff norm and the area of the symmetric difference. For

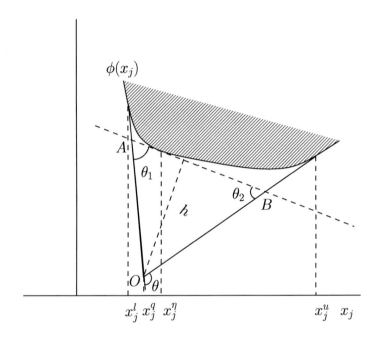

Figure 4.2: Outer-approximation of convex functions

4.2. POLYHEDRAL OUTER-APPROXIMATION

convex functions, with vertical distance as the error measure, the convergence rate of the error of the approximation can be shown to be at best $\Theta(1/n^2)$ by considering a parabola $y = x^2$ (Rote 1992).

The sandwich algorithm is a template of outer-approximation schemes and has been shown to attain the above performance guarantee in a variety of cases (Burkard, Hamacher & Rote 1992, Rote 1992). At a given iteration, this algorithm begins with a number of points at which tangential outer-approximations of the convex function have been constructed. Then, at every iterative step, the algorithm identifies the interval with the maximum outer-approximation error and subdivides it at a suitably chosen point. Various strategies have been proposed for identifying such a point. Five commonly proposed strategies are:

1. *Interval bisection:* bisect the chosen interval (see Figure 4.3).

2. *Slope bisection:* find the supporting line with a slope that is the mean of the slopes at the end points (see Figure 4.4).

3. *Maximum error rule:* Construct the supporting line at the x − ordinate of the point of intersection of the supporting lines at the two end points (see Figure 4.5).

4. *Chord rule:* Construct the supporting line with the slope of the linear overestimator of the function (see Figure 4.6).

5. *Angle bisection:* Construct the supporting line with the slope of the angular bisector of the outer angle θ of ROS (see Figure 4.7).

It was shown by Rote (1992) that, if the vertical distance between the convex function and the outer-approximation is taken as the error measure, then the first four procedures lead to optimal performance guarantees. Interval bisection and slope bisection and, similarly, maximum error rule and chord rule, are dual procedures to each other. If x_j^{l*} is the right-hand derivative at x_j^l and x_j^{u*} is the left-hand derivative at x_j^u then, for $n \geq 2$, the interval bisection and slope bisection schemes produce a largest vertical error ϵ such that

$$\epsilon \leq \frac{9(x_j^u - x_j^l)(x_j^{u*} - x_j^{l*})}{8n^2}$$

136   CHAPTER 4. RELAXATIONS OF FACTORABLE PROGRAMS

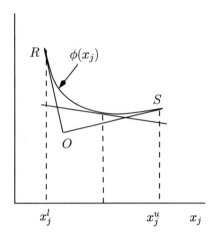

Figure 4.3: Interval bisection in sandwich algorithms

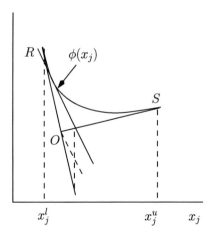

Figure 4.4: Slope bisection in sandwich algorithms

4.2. POLYHEDRAL OUTER-APPROXIMATION

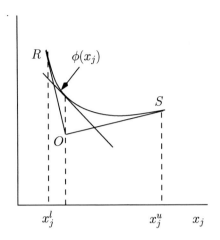

Figure 4.5: Maximum error rule in sandwich algorithms

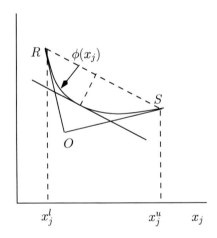

Figure 4.6: Chord rule in sandwich algorithms

138   CHAPTER 4. RELAXATIONS OF FACTORABLE PROGRAMS

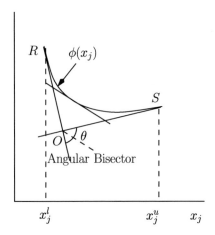

Figure 4.7: Angle bisection in sandwich algorithms

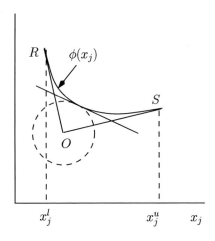

Figure 4.8: Maximum projective error rule in sandwich algorithms

4.2. POLYHEDRAL OUTER-APPROXIMATION

and the maximum error rule and chord rule have a slightly better performance guarantees of

$$\epsilon \leq \frac{(x_j^u - x_j^l)(x_j^{u*} - x_j^{l*})}{n^2}.$$

Note that, since the initial $x_j^{u*} - x_j^{l*}$ can be rather large, the sandwich algorithm may still lead to high approximation errors for low values of n.

Let $\mathcal{G}(f)$ denote the set of points (x, y) such that $y = f(x)$. Projective error is another error measure which is commonly used to describe the efficiency of an approximation scheme. If $\phi^o(x_j)$ is the outer-approximation of $\phi(x_j)$, then

$$\epsilon_p = \sup_{p^{\phi^o} \in \mathcal{G}(\phi^o)} \inf_{p^\phi \in \mathcal{G}(\phi)} \{\|p^\phi - p^{\phi^o}\|\}. \tag{4.2}$$

It was shown by Fruhwirth, Burkard & Rote (1989) that the projective error of the angular bisection scheme decreases quadratically. In fact, if the initial outer angle of the initial function is θ, then

$$\epsilon_p \leq \frac{9\theta(x_j^u - x_j^l)}{8(n-1)^2}.$$

Examples were constructed by Rote (1992) to show that there exist functions for which each of the above outer-approximation schemes perform arbitrarily worse compared to the others. Therefore, the problem of finding the optimal sandwich algorithm for all types of functions is still unanswered.

We suggest another scheme where the supporting line is drawn at the point in $\mathcal{G}(\phi)$ which is closest to the intersection point of the slopes at the two end-points (see Figure 4.8). We call this the *projective error rule*. We are not aware of any function for which the projective error rule will be particularly bad. We show that, using the projective distance as the error measure, the resulting sandwich algorithm also exhibits quadratic convergence. Our proof for this result is similar to the proof for the maximum error case in Rote (1992).

Theorem 4.3. *Consider a convex function $\phi(x_j) : [x_j^l, x_j^u] \mapsto \mathbb{R}$ and the outer-approximation of ϕ formed by the tangents at the x_j^l and x_j^u. Let $R = (x_j^l, \phi(x_j^l))$, $S = (x_j^u, \phi(x_j^u))$ and O be the point where the two tangents meet. Let the angle ROS be $180 - \theta$ and L be the sum of the lengths of RO and OS. Let $k = L\theta/\epsilon_p$. Then, the algorithm needs at most*

$$N(k) = \begin{cases} 0, & k \leq 4; \\ \lceil \sqrt{k} - 2 \rceil, & k > 4, \end{cases}$$

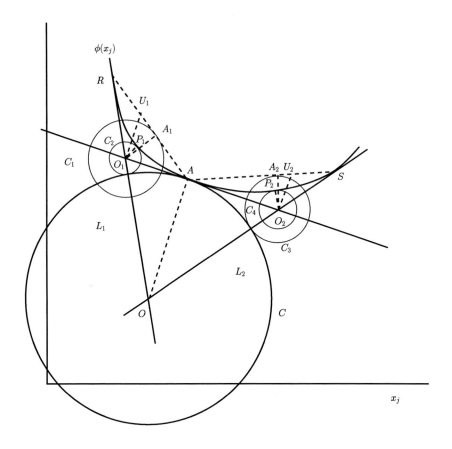

Figure 4.9: Proof of Theorem 4.3

supporting lines.

Proof. We first reason that the supremum in (4.2) is attained at the point O, the intersection of the supporting lines drawn at the end-points of the subinterval under consideration. Since $\phi(x_j)$ is a convex function, as we move from R to S the derivative of ϕ increases, implying that the length of the perpendicular between the point on the curve and RO increases attaining its maximum at A.

Let A and B be any two points. We denote by L_{AB} the length of a line segment that joins points A and B. If A and B lie on the graph of ϕ, and the tangents to ϕ at A and B meet at O, then we define $W_{AB} = L_{AO} + L_{OB}$

4.2. POLYHEDRAL OUTER-APPROXIMATION

and η_{AB} as the angle AOB. Let θ_{AB} denote $180 - \eta_{AB}$.
Note that:

$$(\sqrt{L_{RO}\theta_{RA}} + \sqrt{L_{OS}\theta_{AS}})^2$$
$$\leq (\sqrt{L_{RO}\theta_{RA}} + \sqrt{L_{OS}\theta_{AS}})^2 - (\sqrt{L_{RO}\theta_{AS}} - \sqrt{L_{OS}\theta_{RA}})^2$$
$$= (L_{RO} + L_{OS})(\theta_{RA} + \theta_{AS})$$
$$= W_{RS}\theta_{RS} \tag{4.3}$$

Dividing by ϵ_p and taking square root, we get:

$$\sqrt{L_{RO}\theta_{RA}/\epsilon_p} + \sqrt{L_{OS}\theta_{AS}/\epsilon_p} \leq \sqrt{W_{RS}\theta_{RS}/\epsilon_p}. \tag{4.4}$$

If $[x_j^l, x_j^u]$ is not approximated within ϵ_p, then

$$\epsilon_p < AO \leq L_{O_1 A} \sin(\theta_{RA}) \leq L_{RO}\theta_{RA}. \tag{4.5}$$

Similarly, $\epsilon_p < L_{OS}\theta_{AS}$.

We prove the result by induction. Refer to Figure 4.9. It is easy to show that when $W_{RS}\theta_{RS} \leq 4\epsilon_p$, the curve is approximated within ϵ_p (see (4.4) and (4.5)).

Since $N(\cdot)$ is an increasing function of $(\cdot)$, $W_{RA} \leq L_{RO}$, and $W_{AS} \leq L_{OS}$, it suffices to show that:

$$N(W_{RS}\theta_{RS}/\epsilon_p) \geq \max\{1 + N(L_{RO}\theta_{RA}/\epsilon_p) + N(L_{OS}\theta_{AS}/\epsilon_p)\},$$

where $\theta_{RA} + \theta_{AS} = \theta_{RS}$. Since $\sqrt{L_{RO}\theta_{RA}/\epsilon_p} \geq 1$ and $\sqrt{L_{OS}\theta_{AS}/\epsilon_p} \geq 1$, we can apply the induction hypothesis. We only need to show that:

$$\lceil \sqrt{W_{RS}\theta/\epsilon_p} - 2 \rceil \geq 1 + \max\{\lceil \sqrt{L_{RO}\theta_{RA}/\epsilon_p} - 2 \rceil + \lceil \sqrt{L_{OS}\theta_{AS}/\epsilon_p} - 2 \rceil\}.$$

Since $\lceil c - 2 \rceil \geq 1 + \lceil a - 2 \rceil + \lceil b - 2 \rceil$ when $c \geq a + b$ the result follows. □

The above result shows that the convergence of the approximation gap is quadratic in the number of points used for outer-approximation. The convergence result can be extended to other error-measures such as the area of the symmetric difference between the linear-overestimator and the outer-approximation simply because the area of the symmetric difference grows super-linearly with the projective distance h. Consider, for example, a ring with inner radius r and outer radius $r + h$. Its area is given by $2\pi r h + \pi h^2$. Therefore, as h increases the area grows by $\Theta(h)$. Since any segment of a curve may be treated as a portion of a ring (the curvature of a convex function is always positive), a similar argument shows that the area of the symmetric difference grows by $\Theta(h)$ as h increases. Since h reduces by $O(1/n^2)$, the area of the symmetric difference also reduces by $O(1/n^2)$. However, not much is known about the behavior of these outer-approximation schemes for small values of n.

In order to develop a better understanding of the behavior of outer-approximation schemes for small values of n, we investigate the reduction in the area of the symmetric difference between the curve and the outer-approximation with the introduction of a single supporting line. Our goal is to maximize the area of the triangle AOB (see Figure 4.2), which may be expressed as:

$$\text{Area}(AOB) = \frac{1}{2}h^2(\cot\theta_1 + \cot\theta_2) \qquad (4.6)$$

$$= \frac{1}{2}l(AO)\,l(OB)\operatorname{cosec}(\theta_1 + \theta_2)$$

$$= \frac{1}{2}l(AO)\,l(OB)\operatorname{cosec}(\theta) \qquad (4.7)$$

Note that the area of the triangle is dependent on the square of the projective distance. This motivates us again to employ the projective rule for the selection of the point in the sandwich algorithm. Let us say that, for some line AB, the maximum reduction in area is achieved. Also, let $A_p B_p$ be the line corresponding to the maximum projective distance (see Figure 4.10). Then,

$$\frac{\text{Area}(AOB)}{\text{Area}(A_p O B_p)} = \frac{h_{AB}^2(\cot\theta_1^{AB} + \cot\theta_2^{AB})}{h_{A_p B_p}^2(\cot\theta_1^{A_p B_p} + \cot\theta_2^{A_p B_p})}$$

where h_{AB} and $h_{A_p B_p}$ are the perpendicular distances of AB and $A_p B_p$ with O and θ_1^{AB}, θ_2^{AB}, $\theta_1^{A_p B_p}$, and $\theta_2^{A_p B_p}$ are the angles between AO and AB, AB and OB, $A_p O$ and $A_p B_p$, and $A_p B_p$ and OB_p, respectively. Note that $\theta_1^{AB} + \theta_2^{AB} = \theta_1^{A_p B_p} + \theta_2^{A_p B_p} = \theta$. The minimum value of $\cot\theta_1 + \cot\theta_2$, where $\theta_1 + \theta_2 = \theta$ and θ is a constant less than π, occurs when $\theta_1 = \theta_2 = \theta/2$. This follows from the following fact:

$$\frac{d}{d\theta_1}(\cot\theta_1 + \cot(\theta - \theta_1))$$

4.2. POLYHEDRAL OUTER-APPROXIMATION

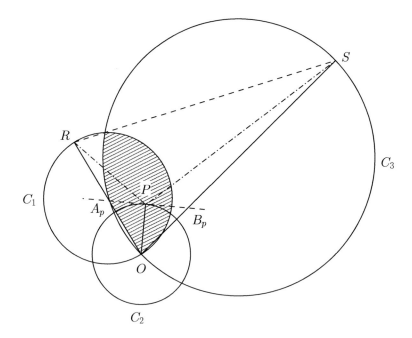

Figure 4.10: Analysis of outer-approximation for projective rule

$$= -\csc^2 \theta_1 + \csc^2(\theta - \theta_1) \begin{cases} < 0, & \text{if } \theta_1 \in (0, \theta/2); \\ = 0, & \text{if } \theta_1 = \theta/2; \\ > 0, & \text{if } \theta_1 \in (\theta/2, \theta). \end{cases}$$

Since $h_{AB} \leq h_{A_p B_p}$, it follows that

$$\frac{\text{Area}(AOB)}{\text{Area}(A_p O B_p)} \leq \frac{\tan \theta}{2 \tan \frac{\theta}{2}} = \frac{1}{1 - \tan \frac{\theta}{2}}.$$

Hence, a simple bound is available for the maximum reduction in the area using any other scheme. The above bound is good for small values of θ. However, as $\theta \to \pi/2$, the bound goes to infinity. A more careful analysis is needed in such a case for a tighter bound (cf. Figure 4.10).

The point identified by the projective rule must lie in the shaded region as shown in Figure 4.10, which is the intersection of the circles C_1 and C_3 and the current outer-approximation ROS. Here, C_1 and C_3 are drawn so that RO and OS are their diameters respectively. Given any point P in the shaded region, it is possible to construct a convex function—like the polyhedral function RPS— with an outer-approximation ROS and P as the point with the minimum Euclidean distance from O.

Given any point P lying on the graph of the convex function, we know that RPS is an upper approximation of the convex function. Hence, any supporting line for the convex function lies below the supporting line with the same slope for RPS. Now consider any line A_1B_1 passing through P (cf. Figure 4.11). If the line is rotated about the point P, then the additional area removed is

$$\text{Area}(B_1PB_2) - \text{Area}(A_1PA_2) = (L^2_{PB_1} - L^2_{A_1P})\,d\theta_a.$$

It is easy to argue that if a small rotation $d\theta_a$ increases the area of A_1OB_1, then further rotation has the same effect. It follows then that the area is a unimodal function of θ_a. Therefore, the maximum area removed by any line is given as below:

$$\text{Area}(ABC) \leq \max\{\text{Area}(ROB_R), \text{Area}(SOA_S)\}.$$

Tighter bounds on the maximum area removed may be derived through the above expression.

A nice property shared by projective rule, chord rule, and angular bisection is that they are invariant under affine transformations of the axes. The projective rule is actually invariant under all projective transformations. If generalized to higher dimensions, the projective rule provides a method for approximation of convex polyhedral sets that is convergent with a simple upper bound on the number of iterations needed (Boyd 1995).

The disadvantage of using the projective rule is that it is not always easy to derive closed form expressions for the point identified by the rule. In fact, it requires the solution of a small convex problem and is, therefore, harder to implement than the other competing strategies in a practical algorithm.

4.2. POLYHEDRAL OUTER-APPROXIMATION

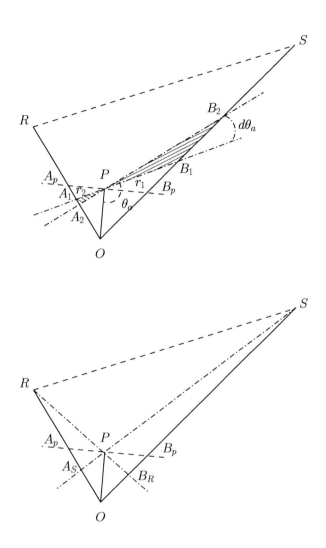

Figure 4.11: Area removed by the supporting line

Chapter 5

Domain Reduction

Synopsis

Domain reduction is the process of eliminating regions from the feasible space if the removal does not affect the convergence of the search process to a global optimum. Domain reduction is also referred to as bounds tightening, domain contraction, and range reduction. Various techniques for domain reduction have been developed by Mangasarian & McLinden (1985), Thakur (1990), Hansen, Jaumard & Lu (1991), Hamed & McCormick (1993), Lamar (1993), Savelsbergh (1994), Andersen & Andersen (1995), Ryoo & Sahinidis (1995), Ryoo & Sahinidis (1996), Shectman & Sahinidis (1998), Zamora & Grossmann (1999). We develop a theory of domain reduction in this chapter and then derive earlier results in the light of these new developments. For example, our results generalize to the nonlinear case, range reduction tests used in the integer linear programming literature.

5.1 Preliminaries

This section overviews the Legendre-Fenchel transform and Lagrangian duality which form the main tools in studies on convex analysis and conjugate duality. For further details, we refer the reader to Rockafellar (1970), Hiriart-Urruty & Lemaréchal (1993a), Hiriart-Urruty & Lemaréchal (1993b), and Rockafellar & Wets (1998).

5.1.1 Legendre-Fenchel Transform

Let $\overline{\mathbb{R}}$ denote the extended real line $(\mathbb{R} \cup +\infty)$. For any $f : \mathbb{R}^n \mapsto \overline{\mathbb{R}}$, the conjugate $f^* : \mathbb{R}^n \mapsto \overline{\mathbb{R}}$ is defined as:

$$f^*(s) = \sup_x \{\langle s, x \rangle - f(x)\}.$$

The mapping $f \mapsto f^*$ is called the Legendre-Fenchel transform. The function $f^{**} = (f^*)^*$ is called the biconjugate of f.

The significance of the Legendre-Fenchel transform becomes evident when a careful analysis of the above definition is made in terms of epigraphical relationships. In the sequel, we shall denote a linear function over x with a slope s and intercept β by $l_{s,\beta}(x)$. Definition of f^* implies

$$\begin{aligned}
(s, \beta) \in \mathrm{epi}(f^*) &\iff \beta \geq \langle s, x \rangle - \alpha \quad \text{for all} \quad (x, \alpha) \in \mathrm{epi}(f) \\
&\iff \alpha \geq \langle s, x \rangle - \beta \quad \text{for all} \quad (x, \alpha) \in \mathrm{epi}(f) \\
&\iff \alpha \geq l_{s,\beta}(x) \quad \text{for all} \quad (x, \alpha) \in \mathrm{epi}(f) \\
&\iff f(x) \geq l_{s,\beta}(x)
\end{aligned}$$

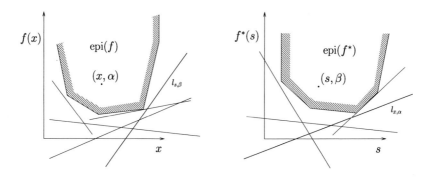

Figure 5.1: Affine minorants of f and f^*

The above relation expresses the epigraph of f^* as the set of all linear minorants of f (see Figure 5.1). Another interpretation may also be derived in the space of linear functionals as follows

$$\begin{aligned}
(s, \beta) \in \mathrm{epi}(f^*) &\iff \beta \geq \langle s, x \rangle - \alpha \quad \text{for all} \quad (x, \alpha) \in \mathrm{epi}(f) \\
\beta \geq f^*(s) &\iff \beta \geq l_{x,\alpha}(v) \quad \text{for all} \quad (x, \alpha) \in \mathrm{epi}(f)
\end{aligned}$$

5.1. PRELIMINARIES

The above relation implies that $f^*(s)$ is the pointwise-supremum of all affine functions $l_{x,\alpha}$ such that $(x,\alpha) \in \text{epi}(f)$. Combining the two interpretations, $f^{**}(x)$ is the pointwise-supremum of all affine functions $l_{s,\beta}$ which are the affine minorants of $f(x)$.

This biconjugate f^{**} characterizes the largest convex function minorizing f under the assumptions that:

> (A1) $f(\cdot) \not\equiv +\infty$ and there is an affine function minorizing f on $\mathbb{R}^n$
> (A2) $f(\cdot)$ is l.s.c. on $\mathbb{R}^n$ ($\liminf_{x^\nu \to x} f(x^\nu) = f(x)$ for all $x \in \mathbb{R}^n$)
> (A3) $f(\cdot)$ is coercive on $\mathbb{R}^n$ ($\liminf_{|x| \to \infty} \frac{f(x)}{|x|} = \infty$)

In order to apply the conjugate transform to functions, it is often easier to work with the functions in "cosmic space." The cosmic space is obtained by adjoining the hyperplane at infinity to an affine space (Rockafellar & Wets 1998). The cosmic space plays a key role in the development of convexity theory. We express the n-dimensional cosmic space as

$$\text{csm}(\mathbb{R}^n) = \mathbb{R}^n \cup \text{hzn}(\mathbb{R}^n)$$

where $\text{hzn}(\mathbb{R}^n)$ denotes the set of directions in $\mathbb{R}^n$.

We provide geometric interpretations of $\text{csm}(\mathbb{R}^n)$ so that the reader is better equipped to appreciate its structure. (For readers familiar with the notion of projective space, the n-dimensional cosmic space is similar to a n-dimensional projective space, the key difference being the lack of orientation in the projective space.)

One way to visualize the cosmic space is to consider the transformation $x \mapsto x/(1+|x|)$. This is equivalent to projecting every point on a unit hemisphere (similar to that in Figure 5.2 modulo scaling). This is sometimes referred to as the celestial model of cosmic space. The dual space can then be adjoined as the other hemisphere. However, it must be noted that the choice of hyperplane at infinity determines the primal and dual space and this choice can be arbitrarily made. Another interpretation is the so called ray space model, which associates with each point $x \in \mathbb{R}^n$, a ray λx, $\lambda > 0$. In Figure 5.2 the celestial representation is given by the projection of the set $C \in \mathbb{R}^n$ on the hemisphere, whereas in the ray space model it is the cone containing C and the origin.

Since each point is transformed into a ray in the cosmic space, the study of convex sets and functions reduces to the study of convex cones.

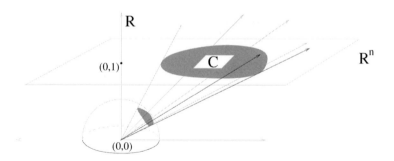

Figure 5.2: The cosmic space

We can now develop an interpretation for the conjugate function in the cosmic space. Consider the cone

$$K_f = \{t(x, \alpha, 1) \mid t > 0, (x, \alpha) \in \text{epi}(f)\}.$$

The polar cone K_f^0 is the set of supporting hyperplanes of K_f. In particular,

$$K_f^0 = \{(s, \lambda, \beta) \mid \langle s, x \rangle + \lambda \alpha + \beta \leq 0 \text{ for all } (x, \alpha) \in \text{epi}(f)\}$$

Imposing $\lambda = -1$, we obtain the hypograph of $-f^*(s)$. For a visual depiction of the relationship between conjugacy and polarity see Figures 5.3 and 5.4.

Next we provide a few results that elucidate important relations between function and its conjugate. Most of these results are classical in the convex analysis literature (Rockafellar 1970).

Proposition 5.1. *For f satisfying (A1) the following are equivalent:*

a. $s^0 \in \partial f(x^0)$

b. $f^*(s^0) + f(x^0) = \langle s^0, x^0 \rangle$

When $\text{cl conv}(f)(x^0) = f(x^0)$ *then the following is also equivalent*

c. $x^0 \in \partial f^*(s^0)$

Also, $\partial f(x^0) \neq \emptyset \Rightarrow \text{cl conv}(f)(x^0) = f(x^0)$.

5.1. PRELIMINARIES

Proof. By rearranging the subgradient inequality, $s^0 \in \partial f(x^0)$ implies that $\langle s^0, x \rangle - f(x) \leq \langle s^0, x^0 \rangle - f(x^0)$. Then, $f^*(s^0) = \langle s^0, x^0 \rangle - f(x^0)$ proves the equivalence of (a) and (b).

Dually, applying the equivalence to $f^*(s)$, we get

$$x^0 \in \partial f^*(s^0) \iff f^*(s^0) + f^{**}(x^0) = \langle s^0, x^0 \rangle$$
$$\iff f^*(s^0) + f(x^0) = \langle s^0, x^0 \rangle$$
$$\iff s^0 \in \partial f(x^0)$$

Also, $s^0 \in \partial f(x^0)$ implies for $l_{s^0}(x) = f(x^0) + \langle s^0, x - x^0 \rangle$, $f(x) \geq l_{s^0}(x)$ and hence cl conv$(f)(x) \geq l_{s^0}(x)$, $(l_{s^0}(x)$ being a convex underestimating function). But, $f(x^0) = l_{s^0}(x^0)$. Therefore, cl conv$(f)(x^0) = l_{s^0}(x^0)$. □

It is not necessarily true that cl conv$(f)(x^0) = f(x^0)$ implies that $\partial f(x^0)$ is a nonempty set. We provide a counterexample in Figure 5.5.

Remark 5.2. *When $f^* = g^*$ and $f(x^0) = g(x^0)$, $\partial f(x^0) = \partial g(x^0)$ since $f(x^0) + f^*(s) = g(x^0) + g^*(s)$. Note that $f^* = g^*$ if and only if cl conv$(f) =$ cl conv(g).*

Proposition 5.3. *Consider a function f satisfying (A1). If f is differentiable at x^0, then $\partial f(x^0) \subseteq \nabla f(x^0)$.*

Proof. The statement is trivially true $\partial f(x^0) = \emptyset$. Assume that $\partial f(x^0) \neq \emptyset$ and let $s \in \partial f(x^0)$. Then,

$$\frac{f(x + td) - f(x)}{t} \geq \langle s, d \rangle$$

Taking the limit as $t \searrow 0$, we get $\langle \nabla f(x), d \rangle \geq \langle s, d \rangle$. Or, $\langle \nabla f(x) - s, d \rangle \geq 0$ for all d. Taking $d = s - \nabla f(x)$, it follows that $\nabla f(x) = s$. □

Proposition 5.1 can be applied to study convex extensions. Consider a function f and $X \subseteq$ ri(dom (f)). Let $x^0 \in X$. We know that $\partial f(x^0) \neq \emptyset$ implies cl conv$(f)(x^0) = f(x^0)$. However, ∂ cl conv$(f)(x^0) \neq \emptyset$ since $X \subset$ ri(conv(dom (f))) (Rockafellar 1970, Theorem 23.4). Also, from Remark 5.2:

$$\text{cl conv}(f)(x^0) = f(x^0) \Rightarrow \partial \text{cl conv}(f)(x^0).$$

Since ∂ cl conv$(f)(x^0) \neq \emptyset$, we have the following equivalence:

$$\partial f(x^0) \neq \emptyset \iff \text{cl conv}(f)(x^0) = f(x^0).$$

$\partial f(x^0) \neq \emptyset$ implies in addition that x^0 is the global minimizer of a linear perturbation of $f(x)$. Hence, Corollary 5.4.

Corollary 5.4. *For a function f satisfying (A1), cl conv$(f)(x^0) = f(x^0)$ if $\partial f(x^0) \neq \emptyset$. If $x^0 \in \text{ri}(\text{conv}(\text{dom}(f)))$, then:*

$$\partial f(x^0) \neq \emptyset \iff \text{cl conv}(f)(x^0) = f(x^0).$$

5.1.2 Lagrangian Relaxation

Consider the *abstract minimization problem* (Rockafellar 1970, Rockafellar & Wets 1998):

$$\inf_{x \in \mathbb{R}^n} \phi(x),$$

where $\phi : \mathbb{R}^n \mapsto \overline{\mathbb{R}}$. Inject $\phi(x)$ in a higher dimensional space by adding perturbation variables $u \in \mathbb{R}^p$ to define a *dualizing parameterization* of ϕ denoted by $\rho(x, u)$ in a manner that $\rho(x, b) = \phi(x)$. We shall mainly be interested in parameterizations that are l.s.c. convex in u (Rockafellar & Wets 1998, pp. 508–509). We further assume that there exists an affine function minorizing $\rho(x, u)$ on $\mathbb{R}^{n+p}$. The *perturbation problem* is defined as $p(u) = \inf_{x \in \mathbb{R}^n} \rho(x, u)$. The Lagrangian function is defined as the negative conjugate of the dualizing parameterization in the u-space through the following relation:

(L) $\qquad l(x, \cdot) = -(\rho(x, \cdot))^* = -\sup_{u}\{\langle \cdot, u \rangle - \rho(x, u)\}.$

The conjugate of the perturbation function and the Lagrangian function are related in a well-defined way.

Theorem 5.5. $p^*(y) = \rho^*(0, y) = -\inf_x l(x, y).$

Proof. The first equality follows from:

$$\begin{aligned}
p^*(y) &= \sup_u\{\langle y, u \rangle - \inf_x \rho(x, u)\} \\
&= \sup_{x,u}\Big\{\langle (0, y), (x, u) \rangle - \rho(x, u)\Big\} \\
&= \rho^*(0, y).
\end{aligned}$$
(5.1)

The second equality follows from:

$$p^*(y) = \sup_u\{\langle y, u \rangle - \inf_x \rho(x, u)\}$$

$$= \sup_{x,u}\{\langle y, u\rangle - \rho(x,u)\}$$
$$= \sup_{x}\left\{\sup_{u}\{\langle y, u\rangle - \rho(x,u)\}\right\}$$
$$= -\inf_{x} l(x, y). \qquad (5.2)$$

$\square$

Figure 5.6 provides a geometric illustration of Lemma 5.5.

5.2 An Iterative Algorithm for Domain Reduction

In this section, we develop an iterative procedure to lower bound the following minimization problem:

$$(\mathrm{M^P}) \qquad \inf_{u}\{h(u) \mid u \in U\}$$

where we do not have an explicit characterization of U available to us. However, we assume that the support function

$$\delta^*(y \mid U) = \sup_{u}\{\langle y, u\rangle \mid u \in U\}$$

is easy to evaluate and the supremum in the above equation is attained. Compactness of U would be adequate to assert the attainment.

Using the Fenchel-Rockafellar duality theorem (Rockafellar 1970), the dual of $\mathrm{M^P}$ is:

$$(\mathrm{M^D}) \qquad \sup_{y}\{-h^*(-y) - \delta^*(y \mid U)\}.$$

It may be noted that there is no duality gap between $\mathrm{M^P}$ and $\mathrm{M^D}$ if $\inf_u h(u + w) + \delta(u \mid U)$ is a lower semicontinuous convex function and $h(u) > -\infty$. We present an algorithm in Figure 5.7 to lower bound $\mathrm{M^P}$.

Theorem 5.6. *Step 3 of Algorithm LinearApproximate in Figure 5.7 lower bounds $\mathrm{M^P}$.*

Proof. The primal relaxed problem constructed in Step 3 of Algorithm LinearApproximate is clearly a relaxation of $\mathrm{M^P}$ since $\langle y^k, u\rangle \leq \delta^*(y^k \mid U)$ for all $u \in U$ by the definition of the support function. $\square$

We shall show later that Algorithm LinearApproximate includes as a special case the Lagrangian lower bounding procedure.

5.3 Theoretical Framework: Abstract Minimization

In this section, we consider the abstract minimization problem:

(A) $$\inf_{x \in \mathbb{R}^n} \phi(x),$$

where $\phi : \mathbb{R}^n \mapsto \overline{\mathbb{R}}$. We assume that an upper bound on the objective function value of A is b_0. Algorithm LinearApproximate and perturbation analysis are used to derive domain reduction schemes for A.

The domain reduction problem is defined in a higher dimensional setting by adding the perturbation variables u_0 and $u \in \mathbb{R}^k$ to the space of problem variables of A as follows:

(R_r^A) $$r^* = \inf_{x, u_0, u} \{r(x, u_0, u) \mid (x, u_0, u) \in P\}.$$

We assume that there exists an x^* that optimizes A and a $u^* \in \mathbb{R}^k$ such that $(x^*, \phi(x^*), u^*) \in P$.

If r_a lower bounds r^*, problem A can clearly be reformulated as:

$$\inf_x \phi(x)$$
$$\text{s.t. } \inf_{u_0, u}\{r(x, u_0, u) \mid (x, u_0, u) \in P\} \geq r_a.$$

In order to follow the development of domain reduction schemes, it may help the reader to consider:

$$P = \{(x, u_0, u) \mid \rho(x, u) \leq u_0 \leq b_0\}$$

where ρ is a dualizing parameterization of A as defined in Section 5.1.2. To construct an effective domain reduction technique, R_r^A should satisfy the following conditions:

- There exists an x^* that optimizes A and a $u^* \in \mathbb{R}^k$ such that

$$(x^*, \phi(x^*), u^*) \in P.$$

5.3. THEORETICAL FRAMEWORK: ABSTRACT MINIMIZATION

- It is easy to optimize r and r^* over linear constraint sets.
- The support function of P can be easily computed.
- $\inf_{u_0,u}\{r(x,u_0,u) \mid (x,u_0,u) \in P\}$ can be explicitly characterized.

We shall mainly restrict our development to functions r that are independent of x and will be denoted as $r(x,u_0,u) = h(u_0,u)$. In this case, R_r^A is rewritten as:
$$h^* = \inf_{x,u_0,u} \{h(u_0,u) \mid (x,u_0,u) \in P\}.$$

Example 5.7. *Consider the optimization problem A with an upper bound of b_0, a dualizing parameterization $\rho(x,y)$ and the corresponding Lagrangian denoted by $l(x,y)$. Problem A can be reformulated as:*

$$\inf_x \phi(x)$$
$$\text{s.t.} \inf_{u_0,u}\{h(u_0,u) \mid \rho(x,u) \leq u_0 \leq \phi(x)\} \geq h^L$$

where h^L is computed by solving the following optimization problem:

$(\overline{R_u^A})$ $\qquad h^L = \inf_{u_0,u}\{h(u_0,u) \mid \langle u,y \rangle + \inf_x l(x,y) \leq u_0 \leq b_0 \text{ for all } y\}.$

Detail. Consider the optimization problem:

(R_e^A) $\qquad h^* = \inf_{x,u_0,u}\{h(u_0,u) \mid \rho(x,u) \leq u_0 \leq \phi(x)\}.$

Since $\phi(x) \leq b_0$, R_e^A is relaxed to:

$$\inf_{u_0,u}\{h(u_0,u) \mid p(u) = \inf_x \rho(x,u) \leq u_0 \leq b_0\},$$

which in further relaxed to:

$$\inf_{u_0,u}\{h(u_0,u) \mid p^{**}(u) \leq u_0 \leq b_0\}. \qquad (5.3)$$

Using Theorem 5.5, (5.3) can be rewritten as:

$$\inf_{u_0,u}\{h(u_0,u) \mid \langle u,y \rangle + \inf_x l(x,y) \leq u_0 \leq b_0 \text{ for all } y\}.$$

□

Let us define a homogenized Lagrangian function as follows:

$$l'_{b_0}(x, y_0, y) = -\sup_{x, u_0, u} \{y_0 u_0 + \langle y, u \rangle \mid \rho(x, u) \leq u_0 \leq b_0\}.$$

Algorithm Reduce presented in Figure 5.8 specializes Algorithm LinearApproximate to lower bound R_r^A when $P = \{\rho(x, u) \leq u_0 \leq b_0\}$. Lemma 5.8 justifies our use of the terminology for the homogenized Lagrangian function.

Lemma 5.8. *If $p^{**}(\cdot)$ is proper, then the support function of $\{(u_0, u) \mid p(u) \leq u_0\}$ is given by $\alpha(y_0, y)$ where α is determined by the following relation:*

$$\alpha(-y_0, y) = \begin{cases} (p^* y_0)(y), & \text{if } y_0 > 0; \\ (p^* 0^+)(y), & \text{if } y_0 = 0; \\ +\infty, & \text{if } y_0 < 0, \end{cases} \qquad (5.4)$$

and $p^* 0^+$ is the recession function of p^*. Alternately,

$$\alpha(y_0, y) = -\inf_x l'_\infty(x, y_0, y).$$

Proof. The first part follows from Corollary 13.5.1 in Rockafellar (1970) which shows that for α defined in (5.4) the following holds:

$$\alpha(y_0, y) = \delta^*((y_0, y) \mid p(u) \leq u_0).$$

Then,

$$\begin{aligned}
\alpha(y_0, y) &= \sup_{y_0, y} \left\{ y_0 u_0 + \langle y, u \rangle - \delta((u_0, u) \mid p(u) \leq u_0) \right\} \\
&= \sup_{y_0, y} \left\{ y_0 u_0 + \langle y, u \rangle - \delta((u_0, u) \mid \inf_x \rho(x, u) \leq u_0) \right\} \\
&= \sup_{y_0, y} \left\{ y_0 u_0 + \langle y, u \rangle - \inf_x \delta((u_0, u) \mid \inf_x \rho(x, u) \leq u_0) \right\} \\
&= \sup_x \sup_{y_0, y} \left\{ y_0 u_0 + \langle y, u \rangle - \inf_x \delta((u_0, u) \mid \inf_x \rho(x, u) \leq u_0) \right\} \\
&= -\inf_x -\sup_{y_0, y} \left\{ y_0 u_0 + \langle y, u \rangle - \delta((u_0, u) \mid \inf_x \rho(x, u) \leq u_0) \right\} \\
&= -\inf_x l'_\infty(x, y_0, y).
\end{aligned}$$

□

5.3. THEORETICAL FRAMEWORK: ABSTRACT MINIMIZATION

It is clear that for a given $x = x^0$:

$$l(x^0, -y_0, y) = \begin{cases} (ly_0)(x^0, y), & \text{if } y_0 > 0; \\ (l0^+)(x^0, y), & \text{if } y_0 = 0; \\ +\infty, & \text{if } y_0 < 0. \end{cases}$$

Therefore, the only difference between the definition of the Lagrangian presented in Section 5.1.2 and $l'_\infty(x, y_0, y)$ is that $l(x, y)$ needs to be computed over a series of points in order to evaluate $\alpha(0, y)$. In the homogenized setting, $\overline{R_u^A}$ in Example 5.3 would be written as:

$(\overline{R_h^A})$ $\displaystyle\inf_{u_0, u} \{h(u_0, u) \mid u_0 \leq b_0,$

$y_0 u_0 + \langle u, y \rangle + \inf_x l'_\infty(x, y_0, y) \leq 0 \text{ for all } (y_0, y)\}.$

The next few results show that $\inf_x l(x, y)$ provides the subgradients of $p^{**}(y)$ even when we relax certain requirements on attainment of supremum over u and infimum over x.

Theorem 5.9. $\bar{y} \in \partial p(\bar{u})$ if and only if there exists a sequence x^i such that

$$-\inf_x l(x, y) = -\lim_{i \to \infty} l(x^i, \bar{y}) = \langle \bar{y}, \bar{u} \rangle - \lim_{i \to \infty} \rho(x^i, \bar{u}).$$

Proof. $\bar{y} \in \partial p(\bar{u})$ if and only if $p^*(\bar{y}) = \langle \bar{y}, \bar{u} \rangle - p(\bar{u})$ (Proposition 5.1), or

$$-\inf_x l(\bar{x}, \bar{y}) = \langle \bar{y}, \bar{u} \rangle - \inf_x \rho(x, \bar{u}). \tag{5.5}$$

($\Rightarrow$) It follows from (5.5) that for any $\epsilon > 0$ there exists a sequence x^i such that

$$-\inf_x l(x, \bar{y}) = \langle \bar{y}, \bar{u} \rangle - \inf_x \rho(x, \bar{u})$$
$$\leq \langle \bar{y}, \bar{u} \rangle - \rho(x^i, \bar{u}) + \frac{\epsilon}{i}$$
$$\leq -l(x^i, \bar{y}) + \frac{\epsilon}{i}$$
$$\leq -\inf_x l(x, \bar{y}) + \frac{\epsilon}{i}.$$

Hence, the equality holds throughout.

(⇐) Conversely,

$$\begin{aligned}
-\inf_x l(x,\bar{y}) &= -\lim_{i\to\infty} l(x^i,\bar{y}) \\
&= \langle \bar{y}, \bar{u}\rangle - \lim_{i\to\infty} \rho(x^i,\bar{u}) \\
&\leq \langle \bar{y}, \bar{u}\rangle - \inf_x \rho(x,\bar{u}) \\
&\leq -\inf_x l(x,\bar{y}).
\end{aligned}$$

Hence, the equality holds throughout. □

Corollary 5.10. *If the infimum in* $\inf_x l(x,y)$ *is attained at* $x = \bar{x}$ *and the supremum in* $\sup_u \{\langle \bar{y}, u\rangle - \rho(\bar{x}, u)\}$ *is attained at* $u = \bar{u}$, *then* $\bar{y} \in \partial p(\bar{u})$.

Proof. Follows easily from Theorem 5.9 if the sequence x^i is chosen such that $x^i = \bar{x}$ for all i. □

Theorem 5.11. *If there exists a sequence* (x^i, u^i) *such that*

$$-\inf_x l(x,y) = -\lim_{i\to\infty} l(x^i,\bar{y}) = \lim_{i\to\infty}\{\langle \bar{y}, u^i\rangle - \rho^{**}(x^i,u^i)\},$$

with $u^i \to \bar{u}$, *then* $\bar{y} \in \partial p^{**}(\bar{u})$.

Proof.

$$\begin{aligned}
-\inf_x l(x,\bar{y}) &= -\lim_{i\to\infty} l(x^i,\bar{y}) \\
&= \lim_{i\to\infty}\{\langle \bar{y}, u^i\rangle - \rho^{**}(x^i,u^i)\} \\
&= \langle \bar{y}, \bar{u}\rangle - \lim_{i\to\infty} \rho^{**}(x^i,u^i) \\
&\leq \langle \bar{y}, \bar{u}\rangle - \lim_{i\to\infty} \inf_x \rho^{**}(x,u^i) \\
&\leq \langle \bar{y}, \bar{u}\rangle - \lim_{i\to\infty} p^{**}(u^i) \\
&\leq \langle \bar{y}, \bar{u}\rangle - p^{**}(\bar{u}) \\
&\leq -\inf_x l(x,\bar{y}).
\end{aligned}$$

Hence, the equality holds throughout. □

5.3. THEORETICAL FRAMEWORK: ABSTRACT MINIMIZATION

Theorem 5.9, Corollary 5.10 and Theorem 5.11 provide schemes for outer-approximating $p^*(y)$ given a Lagrangian relaxation procedure. In particular, if for every $\bar{y}$ the infimum in $\inf_x l(x,y)$ is attained and for every $\bar{x}$ and $\bar{y}$ the supremum in $\sup_u \{\langle \bar{y}, u \rangle - \rho(\bar{x}, u)\}$ is attained, then Corollary 5.10 suffices to generate the epigraph of $p^*(y)$ as it generates, for each $\bar{y}$, a $\bar{u}$ and a corresponding inequality

$$z \geq \langle y, \bar{u} \rangle - p(\bar{u}) = p^*(\bar{y}).$$

We now restrict our attention to the following domain reduction problem:

$$h^* = \inf_{x, u_0, u} \{h(u_0, u) \mid \rho(x, u) \leq u_0\} \qquad (5.6)$$

where

$$h(u_0, u) = \langle a, u \rangle + a_0 u_0 + \delta(u \mid u \leq b) + \delta(u_0 \mid u_0 \leq b_0)$$

with the added restrictions that $(a_0, a) \geq 0$ and $(a_0, a) \neq 0$. Algorithm Reduce can be specialized to lower bound (5.6) by using $l'_\infty(x, y_0, y)$ instead of $l'_{b_0}(x, y_0, y)$. The dual of (5.6) is easily seen to be:

$$h_a^d = \langle a, b \rangle + a_0 b_0 + \sup_{(y_0, y) \leq -(a_0, a)} \{\langle y, b \rangle + y_0 b_0 - \alpha(y_0, y)\}. \qquad (5.7)$$

Define $\beta(y_0, y) = -\langle y, b \rangle - y_0 b_0 + \alpha(y_0, y)$. Let (y_0^i, y^i) be a sequence such that

$$\lim_{i \to \infty} -\beta(y_0^i, y^i) = \sup_{(y_0, y) \leq -(a_0, a)} \{-\beta(y_0, y)\} = \gamma$$

and $(y^i, y_0^i) \leq -(a_0, a)$ for each i. Define $\lambda^i = \max_j -a_j/y_j^i$ and construct the sequence $(\bar{y}_0^i, \bar{y}^i) = \lambda^i (y_0^i, y^i)$. Clearly, $\lambda^i \leq 1$. If $\gamma < +\infty$, then $\beta(y_0^i, y^i) \geq 0$ and (y_0^i, y^i) is not a decreasing direction for $\beta(y_0, y)$ since $(y_0^i, y^i) \leq -(a_0, a) \leq 0$ is a feasible direction and β is a positive homogeneous function. Since $\beta(y_0^i, y^i) \geq 0$ and $\lambda^i \leq 1$,

$$-\beta(\bar{y}_0^i, \bar{y}^i) = -\lambda^i \beta(y_0^i, y^i) \geq -\beta(y_0^i, y^i).$$

Thus:

$$\gamma \geq \lim_{i \to \infty} -\beta(\bar{y}_0^i, \bar{y}^i) \geq \lim_{i \to \infty} -\beta(y_0^i, y^i) = \gamma.$$

Hence, the equality holds throughout. By construction, each $(\bar{y}_0^i, \bar{y}^i)$ has at least one coordinate equal to its corresponding value in $-(a_0, a)$. Then, by taking a subsequence if necessary, it follows when $(a_0, a) \neq 0$ that:

$$\gamma = \sup_{(y_0, y) \leq (a_0, a)} \{-\beta(y_0, y) \mid \exists j \text{ with } a_j > 0 \text{ s.t. } y_j = -a_j\}. \qquad (5.8)$$

Interestingly, if (a_0, a) has only one coordinate nonzero, then (5.7) can be further simplified. For example, if $(a_0, a) = (0, 1)$ then (5.7) simplifies to:

$$h_a^d = b_0 + \sup_{y \leq 0}\{-b_0 - p^*(y)\}.$$

Note that $h_a^d = \infty$ if there exists $y \leq 0$ such that $p^*(y) < -b_0$. This is equivalent to fathoming by inferiority of the node in branch-and-bound. Otherwise, $h_a^d = \sup_{y \leq 0} -p^*(y)$ which will be shown in Section 5.6 to be equivalent to the Lagrangian master problem in the traditional mathematical programming setting.

Using (5.6), the primal problem in Algorithm Reduce can be written as:

$$\min_{(u_0, u)} \quad a_0 u_0 + \langle a, u \rangle$$
$$\text{s.t.} \quad (u_0, u) \leq (b_0, b)$$
$$y_0 u_0 + \langle y, u \rangle + \inf_x l'_\infty(x, y_0, y) \leq 0 \ \forall \, (y_0, y).$$

Using (5.7), the dual problem reduces to:

$$\max_{(y_0, y)} \quad (y_0 + a_0) b_0 + \langle y + a, b \rangle - z$$
$$\text{s.t.} \quad (y_0, y) \leq -(a_0, a)$$
$$z \geq y_0 u_0 + \langle y, u \rangle \ (u_0, u) \in U$$

where $U = \{(u_0, u) \mid \exists x \text{ satisfying } \rho(x, u) \leq u_0\}$.

5.4 Application to Traditional Models

In this section, we apply the results of Section 5.3 to mathematical programming models presented in the traditional form:

(C) $\quad\quad\quad \min \ f(x)$
$\quad\quad\quad\quad \text{s.t.} \ g(x) \leq b$
$\quad\quad\quad\quad\quad\quad x \in X \subseteq \mathbb{R}^n$

where $f : \mathbb{R}^n \mapsto \mathbb{R}$, $g : \mathbb{R}^n \mapsto \mathbb{R}^m$, and X denotes the set of "easy" constraints. Problem C can be viewed as a special case of A using the following definition for ϕ:

$$\phi(x) = f(x) + \delta(x \mid g(x) \leq b) + \delta(x \mid X),$$

5.4. APPLICATION TO TRADITIONAL MODELS

where $\delta(\cdot\,|\,C)$ is the indicator function of the set C (Rockafellar 1970).

The range reduction problem is defined as:

$$(\mathrm{R}_\mathrm{r}^C) \qquad r^* = \inf_{x,u_0,u}\{r(x,u_0,u) \mid f(x) \le u_0 \le b_0, g(x) \le u \le b, x \in X\},$$

where b_0 is an upper bound of $f(x)$. Using a lower bound r_a on r^*, C is reformulated as:

$$(\mathrm{C}') \qquad \begin{array}{l} \min\ f(x) \\ \text{s.t.}\ g(x) \le b \\ \quad\ x \in X \\ \quad\ \inf_{u_0,u}\Big\{r(x,u_0,u) \mid f(x) \le u_0 \le b_0, g(x) \le u \le b\Big\} \ge r_a. \end{array}$$

Problem R_r^C can clearly be rewritten as follows:

$$(\mathrm{R}^C) \qquad r^* = \inf_{x,u_0,u}\ r(x,u_0,u)$$
$$\text{s.t.}\ -y_0(f(x) - u_0) - \langle y, g(x) - u\rangle \le 0 \quad \forall\,(y_0, y) \le 0 \quad (5.9)$$
$$(u_0, u) \le (b_0, b),\ x \in X. \quad (5.10)$$

Note that (y_0, y) can be bounded by any norm without affecting the formulation R^C. It is instructive to note that, when r is a monotonically increasing function of u and u_0, problem C' is equivalent to:

$$\begin{array}{l} \min\ f(x) \\ \text{s.t.}\ g(x) \le b \\ \quad\ x \in X \\ \quad\ r(x, f(x), g(x)) \ge r_a\ \text{or}\ f(x) > b_0. \end{array}$$

We restrict our attention to the functions r that are independent of x and denoted as $r(x,u_0,u) = h(u_0,u)$. Problem R^C is then rewritten as:

$$(\mathrm{R}_\mathrm{u}^C) \qquad h^* = \inf_{x,u_0,u}\ h(u_0,u)$$
$$\text{s.t.}\ -y_0(f(x) - u_0) - \langle y, g(x) - u\rangle \le 0 \quad \forall\,(y_0, y) \le 0 \quad (5.11)$$
$$(u_0, u) \le (b_0, b),\ x \in X. \quad (5.12)$$

R_u^C is relaxed by independently taking an infimum over x in every constraint as follows:

$$(\overline{\mathrm{R}_\mathrm{u}^C}) \qquad h^L = \inf_{u_0,u}\ h(u_0,u)$$
$$\text{s.t.}\ y_0 u_0 + \langle y, u\rangle + \inf_{x \in X}\{-y_0 f(x) - \langle y, g(x)\rangle\} \le 0 \quad \forall\,(y_0, y) \le 0$$
$$(u_0, u) \le (b_0, b),\ x \in X.$$

$\overline{\mathrm{R}^C_u}$ is analogous to $\overline{\mathrm{R}^A_h}$ where we have employed the following Lagrangian subproblem

$$\inf_{x\in X}\{-y_0 f(x) - \langle y, g(x)\rangle\} \text{ where } (y_0, y) \leq 0.$$

The standard definition of the Lagrangian subproblem for C is:

$$\inf_{x\in X}\{f(x) - \langle y, g(x) - b\rangle\} \text{ where } y \leq 0.$$

Note that our definition of the Lagrangian departs slightly from the standard definition in that it includes a homogenization with y_0 and removes the shift by b. The relaxation of R^C_r can be generated by defining any easily computable support function as the Lagrangian function. As an example, consider the homogenized Lagrangian below:

$$l'(x, y_0, y) = \sup_{u_0, u}\{y_0 u_0 + \langle y, u\rangle \mid f(x) \leq u_0 \leq b_0,\ g(x) \leq u \leq b, x \in X\}.$$

The optimum in the definition of the homogenized Lagrangian above is attained at $(u_0^*, \ldots, u_m^*)$ below:

$$u_i^* = \begin{cases} b_i, & y_i > 0; \\ f(x), & i = 0,\ y_0 \leq 0; \\ g_i(x), & i \neq 0,\ y_i \leq 0, \end{cases}$$

where g_i denotes the ith constraint. Let $u^* = (u_1^*, \ldots, u_m^*)$. The following constraint is valid for R^C_r:

$$-y_0(u_0^* - u_0) - \langle y, u^* - u\rangle \leq 0.$$

The relaxation can then be constructed easily:

$(\overline{\mathrm{R}^C_r})$ $\quad\inf_{u_0, u} h(u_0, u)$
$\qquad\quad$ s.t. $y_0 u_0 + \langle y, u\rangle + \inf_x l'(x, y_0, y) \leq 0\quad (y_0, y) \in \mathbb{R}^{m+1}$
$\qquad\qquad (u_0, u) \leq (b_0, b),\ x \in X.$

An astute reader may realize that l' is hard to compute. Thus, $(\overline{\mathrm{R}^C_r})$ has limited use beyond demonstration of the principles involved.

5.5 Geometric Intuition

The theory presented in Sections 5.3 and 5.4 can be understood easily in a geometric setting. The purpose of this section is to provide the geometric intuition behind our domain reduction theory.

We illustrate the procedure through an example shown in Figure 5.9. The objective is to find the point on the value function of the perturbation problem $p^{**}(u) \leq b_0$ that is nearest to u^0 (see Figure 5.9).

We first distinguish between hard and easy problems. Consider a point u^1 such that the slope of $p(u)$ at $u = u^1$ is given by $y = y^1$. Given u^1, it is hard to compute $p(u^1)$ and y^1 (see Figure 5.10). However, given y^1, the point u^1 can be retrieved by solving the Lagrangian dual problem and $p(u^1)$ can also be easily computed (see Figure 5.11).

We now illustrate an iterative step of Algorithm Reduce. Start with an inner-approximation and an outer-approximation of $p^{**}(u) \leq b_0$ as in Figure 5.12. An upper bound on the domain reduction problem is obtained in Step 1 of the algorithm by locating the closest point to u^0, say $\bar{u}_1$, that belongs to the epigraph of the current inner-approximation. Similarly, a lower bound on the problem is obtained by finding the closest point to u^0 in the epigraph of the outer-approximation of $p^{**}(u) \leq b_0$. The upper bounding problem is easier to solve in the dual space while the lower bounding problem is more easily solved in the primal space.

Let $\bar{y}_1$ be the slope of the inner-approximation at $\bar{u}_1$. Then, solving the Lagrangian relaxation we obtain u_1 as shown in Figure 5.14. As a result, improved inner- and outer-approximations are constructed and better bounds derived on the domain reduction problem (see Figure 5.13).

5.6 Domain Reduction Problem: Motivation

In the previous sections, we developed a theoretical framework and claimed that it unifies many domain reduction results in the literature. However, we have not yet provided much insight into the application of the developed theory for range reduction purposes. The purpose of this section is to provide such an application.

Much insight into our definition of the domain reduction problem and its potential uses is gained by restricting to $h(u_0, u) = a_0 u_0 + \langle a, u \rangle$ where $(a_0, a) \geq 0$ and $(a_0, a) \neq 0$. In fact, the most important applications set

(a_0, a) to one of the principal directions in $\mathbb{R}^{m+1}$. We employ the setting of Section 5.4 for our exposition in this section.

In Figure 5.16, we specialize Algorithm Reduce using the Lagrangian function defined below:

$$l(x, y_0, y) = \begin{cases} \infty, & x \notin X; \\ y_0 f(x) - \langle y, g(x) \rangle, & x \in X, (y_0, y) \leq 0; \\ -\infty, & \text{otherwise.} \end{cases} \quad (5.13)$$

$l(x, y_0, y)$ was implicitly used in the construction of $\overline{R_u^C}$ in Section 5.4.

The reader should notice that Algorithm SpecialReduce reduces to the standard Lagrangian lower bounding procedure for problem C when $(a_0, a) = (1, 0)$ (see discussion following Theorem 5.11). Similarly, Algorithm SpecialReduce reduces to the standard Lagrangian lower bounding procedure for the problem:

$$(\text{R}_{g_i}) \qquad \begin{aligned} \min \quad & g_i(x) \\ \text{s.t.} \quad & g(x) \leq b \\ & f(x) \leq b_0 \\ & x \in X \subseteq \mathbb{R}^n \end{aligned}$$

when $a_i = 1$ and $a_j = 0$ for all $j \neq i$.

Let us assume that we use the Lagrangian procedure (or any algorithm that provides dual solutions) for deriving a lower bound on C in the context of branch-and-bound. In the process of doing so, we solve multiple Lagrangian subproblems and generate cuts for the primal master problem in Step 3 of Algorithm SpecialReduce. The generated cuts take the following form:

$$y_0^k u_0 + \langle y^k, u \rangle + \inf_x l(x, y_0^k, y^k) \leq 0. \quad (5.14)$$

Clearly, these constraints are also valid for the primal master problem associated with the Lagrangian lower bounding procedure for R_{g_i}. This central observation can be used to derive most of the domain reduction strategies in the current literature which derive lower bounds for $g_i(x)$. The observation rests on the fact that all the Lagrangian bounding schemes produce inner- and outer-approximations of $\{(u_0, u) \mid p(u) \leq u_0 \leq b_0\}$ (see Figure 5.15).

5.7 Relation to Earlier Works

We have developed a theoretical framework for domain reduction. In this section, we will choose some prevalent domain reduction strategies and derive

5.7. RELATION TO EARLIER WORKS

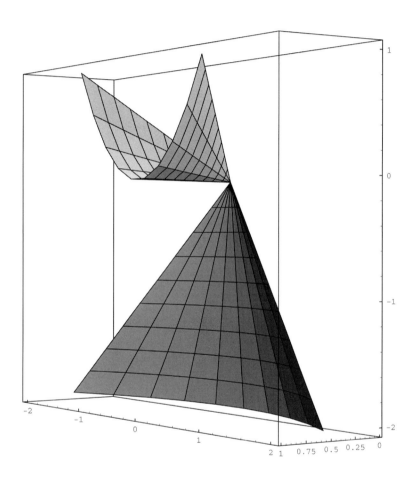

Figure 5.3: x^2 and conjugate $x^2/4$

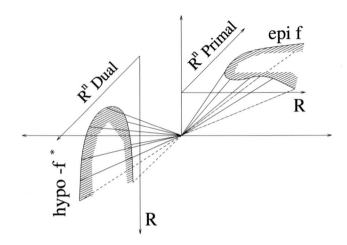

Figure 5.4: Interpreting f^* in cosmic space

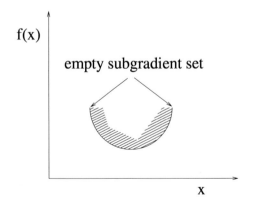

Figure 5.5: $\partial f(x) = \emptyset$ for a closed convex function

5.7. RELATION TO EARLIER WORKS

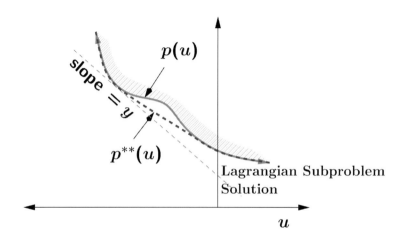

Figure 5.6: Perturbation problem and Lagrangian duality

Algorithm LinearApproximate

Step 0. Set $K = 0$.

Step 1. Solve the dual:
$$h_U^K = \max_{y,z} \quad -h^*(y) - z$$
$$\text{s.t.} \quad z \geq \langle y, u^k \rangle \quad k = 1, \ldots, K-1$$

Let the optimum be denoted by y^K.

Step 2. Evaluate the support function of U:
$$\delta^*(y^K \,|\, U) = \sup_u \{\langle y^K, u \rangle \,|\, u \in U\}.$$

Let the optimum be attained at u^K.

Step 3. Augment and solve the relaxed primal problem:
$$h_L^K = \min_{(u)} \quad h(u)$$
$$\text{s.t.} \quad \langle y^k, u \rangle \leq \delta^*(y^k \,|\, U) \quad k = 1, \ldots, K$$

Step 4. Termination check: If $h_U^K - h_L^K \leq \epsilon$, stop; otherwise, set $K = K+1$ and goto Step 1.

Figure 5.7: Algorithm to lower bound M^P

Algorithm Reduce

Step 0. Set $K = 0$.

Step 1. Solve the relaxed dual:

$$h_U^K = \max_{(y_0, y)} \; -\bar{h}^*(-y_0, y) - z$$
$$\text{s.t.} \quad y_0 \leq 0$$
$$z \geq y_0 u_0^k + \langle y, u^k \rangle \quad k = 1, \ldots, K-1$$

Let the solution be (y_0^K, y^K).

Step 2. Solve the Lagrangian subproblem:

$$\inf_x l'_{b_0}(x, y_0^K, y^K) = -\max_{u_0, u} \; y_0^K u_0 + \langle y^K, u \rangle$$
$$\text{s.t.} \quad \inf_x \rho(x, u) \leq u_0 \leq b_0$$

Let the solution be (x^K, u_0^K, u^K).

Step 3. Augment and solve relaxed primal problem:

$$h_L^K = \min_{(u_0, u)} \; \bar{h}(u_0, u)$$
$$\text{s.t.} \quad y_0^k u_0 + \langle y^k, u \rangle + \inf_x l'_{b_0}(x, y_0^k, y^k) \leq 0 \quad k = 1, \ldots, K$$

Step 4. Termination check: If $h_U^K - h_L^K \leq \epsilon$, stop; otherwise, set $K = K+1$ and goto Step 1.

Figure 5.8: Algorithm to solve R_c^A

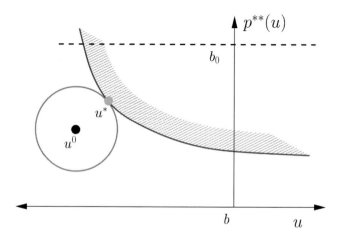

Figure 5.9: Domain reduction problem

5.7. RELATION TO EARLIER WORKS

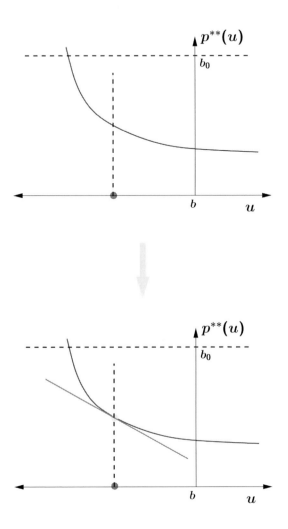

Figure 5.10: Hard problem

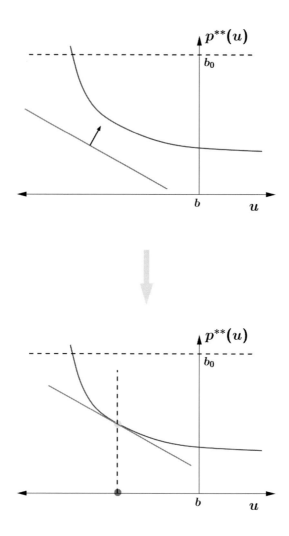

Figure 5.11: Easy problem

5.7. RELATION TO EARLIER WORKS

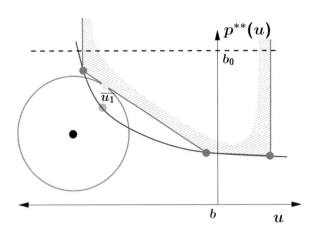

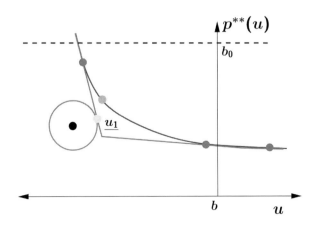

Figure 5.12: Bounds with initial approximations

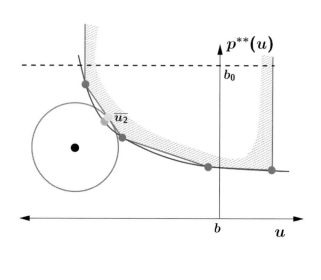

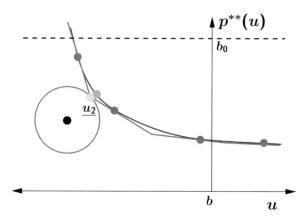

Figure 5.13: Bounds with improved approximations

5.7. RELATION TO EARLIER WORKS

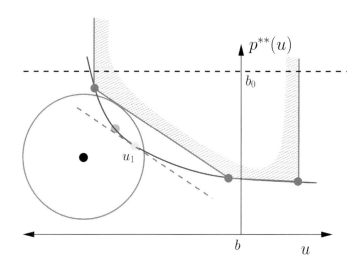

Figure 5.14: The Lagrangian subproblem

176   CHAPTER 5. DOMAIN REDUCTION

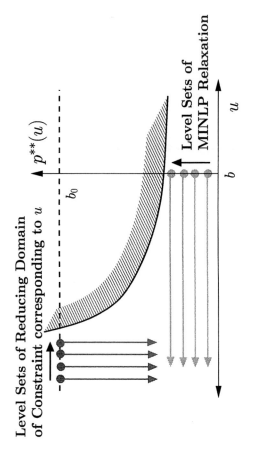

Figure 5.15: Closely related lower bounding procedures

5.7.1 Bounds Via Monotone Complementarity

A range reduction scheme was derived by Mangasarian & McLinden (1985) using an argument that relied on the monotonicity property of the "twisted" derivative. We show in this section that the same result can be derived using the proposed domain reduction framework.

Consider the following differentiable convex program:

$$\min_x \{f(x) \mid y = -g(x) \geq 0, x \geq 0\} \quad (5.15)$$

where $f : \mathbb{R}^n \mapsto \mathbb{R}$, $g : \mathbb{R}^n \mapsto R^m$ are convex and differentiable. Also, consider the following dual of the above program:

$$\max_{x,u} \{L(x, u) - \langle v, x \rangle \mid v = \nabla_x L(x, u) \geq 0, u \geq 0\} \quad (5.16)$$

where $L(x, u) = f(x) - \langle u, y \rangle$.

Theorem 5.12 (Mangasarian & McLinden 1985). *Let f and g be differentiable and convex in $\mathbb{R}^n$. Each primal-dual feasible point of (5.15)-(5.16) bounds any $(\bar{x}, \bar{y}, \bar{u}, \bar{v})$ which solves (5.15)-(5.16) as follows:*

1. $\sum_{i \in I_1} \bar{x}_i \leq (\langle x, v \rangle + \langle u, y \rangle)/\min v_{i \in I_1}$
2. $\sum_{j \in J_2} \bar{y}_j \leq (\langle x, v \rangle + \langle u, y \rangle)/\min u_{j \in J_2}$
3. $\sum_{i \in I_2} \bar{u}_i \leq (\langle x, v \rangle + \langle u, y \rangle)/\min y_{i \in I_2}$
4. $\sum_{j \in J_1} \bar{v}_j \leq (\langle x, v \rangle + \langle u, y \rangle)/\min x_{j \in J_1}$

Proof. Since (x, y, u, v) is a feasible primal-dual pair, it satisfies the KKT conditions for optimality associated with the Lagrangian subproblem (see (5.13)). As a result, (5.14) reduces to:

$$-b_0 + \sum_{j=1}^{m} u_j \bar{y}_j + \sum_{i=1}^{n} v_i \bar{x}_i + f(x) - \langle u, y \rangle - \langle v, x \rangle \leq 0 \quad (5.17)$$

where $(u, v) \geq 0$. Since (x, y) is feasible to (5.15), $b_0 \leq f(x)$. Optimizing $\sum_{i \in I_1} \bar{x}_i$ and $\sum_{j \in J_2} \bar{y}_j$ over (5.17), yields the first two relations. Assume that the inner-approximation consists of only $(f(x), y, x)$. Then, the dual solutions are bounded as below:

$$f(x) - \langle \bar{u}, y \rangle - \langle \bar{v}, x \rangle \geq h_L^1 \geq f(x) - \langle u, y \rangle - \langle v, x \rangle.$$

The remaining inequalities are easily derived from the above relation. □

5.7.2 Tightening using Reduced Costs

Many domain reduction schemes take advantage of the reduced costs of variables and/or marginals of the constraints to develop tighter bounds on them. The following result appears as Theorem 2 in Ryoo & Sahinidis (1996).

Theorem 5.13 (Ryoo & Sahinidis 1996). *Consider an optimization problem:*

(P) $\qquad\qquad\qquad \min\ f(x)$
$\qquad\qquad\qquad\ \text{s.t.}\ g(x) \leq 0$
$\qquad\qquad\qquad\qquad\ x \in X \subseteq \mathbb{R}^n$

Let R be a convex relaxation of P and L be the optimal objective function value of R. Consider a constraint $\bar{g}_i(x) \leq 0$ that is active at the solution of problem R with a dual multiplier value of $\mu_i > 0$. Let b_0 be a known upper bound for problem C. Then, the following constraint is valid for C:

$$\bar{g}_i(x) \geq -(b_0 - L)/\mu_i.$$

Proof. In light of the developments in Section 5.6, any dual feasible solution y^k of R can be used to develop the constraint:

$$-u_0 + \langle y^k, u \rangle + \inf_x l(x, -1, y^k) \leq 0$$

for the primal master problem of R_{gi} where the Lagrangian subproblem is defined by dualizing the constraints of R. If $-y^*$ is the optimal dual solution to R, then $\inf_x l(x, -1, y^*) = L$. It is given that $y_i^* = \mu_i$. The following relaxed master problem lower bounds the minimum of R_{gi}:

$$\min\{u_i \mid -\langle y^*, u \rangle \leq u_0 - L,\ u \leq 0,\ u_0 \leq b_0\}.$$

The optimal solution to the above problem is

$$u_i = -(b_0 - L)/\mu_i,$$

which is attained when each u_j for $j \neq i$ is set to its upper bound. □

5.7. RELATION TO EARLIER WORKS

Theorem 3, Corollaries 1, 2 and 3 from Ryoo & Sahinidis (1996), as well as marginals-based reduction techniques in Thakur (1990), Lamar (1993), and the early integer programming literature (cf. Nemhauser & Wolsey 1988) follow similarly.

The following appears in a slightly restricted form in Zamora & Grossmann (1999).

Theorem 5.14. *Consider the optimization problem:*

$$(R_z) \qquad \begin{aligned} l_i = \min \quad & g_i(x) \\ \text{s.t.} \quad & g(x) \leq b \\ & f(x) \leq b_0 \\ & x \in X \subseteq \mathbb{R}^n \end{aligned}$$

where $b_j = 0$ for all $j \in \{1, \ldots, m\} \setminus \{i\}$. Let $y_0^* \leq 0$ be the optimal multiplier corresponding to $f(x) \leq b_0$. Then,

$$f(x) \geq b_0 + (b_i - l_i)/y_0^*.$$

Proof. Let y^* be the vector of optimal dual multipliers corresponding to $g(x) \leq b$. Note that the Lagrangian subproblem solution is $l_i - y_0^* b_0 - y_i^* b_i$. It can be easily argued that, when $y_i^* < 0$, $l_i \geq b_i$. Therefore, we may restrict our attention to $y_i^* = 0$. The following relaxed master problem lower bounds R_z:

$$\min \left\{ u_i \;\middle|\; \sum_{j=0}^{m} y_j^* u_j \leq u_i - l_i + y_0^* b_0,\; u \leq b,\; u_0 \leq b_0 \right\}.$$

□

5.7.3 Linearity-based Tightening

Linearity-based tightening methods have been popular in integer programming literature (Savelsbergh 1994, Andersen & Andersen 1995) and have also been proposed for various nonlinear programming problems (Shectman & Sahinidis 1998).

The linearity-based tightening procedure works with a constraint set of the form $\sum_{j=1}^{n} a_{ij} x_j \leq b_i$, $i = 1, \ldots, m$. Processing one constraint at a time, the following bounds are derived on each variable:

$$\begin{cases} x_h \leq \dfrac{1}{a_{ih}} \left(b_i - \sum_{j \neq h} \min\{a_{ij} x_j^u, a_{ij} x_j^l\} \right), & a_{ih} > 0; \\ \\ x_h \geq \dfrac{1}{a_{ih}} \left(b_i - \sum_{j \neq h} \min\{a_{ij} x_j^u, a_{ij} x_j^l\} \right), & a_{ih} < 0, \end{cases} \quad (5.18)$$

where x_j^u and x_j^l are the tightest upper and lower bounds available for x_j at the current node.

Linearity-based tightening uses the Fourier-Motzkin algorithm to express the solution to the following linear program in closed-form:

$$\begin{aligned} \min\ & x_h \\ \text{s.t.}\ & a_i x \leq b_i \\ & x \leq x^u \\ & x \geq x^l. \end{aligned}$$

Assuming $a_{ih} < 0$ and x_h is at not at its lower bound, the optimum dual solution can easily be gleaned from the above linear program to be:

$$\begin{aligned} y &= 1/a_{ih} \\ r_j &= -\max\{a_{ij}/a_{ih}, 0\} \text{ for all } j \neq h \\ s_j &= \min\{a_{ij}/a_{ih}, 0\} \text{ for all } j \neq h \\ r_h &= s_h = 0 \end{aligned}$$

where y, r_j, and s_j are the dual multipliers corresponding to $a_i x \leq b_i$, $x_j \leq x_j^u$, and $x_j \geq x_j^l$, respectively. Then, the following relaxed primal master problem is constructed:

$$\min\{x_h \mid -x_h + yu + \langle r, v \rangle - \langle s, w \rangle \leq 0, u \leq b, v \leq x^u, w \geq x^l\}. \quad (5.19)$$

The bounds presented in (5.18) follow easily from (5.19).

The reader may notice that we have implicitly used Fourier-Motzkin elimination to solve the relaxed master problems (5.19) and also other relaxations developed in Section 5.7.1 and Section 5.7.2 to derive closed form expressions much in the same way as Fourier-Motzkin elimination is used to derive (5.18). All the above problems minimize a variable over a constraint set consisting of one linking constraint and various bound constraints. The resulting problem is easily solved using Fourier-Motzkin elimination. From this standpoint, linearity-based tightening can be regarded as a procedure for solving some of the linear programs that arise from our domain reduction framework rather than a consequence of our framework.

5.8 Probing

In this section, we describe a probing technique for domain reduction. Probing has been a popular domain reduction technique for integer programming problems (cf. Savelsbergh 1994) and has been extended to the context of nonlinear global optimization problems by Ryoo & Sahinidis (1996). We present and generalize some of these results using our domain reduction techniques.

Consider the following mathematical program:

(P)
$$\begin{aligned} \min \ & g_1(x) \\ \text{s.t.} \ & g(x) \leq b \\ & x \in X \subseteq \mathbb{R}^n \end{aligned}$$

where $g : \mathbb{R}^n \mapsto \mathbb{R}^m$. Define the following probing problems P_i for $i = 1, \ldots, m$ as:

(P^i)
$$\begin{aligned} \min \ & g_i(x) \\ \text{s.t.} \ & g(x) \leq p + \infty \cdot e_i \\ & x \in X \subseteq \mathbb{R}^n \end{aligned}$$

Here, e_i is the ith unit vector and has been introduced as a notational convenience for removing the upper bounding constraint on the objective from the probing problem. The standard Lagrangian subproblem for P^i is:

$$\inf_x g_i(x) + \sum_{\substack{j=1 \\ j \neq i}}^{m} (y_j^i g_j(x) - p_j) \tag{5.20}$$

Let $y^i = [y_1^i, \ldots, y_m^i]$ be a dual feasible vector for P^i where $y_i^i = -1$ and let L^i be a lower bound on the corresponding optimal solution of the Lagrangian subproblem presented in (5.20). Then, the following constraints are valid:

$$-u_i + \sum_{\substack{j=1 \\ j \neq i}}^{m} y_j^i (u_j - p_j) + L^i \leq 0,$$

where $u_j \leq b_j$. Domain reduction can then be performed by minimizing some function, say $h(u)$, over the generated constraints. A simple consequence of this fact appears as Theorem 3 in Ryoo & Sahinidis (1996). Our proof of this result provides an example of the applicability of probing in range reduction.

Theorem 5.15 (Ryoo & Sahinidis 1996). *Let R be a convex relaxation of a mathematical programming problem P and consider a constraint $a_k^t x - b_k \leq 0$ that is not active at the solution of R. Let b_0 be a known upper bound for problem P. Solve R by fixing $a_k^t x$ at b_k, i.e., after adding the constraint $b_k \leq a_k^t x$ in the formulation. Let Z be the optimal objective value of this partially restricted relaxed problem. If a positive dual multiplier μ_k is obtained in the solution of the new problem, then the following constraint is valid for P*

$$a_k^t x \leq b_k + (b_0 - Z)/\mu_k.$$

Proof. We introduce $-a_k^t x \leq \infty$ as the $m+1$ constraint in the definition of P and restrict our discussion before the statement of the theorem to the probing problem P_1, where $\mu_k = -y_{m+1}^1$, $L^i = Z$, $p_j = b_j$ for all $j \in \{2,\ldots,m\}$ and $p_{m+1} = -b_k$. Then, minimizing u_{m+1} over

$$-u_1 + \sum_{j=2}^{m} y_j^1(u_j - b_j) - \mu_k(u_{m+1} + b_k) + Z \leq 0 \qquad (5.21)$$

where $u_i \leq b_i$ for all $i \in \{1,\ldots,m\}$, we get

$$-a_i^t x \geq -b_k - (b_0 - Z)/\mu_k.$$

□

It may be noted that the above result probed only one constraint to derive an improved lower bound on the same constraint. However, in the proof we argued the validity of (5.21). Constraint (5.21) can therefore be used to minimize any of u_1 through u_m if an upper bound is available for the probed constraint. An interesting application of this observation appears in branch-and-bound methods. Branch-and-bound relaxes P over a hypercube at every node of the search tree. When such a relaxation is probed to identify the bounds on a critical variable, additional tightening occurs in other variable axes as a bonus since upper bounds are always available on all variables contributing to nonconvexity in the model.

5.9 Learning Reduction Procedure

In this section, we develop the *learning reduction procedure*, a procedure that takes advantage of the proposed domain reduction framework. We first motivate the procedure. Branch-and-bound for global optimization problems can be shown to be convergent under the assumption that the branching scheme is exhaustive. Therefore, in order to guarantee exhaustiveness, most rectangular partitioning schemes resort to periodic bisection of the feasible space along each variable axis. Let us assume that, after branching, one of the child nodes turns out to be inferior. Then, it is highly probable that a larger region could have been fathomed if a proper branching point was chosen initially. It is in cases of this sort that the learning procedure introduces an error-correcting capability by expanding the region that is provably inferior.

We now give a formal description of how this is achieved. Consider a node N where the following relaxation is generated:

$$\begin{aligned}\min_x \quad & (f^N(x^L, x^U))(x) \\ \text{s.t.} \quad & (g_i^N(x^L, x^U))(x) \leq 0 \quad i = 1, \ldots, m \\ & x^L \leq x \leq x^U\end{aligned}$$

where $(f(x^L, x^U))(\cdot)$ and $(g_i(x^L, x^U))(\cdot)$, for $i = 1, \ldots, m$ are convex functions. We assume that the relaxation strategy is such that, when a variable is at its bound, the dependent functions are represented exactly.

Let us assume that the node N is partitioned into N_1 and N_2 by branching along x_j at x_j^B. Define

$$x^a = [x_1^U, \ldots, x_{j-1}^U, x_j^B, x_{j+1}^U, \ldots, x_n^U]$$

and

$$x^b = [x_1^L, \ldots, x_{j-1}^L, x_j^B, x_{j+1}^L, \ldots, x_n^L].$$

N_1 and N_2 are constructed such that $x^L \leq x \leq x^a$ for every $x \in N_1$ and $x^b \leq x \leq x^U$ for every $x \in N_2$. Assume further that the node N_1 is found to be inferior with y^* as the vector of optimal dual multipliers for the relaxation generated at node N_1. The constraint $x \leq x_j^B$ is assumed to be active in the solution of N_1. Note that, even though $x \leq x_j^B$ is often active in practice, it is not provably so (see Figure 5.17 for a counterexample).

Most nonlinear programming relaxation techniques are such that the relaxation at $x = x_j^B$ is the same in both the nodes N_1 and N_2. If we can construct the optimal dual solution $y_{N_2^P}^*$ for the following program:

$$(\text{P}_{N_2}) \qquad \min_x \ (f^{N_2}(x^L, x^U))(x)$$
$$\text{s.t.} \ (g_i^{N_2}(x^L, x^U))(x) \leq 0 \quad i = 1, \ldots, m$$
$$x^L \leq x \leq x^U$$
$$x_j \geq x_j^B$$

then we can apply the techniques developed in Section 5.8 for domain reduction. Construction of $y_{N_2^P}^*$ can be viewed as a special case of the problem we recognized in Section 5.5 as being as hard as the problem of lower bounding the nonconvex problem at any node of the tree. However, when the relaxation technique obeys certain restrictions, $y_{N_2^P}^*$ can be computed efficiently. We assumed that the relaxations at node N_1 and node N_2 are identical when $x_j = x_j^B$ and that $x \leq x_j^B$ is active. These assumptions imply that not only is the optimal solution of the relaxation at N_1 optimal to P_{N_2}, but the geometry of the KKT conditions at optimality is identical apart from the activity of the constraint $x_j \leq x_j^B$. Therefore, it is possible to identify the set of active constraints for P_{N_2}. Once this is done, $y_{N_2^P}^*$ can be computed by solving a set of linear equations. In fact, the only change required to $y^* N_1$ is in the multiplier corresponding to $x_j \leq x_j^B$.

There is, however, a subtle point that must be realized while identifying the set of active constraints. A function $r(x)$ in the original nonconvex program may be relaxed to $\max\{r_1(x), \ldots, r_k(x)\}$ where each $r_i(\cdot)$ is separable and convex. In this case, the function attaining the supremum can change between the nodes N_1 and N_2. We illustrate this through the example of the epigraph of a bilinear function $z \leq x_j w$. z is often relaxed using the bilinear envelopes as follows (McCormick 1976):

$$z \leq \min\{x_j^L w + w^U x_j - x_j^L w^U, x_j^U w + w^L x_j - x_j^U w^L\}.$$

In particular, at node N_1:

$$z \leq \min\{x_j^L w + w^U x_j - x_j^L w^U, x_j^B w + w^L x_j - x_j^B w^L\}$$

and at node N_2:

$$z \leq \min\{x_j^B w + w^U x_j - x_j^B w^U, x_j^U w + w^L x_j - x_j^U w^L\}.$$

5.9. LEARNING REDUCTION PROCEDURE

When $x = x_j^B$ in the relaxation at node N_1, the dominating constraint is $z \leq x_j^B w + w^L x_j - x_j^B w^L = x_j^B w$ whereas in P_{N_2}, the dominating constraint is $z \leq x_j^B w + w^U x_j - x_j^B w^U = x_j^B w$. If μ is the dual multiplier for the constraint $z \leq x_j^B w + w^L x_j - x_j^B w^L$ in $y_{N_1}^*$, then μ should be the dual multiplier of $z \leq x_j^B w + w^U x_j - x_j^B w^U$ in $y_{N_2^P}^*$. Adding $(w^U - w^L)i(x_j^B - x_j) \geq 0$, results in the desired constraint:

$$z \leq x_j^B w + w^L x_j - x_j^B w^L.$$

Thus, it is possible that the multiplier corresponding to the upper bounding constraint $x_j \leq x_j^B$ increases when a dual solution is constructed for P_{N_2} from a dual solution of N_1. Such an increase would not occur if the only difference between the relaxations at N_1 and N_2 is the bound on x_j, which is most often the case with relaxations used for integer linear programs. A higher value of the multiplier implies smaller reduction and, therefore, it may seem that the learning reduction procedure will work better for integer linear programming problems. However, in integer linear programming, it is not necessary to bisect the axis of the branching variable from time to time to induce exhaustiveness. As a result, the learning reduction procedure may not be of much use for integer linear programs.

Algorithm SpecialReduce

Step 0. Set $K = 0$.

Step 1. Solve the relaxed dual:

$$h_U^K = \max_{(y_0,y)} \; (y_0 + a_0)b_0 + \langle y + a, b \rangle - z$$
$$\text{s.t.} \quad z \geq y_0 u_0^k + \langle y, u^k \rangle \quad\quad k = 1, \ldots, K-1$$
$$(y_0, y) \leq -(a_0, a)$$

Let the solution be (y_0^K, y^K).

Step 2. Solve the Lagrangian subproblem:

$$\inf_x l(x, y_0^K, y^K) = -\max_{x,u_0,u} \; y_0^K u_0 + \langle y^K, u \rangle$$
$$\text{s.t.} \quad f(x) \leq u_0$$
$$g(x) \leq u$$
$$x \in X$$

Let the solution be (x^K, u_0^K, u^K). Note that $(y_0, y) \leq 0$ implies that $u_0^K = f(x^K)$ and $u^K = g(x^K)$.

Step 3. Augment and solve relaxed primal problem:

$$h_L^K = \min_{(u_0,u)} \; a_0 u_0 + \langle a, u \rangle$$
$$\text{s.t.} \quad y_0^k u_0 + \langle y^k, u \rangle + \inf_x l(x, y_0^k, y^k) \leq 0$$
$$\quad\quad k = 1, \ldots, K$$
$$(u_0, u) \leq (b_0, b)$$

Step 4. Termination check: If $h_U^K - h_L^K \leq \epsilon$, stop; otherwise, set $K = K + 1$ and goto Step 1.

Figure 5.16: Equivalence with standard Lagrangian lower bounding

5.9. LEARNING REDUCTION PROCEDURE

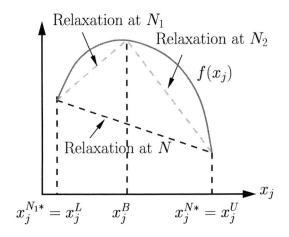

Figure 5.17: Branching constraint is not active

Chapter 6
Node Partitioning Schemes

Synopsis

In this chapter, we concentrate on branching strategies for mixed-integer nonlinear programs. We introduce the notion of an ideal violation and use it to develop a partitioning technique for factorable programs. Not only does this partitioning scheme lead to a convergent branch-and-bound algorithm but it is found to be practically efficient as well. In the second part of this chapter, we study finiteness issues for branch-and-bound. In particular, we develop a finite branching scheme for stochastic two-stage integer programs with pure-integer recourse.

6.1 Introduction

Rectangular partitioning—to which we restrict our attention—splits the feasible region at every branch-and-bound iteration into two parts by intersecting it with the half-spaces of a hyperplane parallel to one of the principal hyperplanes. In order to define such a hyperplane uniquely, the coordinate axis perpendicular to the hyperplane as well as the intercept of the hyperplane on the axis must be specified. The choice of the coordinate axis and the intercept are typically referred to as branching variable selection and branching point selection, respectively.

In Section 6.2 we develop a partitioning technique for factorable programs. We have found that this scheme works efficiently in practice. However, there does not seem to be any framework for understanding what constitutes a good branching decision in factorable programming. We make a systematic analysis by formalizing our intuitive judgment of a good branching scheme and then justifying our design choices in formulating the branching scheme in light of the identified characteristics.

In a related issue, we study the finiteness properties of branching schemes in stochastic programs in Section 6.3 where we develop the first branch-and-bound algorithm in the space of the first-stage variables that exhibits finite termination and is practically efficient. Our analysis in this section highlights the importance of research geared towards developing finite branching schemes for a larger class of mixed-integer nonlinear programs where such guarantees are virtually absent in current literature with the exception of a few results (Shectman & Sahinidis 1998, Al-Khayyal & Sherali 2000, Ahmed et al. 2000).

6.2 Partitioning Factorable Programs

In this section, we develop a rectangular partitioning scheme for factorable programs. Node partitioning is, at the time of this writing, the component of branch-and-bound that is least amenable to analysis. The schemes employed in the current implementations are mostly heuristically designed. Even for integer programming algorithms, which have been studied for decades, no thorough analysis of branching schemes has been done. For the larger class of factorable integer programs, the situation is much worse. The proposed schemes are not only rarely analyzed, but also do not seem to work well in practice. In this section, we take some steps towards remedying the current situation by developing a branching scheme for factorable programs and, whenever possible, providing insights and justifying our design choices.

6.2.1 Branching Variable Selection

Branching decisions are made either by computing a *violation* that measures the error introduced when a nonconvex function is underestimated by convex functions over a finite interval, or by estimating the improvement in the lower bound, *pseudo-cost*, as a result of branching on a certain variable. In essence, both estimates try to capture the effect of branching on the relaxation construction. In order to account for both feasibility errors and improvements in objective function value, researchers have often proposed a

6.2. PARTITIONING FACTORABLE PROGRAMS

branching strategy relying on a combination of the two estimates.

The challenge in devising branching schemes for mixed-integer nonlinear programs is amplified by the fact that the relaxations are often constructed in a higher-dimensional space than the original nonconvex problem. It is adequate, for the purpose of convergence, to branch on the space of original problem variables. However, it is not an effective strategy in practice. On the other hand, it is quite challenging to *meaningfully* project violations and/or pseudo-costs to the space of original variables. In this section, we introduce a scheme for performing such a projection.

Consider the following mixed-integer nonlinear program:

(P)
$$\begin{aligned}\min\ & f(x,y)\\ \text{s.t.}\ & g(x,y)\leq 0\\ & x\in X\subseteq \mathbb{Z}^p\\ & y\in Y\subseteq \mathbb{R}^n\end{aligned}$$

where $f:(X,Y)\mapsto \mathbb{R}$ and $g:(X,Y)\mapsto \mathbb{R}^m$ and its relaxation given below:

(R)
$$\begin{aligned}\min\ & \bar{f}(\bar{x},\bar{y})\\ \text{s.t.}\ & \bar{g}(\bar{x},\bar{y})\leq 0\\ & \bar{x}\in \bar{X}\subseteq \mathbb{Z}^{\bar{p}}\\ & \bar{y}\in \bar{Y}\subseteq \mathbb{R}^{\bar{n}}\end{aligned}$$

where $\bar{f}:(\bar{X},\bar{Y})\mapsto \mathbb{R}$ and $\bar{g}:(\bar{X},\bar{Y})\mapsto \mathbb{R}^{\bar{m}}$. We assume that there exists a mapping, $M:\mathbb{R}^{p+n}\mapsto \mathbb{R}^{\bar{p}+\bar{n}}$ such that for all $(x,y)\in(X,Y)$ with $g(x,y)\leq 0$, $M(x,y)\in (\bar{X},\bar{Y})$, $\bar{g}(M(x,y))\leq 0$ and $\bar{f}(M(x,y))=f(x,y)$. We assume that there exists a series of functional evaluations as below:

(E)
$$\begin{aligned}u_1 &= \gamma_1(x,y)\\ u_2 &= \gamma_2(x,y,u_1)\\ &\vdots\\ u_n &= \gamma_n(x,y,u_1,\ldots,u_{n-1})\end{aligned}$$

such that $M(x,y)$ can be easily retrieved from $(x,y,u_1,\ldots,u_n)$ and vice-versa.

As an example, consider the set $y\geq \log(x)-x^2$. The relaxation scheme might add the variables u_1 and u_2 and relax the following set:

$$\begin{aligned}y &\geq u_1+u_2\\ u_1 &\geq \log(x)\\ u_2 &\geq -x^2.\end{aligned}$$

The series of evaluations is given by:

$$u_1 = \log(x)$$
$$u_2 = x^2$$

and $M(x, y)$ is given by (x, y, u_1, u_2).

Definition 6.1. *Let $F(P)$ denote the feasible region of P and*

$$(x^*, y^*) \in \operatorname*{argmin}_{(x,y) \in F(P)} \|(\overline{x}, \overline{y}) - M(x, y)\|,$$

where $\|\cdot\|$ denotes some norm in $\mathbb{R}^{\overline{p}+\overline{n}}$. Then, the ideal violation is a vector $v(\overline{x}, \overline{y}) \in \mathbb{R}^{\overline{p}+\overline{n}}$ such that $v_j(\overline{x}, \overline{y}) = |(M(x^*, y^*))_j - (\overline{x}, \overline{y})_j|$. The ideal violation of a set S is defined by $v_j^S(\overline{x}, \overline{y}) = \max_{(\overline{x}, \overline{y}) \in S} v_j(\overline{x}, \overline{y})$.

The motivation for our definition should be clear. If $(\overline{x}^*, \overline{y}^*)$ is the optimal solution to the relaxation of P over a partition element Π and $v(\overline{x}^*, \overline{y}^*) = 0$, then Π can be fathomed since there exists an (x^*, y^*) in $F(P)$ such that $f(x^*, y^*) = \overline{f}(\overline{x}^*, \overline{y}^*)$. The ideal violation is indeed hard to compute. Therefore, our attempt will only be to compute a crude approximation of it.

We start with a degenerate hypercube, $H = (\overline{x}^*, \overline{y}^*)$, where $(\overline{x}^*, \overline{y}^*)$ is the optimal solution of the relaxation over the partition element Π in consideration. Then, we traverse the evaluation procedure E in reverse order, and enlarge H as necessary to accommodate at least one point that satisfies the relation in consideration. For example, let us say there is a bilinear relation $\overline{w} = \overline{x}\,\overline{y}$ in the evaluation procedure. We choose the point $(\hat{x}, \hat{y}, \hat{w})$ in H that is farthest from $\overline{w} = \overline{x}\,\overline{y}$. Then, we identify a point (preferably the closest to $(\hat{x}, \hat{y}, \hat{w})$) on the surface $\overline{w} = \overline{x}\,\overline{y}$ and enlarge H to include this point. At every step, we make sure that H does not cross the bounds available at the current partition element. Finally, the width of H along each coordinate direction provides the violation in that direction. Note that integrality restrictions are automatically handled by the proposed scheme since the minimum violation of a variable that is not integer-valued in the relaxation solution is its distance from the nearest integral point.

The procedure presented in the preceding paragraph is still fairly complex and an implementation should make further simplifications. We illustrate the nature of the possible simplifications using the example of a bilinear function $\overline{w} = \overline{xy}$. It is well-known that the optimum (maximum/minimum) of $\overline{w} - \overline{xy}$ over a hypercube is attained at one of the eight extreme points. Therefore,

6.2. PARTITIONING FACTORABLE PROGRAMS

given H we identify the extreme point, say $(\hat{x}, \hat{y}, \hat{w})$, that maximizes $|\bar{w} - \overline{xy}|$. We would now like to solve the following problem:

$$\min_{x,y,w}\{(x-\hat{x})^2 + (y-\hat{y})^2 + (w-\hat{w})^2 \mid w = x\,y\}. \quad (6.1)$$

The solution of the above mathematical program can be computed by analyzing the KKT conditions. However, the resulting solution involves finding roots of certain univariate polynomials, which is a nontrivial task. Therefore, we overestimate $x^* - \hat{x}$, $y^* - \hat{y}$ and $w^* - \hat{w}$, where (x^*, y^*, z^*) is the optimal solution of (6.1). We simply overestimate $w^* - \hat{w}$ by $\hat{x}\hat{y} - \hat{w}$. Then, we add the following constraints to (6.1):

$$\|x^* - \hat{x}\|_2 = \|y^* - \hat{y}\|_2$$
$$w^* = \hat{w}$$

and solve the augmented mathematical program by substituting $a = \|x - \hat{x}\|$. It is easy to see that the optimal a is the minimum norm solution to:

$$l(a) = \hat{w} - (\hat{x} + a)(\hat{y} \pm a) = 0.$$

Once the optimal value of a is known, x^* and y^* can be easily evaluated. Then, H is expanded to include the point $(x^*, y^*, \hat{x}\hat{y})$.

The violations provide, at least intuitively, the range in which the variable is likely to vary as the relaxation gap vanishes to zero. The violation is still to be translated into the change in the lower bound generated as a result of solving the relaxation. It should be noted that the variables are not allowed to change independent of each other. Therefore, it is not adequate to multiply the violation with the slope of the objective in the same direction. Instead, we recommend the following strategy.

Let the projection of the hypercube H on the jth axis be $[l_j, u_j]$. If $(\overline{x}^*, \overline{y}^*)$ is the optimal solution to the current relaxation, then our construction of H guarantees that $(\overline{x}^*, \overline{y}^*)_j \in [l_j, u_j]$. Assume further that we have successfully constructed H so that $M(x^*, y^*) \in H$, where (x^*, y^*) is the optimal solution of P restricted to Π. Now, for $M(x^*, y^*)$ to be optimal to any ensuing relaxation, the constraints must alter in an appropriate way to satisfy the KKT conditions. In order to account for the movement of the constraints, we take the following approach. We perturb the right-hand-sides of the constraints that are active at $(\overline{x}^*, \overline{y}^*)_j$ and inactive at $(M(x^*, y^*))_j$ in a way that they now become active at $(M(x^*, y^*))_j$. Then, we assume that the local geometry of KKT conditions does not change. The assumption is justified for linear relaxations whereas for nonlinear relaxations, the assumption is reasonable only if the gradients at $M(x^*, y^*)$ do not change drastically. If the local geometry of the KKT conditions does not change, the dual solution to the relaxation of P remains optimal and can be used to evaluate a Lagrangian lower bound for the perturbed problem. In the case of linear relaxations, if π^* is the optimal dual solution to the current relaxation, the process of converting violations to change in relaxation solution is approximated by $|\pi^{*t}A|v(\overline{x}^*, \overline{y}^*)$ where A is the matrix of the linear constraints. If we assume that the relaxation variables include as a subset, say S, the variables of P, then the branching variable k is chosen so that:

$$k \in \operatorname*{argmax}_{j \in S} \left\{ (|\pi^{*t}A|v(\overline{x}^*, \overline{y}^*))_j \right\}.$$

In order to ensure that the branching scheme is not polarized towards a few variables, we occasionally choose a variable whose bounds are the furthest amongst all the variables that contribute a nonzero violation.

6.2.2 Branching Point Selection

Contrary to integer linear programming, branching at the current relaxation solution does not exhaustively partition the feasible space and, as a consequence, branch-and-bound with such a partitioning scheme is not guaranteed to converge to the optimal solution (Horst & Tuy 1996). Another point of departure between relaxation schemes for integer nonlinear programs and integer linear programs is that for integer nonlinear programs the choice of branching point affects not only the right-hand-sides of the constraints but also the constraints themselves. For example, the slopes of the bilinear underestimators are dependent on the bounds of the independent variables. As a result, the choice of the branching point for integer nonlinear programming problems is more difficult.

The branching point is typically chosen by an application of the bisection rule, omega rule, or their modified versions (Liu, Sahinidis & Shectman 1996, Shectman & Sahinidis 1998). In BARON, the branching point is chosen as a convex combination of the current relaxation solution (omega rule) and the midpoint (bisection rule). The convex combination is taken to induce exhaustiveness in the partitioning scheme while biasing it towards the relaxation

6.2. PARTITIONING FACTORABLE PROGRAMS

solution. In case the primary cause for violation is integrality restrictions, the node is partitioned using the omega branching rule.

In the remainder of this section, we mathematically formulate the problem of identifying the branching point which maximizes the improvement of the relaxation lower bound. To keep our presentation simple, we limit our discussion to linear relaxations. Consider the following relaxation problem:

$$\min_{x} \quad cx$$
$$\text{s.t.} \quad A(x^L, x^U)x \geq p(x^L, x^U)$$

Note that A and p are functions of the bounds x^L and x^U on x. Let x_j be the branching variable and x_j^B be the branching point to be chosen. Let

$$x^a = [x_1^U, \ldots x_{j-1}^U, x_j^B, x_{j+1}^U, \ldots, x_n^U]$$

and

$$x^b = [x_1^L, \ldots x_{j-1}^L, x_j^B, x_{j+1}^L, \ldots, x_n^L].$$

Let L_a and L_b be the lower bounds on the optimization problem over $[x^L, x^a]$ and $[x^b, x^U]$ respectively. Then, $L = \min\{L_a, L_b\}$ is a valid lower bound on $[x^L, x^U]$ obtained by the solution of the following mathematical program:

$$L(x_j^B) = \min_{x, x_1, \lambda} \quad cx$$
$$\text{s.t.} \quad A(x^L, x^a)x_1 \geq p(x^L, x^a)\lambda$$
$$A(x^b, x^U)(x - x_1) \geq p(x^b, x^U)(1 - \lambda)$$
$$0 \leq \lambda \leq 1.$$

By LP duality,

$$L(x_j^B) = \max_{u,v} \quad vp(x^b, x^U) - s$$
$$\text{s.t.} \quad vA(x^b, x^U) = c$$
$$uA(x^L, x^a) = c$$
$$up(x^L, x^a) \geq vp(x^b, x^U) - s$$
$$u, v, s \geq 0.$$

Since we are interested in x_j^B that yields the highest lower bound L, we solve the following optimization problem

$$L = \max_{u,v,x_j^B} \; vp(x^b, x^U) - s$$

$$\text{s.t.} \quad vA(x^b, x^U) = c$$

$$uA(x^L, x^a) = c$$

$$up(x^L, x^a) \geq vp(x^b, x^U) - s$$

$$u, v, s \geq 0.$$

Note that, as soon as x_j^B is fixed, the above program decomposes into two linear programs which are the duals of the linear relaxations generated at the resulting children of the current node. Typically, A and p are linear functions of x_j^B. In such a case, the above optimization problem is a bilinear programming problem. Since it is generally hard to solve bilinear programming problems to optimality, either a local solution or some approximation thereof can be used to identify x_j^B. The discussion on the *learning reduction heuristic* in Section 5.9 provides some insights into approximating x_j^B.

6.3 Finiteness Issues and Stochastic Integer Programs

Contrary to its application for integer programming, branch-and-bound is not a finite algorithm for integer nonlinear programming. Finiteness can be easily retrieved by allowing an absolute tolerance on the objective function. Even so, the convergence often becomes painstakingly slow during the final phase of the algorithm. Therefore, it seems even from a practical standpoint, that research needs to be done to formulate finitely terminating schemes for global optimization problems. Some attempts at analyzing partitioning schemes have resulted in finite algorithms for certain special classes of global optimization problems including concave programming (Shectman & Sahinidis 1998, Al-Khayyal & Sherali 2000) and stochastic integer programming (Ahmed et al. 2000).

In this section, we construct a finite branch-and-bound algorithm for stochastic integer programs with pure-integer recourse by developing spe-

6.3.1 Stochastic Integer Programs

Traditional mathematical programming models assume that information affecting the decision is available before the optimization model is solved. However, many practical situations demand that important planning decisions be taken in uncertain environments. When the uncertainties unravel, it is often possible to take additional corrective and/or predictive measures. As an example, consider the textile industry, where a textile company typically orders raw materials and reserves production capacities long before the demand realizes. Once the demand realizes, the company can make better predictions for the future, and at the same time perform corrective actions by transferring raw-materials/production capacities from one product to another when such a transfer is technologically feasible and profitable. The stochastic programming approaches address situations of this sort by solving an optimization model that identifies decisions that are *expected* to maximize the utility of the decision maker over all possible realizations of the unforeseen events.

A two-stage stochastic program with pure-integer recourse is formulated as:

(2SSIP):
$$z = \min_{x} \left\{ c^t x + \mathrm{E}_{\omega \in \Omega}[Q(x, \omega)] \mid x \in X \right\}$$
$$Q(x, \omega) = \min_{y} \; f(\omega)^t y(\omega)$$
$$D(\omega) y(\omega) \geq h(\omega) + T(\omega) x$$
$$y(\omega) \in Y \subseteq \mathbb{R}^{n_2}$$

where $X \subseteq \mathbb{R}^{n_1}$, $c \in \mathbb{R}^{n_1}$ and $Y \subseteq \mathbb{R}^{n_2}$. For each $\omega \in \Omega$, $f(\omega) \in \mathbb{R}^{n_2}$, $h(\omega) \in \mathbb{R}^{m_2}$, and $D(\omega) \in \mathbb{R}^{m_2 \times n_2}$ and $T(\omega) \in \mathbb{R}^{m_2 \times n_1}$. In addition, we make the following additional assumptions:

(A1) The uncertain parameter ω follows a discrete distribution with finite support $\Omega = \{\omega^1, \ldots, \omega^S\}$ with $\Pr(\omega = \omega^s) = p^s$.

(A2) The second-stage variables y are purely integer, i.e., $y \in \mathbb{Z}^{n_2}$.

(A3) The technology matrix T linking the first- and second-stage problems is deterministic, i.e., $T(\omega) = T$.

198   CHAPTER 6. NODE PARTITIONING

(A4) The first-stage constraint set X is nonempty and compact.

(A5) $Q(x,\omega) < \infty$ for all $x \in \Re^{n_1}$ and all s.

(A6) For each s, there exists $u(\omega) \in \mathbb{R}_+^{m_2}$ such that $u(\omega)D(\omega) \leq f(\omega)$.

(A6) For each s, the second-stage constraint matrix is integral, i.e., $D(\omega) \in Z^{m_2 \times n_2}$.

Stochastic integer programs arise when the second stage involves, for example, scheduling decisions (Dempster 1982), routing decisions (Spaccamela, Rinnooy Kan & Stougie 1984), resource acquisition decisions (Bienstock & Shapiro 1988), fixed-charge costs (Bitran, Haas & Matsuo 1986), and changeover costs (Carøe, Ruszczyński & Schultz 1997). In addition, binary variables in the second stage can also arise in the modeling of risk objectives in stochastic linear programming (King, Takriti & Ahmed 1997).

Various algorithms for stochastic integer programming have been proposed by Klein Haneveld, Stougie & van der Vlerk (1995), Klein Haneveld, Stougie & van der Vlerk (1996), Laporte & Louveaux (1993), Carøe & Tind (1998), Schultz, Stougie & van der Vlerk (1998), and Carøe & Schultz (1999). For a detailed discussion on various algorithms for stochastic integer programming, refer to the recent surveys of van der Vlerk (1995), Schultz, Stougie & van der Vlerk (1996), and Stougie & van der Vlerk (1997). To the best of our knowledge, this literature provides no algorithms for 2SSIP which are finite and do not resort to explicit enumeration.

6.3.2 The Question of Finiteness

When viewed in the space $(x, y(\omega_1), \ldots, y(\omega_S))$, 2SSIP is an integer linear programming problem and in that form finite algorithms for it have been available for decades. However, it can be easily seen that, as the number of scenarios grows large, the number of variables in the problem grows rapidly. Since the time-complexity of algorithms for integer programming is typically exponential in the number of variables, such algorithms turn out to be inefficient in practice.

2SSIP can also be viewed as a global optimization problem in the space of the first-stage variables. $Q(\cdot, \omega)$ is a nonlinear function which is neither convex nor upper semicontinuous. It still differs from most integer nonlinear

6.3. FINITENESS ISSUES

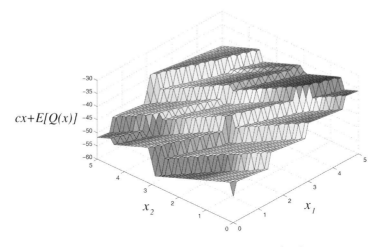

Figure 6.1: Objective function of (Ex)

programming problems addressed in this book in that the nonlinear formulation is not known explicitly and can only be evaluated at any point by solving a sequence of integer programs. In Figure 6.1, we illustrate the nonconvex nature of the value function by plotting the value function for the following example from Schultz et al. (1998):

$$
\begin{aligned}
\text{(Ex)} \quad z = \min \quad & -1.5x_1 - 4x_2 + E[Q(x_1, x_2, \omega_1, \omega_2)] \\
\text{s.t.} \quad & 0 \leq x_1, x_2 \leq 5 \\
Q(x_1, x_2, \omega_1, \omega_2) = \min \quad & -16y_1 - 19y_2 - 23y_3 - 28y_4 \\
\text{s.t.} \quad & 2y_1 + 3y_2 + 4y_3 + 5y_4 \leq \omega_1 - \frac{1}{3}x_1 - \frac{2}{3}x_2 \\
& 6y_1 + y_2 + 3y_3 + 2y_4 \leq \omega_2 - \frac{2}{3}x_1 - \frac{1}{3}x_2 \\
& y_1, y_2, y_3, y_4 \in \{0, 1\}.
\end{aligned}
$$

In the example above, $(\omega_1, \omega_2) \in \{5, 15\} \times \{5, 15\}$ with a uniform probability distribution.

6.3.3 Key to Finiteness

Instead of solving 2SSIP in the x-space, we view the problem in the space of the *tender* variables, $\chi = Tx$. For notational convenience, we define:

$$\Psi(Tx, \omega) = Q(x, \omega),$$

$$\zeta(x,\chi) = c^t x + \mathrm{E}_{\omega\in\Omega}[P(\chi,\omega)],$$
$$\mathcal{X} = \{\chi \mid \chi = Tx, x \in X\}.$$

It follows easily from perturbation analysis that for a given ω, $\Psi(\cdot,\omega)$ is lower semicontinuous, nondecreasing and piecewise constant. Further, Ψ exhibits a step only if $h_j(\omega) + \chi_j \in \mathbb{Z}$ for some $j \in \{1,\ldots,n_2\}$. Therefore, there exists a function $\gamma : \mathbb{Z}^{n_2} \times \Omega \mapsto \mathbb{R}$, nondecreasing in a, such that Ψ can be expressed as:

$$\Psi(\chi,\omega) = \sum_{a\in\mathbb{Z}^{n_2}} \gamma(a,\omega) U(a, \chi + h(\omega)), \qquad (6.2)$$

where

$$U(i,j) = \begin{cases} 1, & \text{if } i-1 < j \le i; \\ 0, & \text{otherwise.} \end{cases}$$

It is easy to compute $\gamma(a,\omega)$ since $\Psi(a - h(\omega),\omega) = \gamma(x,\omega)$. If, during the course of branch-and-bound, we isolate a rectangular partition element Π and there exists for every $\omega \in \Omega$ an $a_\omega \in \mathbb{Z}^{n_2}$ such that $a_\omega - 1 < \chi + h(\omega) \le a_\omega$ for all $\chi \in \Pi$, then $\Psi(\chi,\omega)$ is independent of χ over Π and equals $\gamma(a_\omega,\omega)$. We can therefore construct a relaxation that is gapless using the relation:

$$E_{\omega\in\Omega}[\psi(\chi,\omega)] = \sum_{\omega\in\Omega} p(\omega)\gamma(a_\omega,\omega),$$

where $p(\omega)$ is the probability associated with the occurrence of ω. Assumption (A4) guarantees the existence of a^L and a^U such that $a^L(\omega) < h(\omega) + Tx \le a^U(\omega)$ for every $x \in X$. Therefore, in order to explicitly characterize Ψ, we only need to compute $\gamma(a,\omega)$ over the finite set A given by:

$$A = \left\{ (a,\omega) \mid a^L(\omega) < a \le a^U(\omega), \omega \in \Omega \right\}.$$

6.3.4 Lower Bounding Problem

In this section, we construct the lower bounding problem over a partition element $\Pi = \prod_{j=1}^{m_2}(\chi_j^L, \chi_j^U]$. We assume that:

(A8) there exists an $\bar{\omega}_j \in \Omega$ such that $\chi_j^L + h_j(\bar{\omega}_j)$ is integral,

i.e., χ_j^L is a point with a potential discontinuity. The lower bounding problem is constructed as:

(LB): $$z^L(\Pi) = \min_{x,\chi} \; cx + \sum_{\omega\in\Omega} p(\omega)\Psi(\chi^L + \epsilon \cdot \mathbf{1}, \omega) \qquad (6.3)$$

6.3. FINITENESS ISSUES

Algorithm EpsilonCalculate

Do for $j = 1, \ldots, m_2$:

    Choose any $\omega_1 \in \Omega$

    For all $\omega \in \Omega$, compute:

$$\delta_j(\omega, \omega_1) = \lceil h_j(\omega) - h_j(\omega_1) \rceil + h_j(\omega_1) - h_j(\omega)$$

    Order $\{\delta_j(\omega, \omega_1) \mid \omega \in \Omega\}$ and rewrite as $d_1 \leq \cdots \leq d_{S-1}$

    Let $d_S = 1$

    Define the index set $Z = \{i \mid d_i \neq 0, d_{i+1} \neq d_i, i \in \{1, \ldots, S\}\}$

    Compute $\epsilon_j = \min_{i \in Z} d_i$

End Do

Set $\epsilon = 0.5 \times \min_{j=1,\ldots,m_2} \{\epsilon_j\}$

Figure 6.2: Procedure to calculate ϵ

$$\begin{aligned} \text{s.t.} \quad & x \in X \\ & Tx = \chi \\ & \chi^L \leq \chi \leq \chi^U \end{aligned}$$

where

$$\begin{aligned} \Psi(\chi, \omega) = \min_{y(\omega)} \quad & f(\omega)^t y(\omega) \\ \text{s.t.} \quad & D(\omega) y(\omega) \geq h(\omega) + \chi \\ & y(\omega) \in Y \cap \mathbb{Z}^{n_2}. \end{aligned} \quad (6.4)$$

In problem LB, ϵ is sufficiently small such that $\Psi(\cdot, \omega)$ is constant over $(\chi^L, \chi^L + \epsilon \cdot \mathbf{1}]$ for all ω where $\mathbf{1}$ is a vector of all ones. Such an ϵ can be computed by the procedure in Figure 6.2.

Theorem 6.2. *Algorithm EpsilonCalculate computes an ϵ which guarantees that $\Psi(\chi, \omega)$ is constant for all $\omega \in \Omega$ when $\chi \in (\chi^L, \chi^L + \epsilon]$ assuming χ_j^L satisfies (A8).*

Proof. We assumed that there exists an $\bar{\omega}_j$ for every $j \in \{1, \ldots, m_2\}$ such that $a_j = \chi_j^L + h(\bar{\omega}_j)$ is integral. Define:

$$\delta_j(\omega_1, \omega_2) = \lceil h_j(\omega_1) - h_j(\omega_2) \rceil + h_j(\omega_2) - h_j(\omega_1).$$

and η by the following relation:

$$\eta(j, \omega) = \lceil \chi_j^L + h(\omega) \rceil - \chi_j^L - h(\omega).$$

If $\epsilon < \eta(j, \omega)$ when $\eta(j, \omega) \neq 0$ and $\epsilon < 1$ when $\eta(j, \omega) = 0$, then there is no discontinuity perpendicular to χ_j at any point in $\chi_j \in (\chi_j^L, \chi_j^L + \epsilon]$. Also,

$$\begin{aligned}\eta(j, \omega) &= \lceil \chi_j^L + h_j(\omega) \rceil - \chi_j^L - h_j(\omega) \\ &= \lceil a_j - h_j(\bar{\omega}_j) + h_j(\omega) \rceil - a_j + h_j(\bar{\omega}_j) - h_j(\omega) \\ &= \lceil h_j(\omega) - h_j(\bar{\omega}_j) \rceil + h(\bar{\omega}_j) - h_j(\omega) \\ &= \delta_j(\omega, \bar{\omega}_j).\end{aligned}$$

Since Algorithm EpsilonCalculate computes ϵ satisfying the following relation:

$$\epsilon = \frac{1}{2} \min\{1, \min\{\delta_j(\omega_1, \omega_2) \mid \delta_j(\omega_1, \omega_2) \neq 0, j \in \{1, \ldots, m_2\}, \omega_1, \omega_2 \in \Omega\}\},$$

Ψ is constant over $(\chi^L, \chi^L + \epsilon \cdot \mathbf{1}]$. $\square$

Proposition 6.3. *For any partition element $\Pi = (\chi^L, \chi^U]$, satisfying (A8),*

$$z^L(\Pi) \leq \inf_{x, \chi} \left\{ c^t x + \mathrm{E}_{\omega \in \Omega}[\Psi(\chi, \omega)] \mid x \in X, \chi \in \Pi \cap \mathcal{X} \right\}.$$

Proof. Since for a given ω, $\Psi(\chi, \omega)$ is a nondecreasing function and constant over $(\chi^L, \chi^L + \epsilon \cdot \mathbf{1}]$ (Theorem 6.2), it follows that $\Psi(\chi^L + \epsilon \cdot \mathbf{1}, \omega) \leq \Psi(\chi, \omega)$ for all $\chi \in \Pi \cap \mathcal{X}$. $\square$

6.3.5 Upper Bounding

The reader may be caught off-guard upon realizing that LB is not a lower bounding scheme over its feasible set $F = \{x \in X, \chi \in \mathrm{cl}\,(\Pi) \cap \mathcal{X}\}$. On the contrary, coupled with Proposition 6.3, our next result shows that, when Π is small enough, LB provides a natural upper bounding solution for 2SSIP.

6.3. FINITENESS ISSUES

Proposition 6.4. *For any partition element* $\Pi = (\chi^L, \chi^U]$, *satisfying (A8),* $\Psi(\cdot, \omega)$ *is constant for all* $\omega \in \Omega$, *if* $(\bar{x}, \bar{\chi})$ *is the optimal solution to LB then:*

$$\min_{x,\chi}\left\{c^t x + E_{\omega \in \Omega}[\Psi(\chi, \omega)] \mid x \in X, \chi \in \mathrm{cl}\,(\Pi) \cap \mathcal{X}\right\} \leq \zeta(\bar{x}, \bar{\chi}) \leq z^L(\Pi).$$

Proof. Let

$$(x^*, \chi^*) \in \operatorname*{argmin}_{x,\chi}\left\{c^t x + E_{\omega \in \Omega}[\Psi(\chi, \omega)] \mid x \in X, \chi \in \mathrm{cl}\,(\Pi) \cap \mathcal{X}\right\}.$$

If $\bar{\chi}^* \in \Pi$, Ψ is independent of χ. Therefore,

$$z(x^*, \chi^*) \geq z^L(\Pi) = \zeta(\bar{x}, \chi^L + \epsilon \cdot \mathbf{1}) = \zeta(\bar{x}, \bar{\chi}) \geq z(x^*, \chi^*).$$

Hence the equality holds throughout. If $\bar{\chi}^* \notin \Pi$, then

$$z(x^*, \chi^*) \leq \zeta(\bar{x}, \bar{\chi}) \leq \zeta(\bar{x}, \chi^L + \epsilon \cdot \mathbf{1}),$$

where the second-inequality follows from the nondecreasing nature of Ψ and can be strict. □

If $(\bar{x}, \bar{\chi})$ is a solution to LB, an upper bound on 2SSIP is computed by evaluating $\zeta(\bar{x}, \bar{\chi})$.

Proposition 6.5. *If for a partition element* $\Pi = (\chi^L, \chi^U]$ *that satisfies (A8)* $\Psi(\cdot, \omega)$ *is constant for all* $\omega \in \Omega$, *then the partition element is fathomed.*

Proof. Follows directly from Proposition 6.4 since $\zeta(\bar{x}, \bar{\chi}) \leq z^L(\Pi)$. □

6.3.6 Branching Scheme

We assumed while constructing our lower bounding and upper bounding schemes that the branching scheme constructs partition elements which satisfy (A8). We can easily construct such a branching scheme by choosing j and ω, such that $\lfloor \chi_j^U + h_j(\omega) \rfloor - h_j(\omega)$, lies in $(\chi_j^L, \chi_j^U]$. Note that, if no such (j, ω) pair exists, then Ψ is constant over the current partition element and should be fathomed by Proposition 6.5. Figure 6.3 provides a somewhat more clever scheme which still obeys (A8). We assume in the description of Algorithm SipBranch that $y^*(\omega)$ attains the optimal solution to the second-stage problem in LB that corresponds to $\Psi(\chi + \epsilon \cdot \mathbf{1}, \omega)$.

Algorithm SipBranch

Compute
$$p_j = \min_{\omega \in \Omega}\left\{(D(\omega)y^*(\omega))_j - h_j(\omega)\right\}$$

Let $j' \in \operatorname{argmax}\{\min\{p_j - \chi_j^L, \chi_j^U - p_j\}\}$.

Choose j' as the branching variable and $p_{j'}$ as the branching point and construct
$$\Pi_1 = (\chi_{j'}^L, p_{j'}] \prod_{j \neq j'} (\chi_j^L, \chi_j^U)$$

and
$$\Pi_2 = (p_{j'}, \chi_{j'}^U] \prod_{j \neq j'} (\chi_j^L, \chi_j^U)$$

Figure 6.3: Branching scheme for 2SSIP

Theorem 6.6. *Algorithm SipBranch successfully partitions every unfathomed partition element Π that satisfies (A8) into two partition elements Π_1 and Π_2 that satisfy (A8).*

Proof. If the partition element $\Pi = (\chi^L, \chi^U]$ is unfathomed, Ψ is not constant over Π (Proposition 6.3). We prove by contradiction that there exists some k such that $p_k > 0$. Certainly $p_j > \chi_j^L + \epsilon$ since $y^*(\omega)$ is feasible when $\chi = \chi_j^L + \epsilon$. Therefore, p_j is nonpositive for every j if and only if $D(\omega)y^*(\omega) - h(\omega) \geq \chi^U$. In other words, $y^*(\omega)$ is feasible over the entire partition element. Since $\Psi(\chi^L + \epsilon \cdot \mathbf{1}, \omega) = f(\omega)^t y^*(\omega)$, $\Psi(\chi, \omega) \leq f(\omega)^t y^*(\omega)$. We know from Theorem 6.2 and the monotonicity of Ψ that $\Psi(\chi, \omega) \geq f(\omega)^t y^*(\omega)$. Hence, $\Psi(\chi, \omega) = f(\omega)^t y^*(\omega)$ over Π contradicting our assumption. Clearly, every point in Π belongs to either Π_1 or Π_2. If $p_{j'}$ is chosen as the branching point, the partition element containing $(\chi_{j'}^L, p_{j'}]$ satisfies (A8) trivially, while the partition element containing $(p_{j'}, \chi_{j'}^U]$ also does so because there exists an $\bar{\omega}$ such that $p_{j'} + h_{j'}(\bar{\omega}) = (D(\omega)y^*(\omega))_{j'}$ and the right-hand-side is clearly integral. □

6.3.7 Finiteness Proof

Theorem 6.7. *Consider the branch-and-bound algorithm with: LB as the lower bounding scheme, ϵ computed as in Algorithm EpsilonCalculate, an upper bounding scheme that evaluates ζ at the optimal point in LB, and the branching scheme presented in Algorithm SipBranch . This algorithm solves 2SSIP and terminates finitely.*

Proof. During the course of the algorithm, either a partition element is fathomed by Proposition 6.5 or refined into smaller partition elements by Theorem 6.6 at a point of potential discontinuity. Since X is compact, so is $\mathcal{X}$, and consequently there are only a finite number of potential discontinuities in the feasible region and therefore, the algorithm is finite. If x^* is the optimal solution to 2SSIP, then (x^*, Tx^*) is not fathomed until a $(\bar{x}, \bar{\chi})$ is located such that $\zeta(\bar{x}, \bar{\chi}) \leq \zeta(x^*, Tx^*)$ and $\bar{\chi} = T\bar{x}$. Therefore, $\bar{x}$ is also optimal to 2SSIP. □

6.3.8 Enhancements

The branch-and-bound algorithm proposed can be enhanced in a number of ways. We present some of the improvements in this section.

An alternative upper bounding scheme avoids the need to solve integer programs for upper bounding 2SSIP. Construct the vector p such that

$$p_j = \min_{\omega \in \Omega} \left\{ (D(\omega) y^*(\omega))_j - h_j(\omega) \right\}$$

where $y^*(\omega)$ attains the optimal solution to the second-stage problem in LB that computes $\Psi(\chi + \epsilon \cdot \mathbf{1}, \omega)$. Let $\bar{x} \in \mathrm{argmin}\, \{cx \mid x \in X, \chi \in [\chi^L, p^j] \cap \chi\}$. Then, $c\bar{x} + E_{\omega \in \Omega}[f(\omega)^t y^*(\omega)]$ is an upper bound on 2SSIP since $\Psi(T\bar{x}, \omega) \leq f(\omega)^t y^*(\omega)$ when $T\bar{x} \in [\chi^L, p^j]$.

The proposed lower bounding scheme can be used with any other lower bounding scheme that dominates LB. In particular, Benders cuts generated from the linear relaxation of the second-stage problems can be introduced to lower bound Ψ.

Another lower bounding scheme can be constructed my modifying the scenario disaggregation approach used by Carøe & Schultz (1999). The main idea is to create copies of the first-stage variables for each scenario as $x(\omega)$ with the additional constraint that $x(\omega_i) = x(\omega_j)$ for all $\omega_i, \omega_j \in \Omega$.

It may be noticed that it is not necessary to enforce all the equivalence constraints. Clearly, any spanning tree of a complete graph with Ω nodes, where each node is labelled with $x(\omega)$ and each arc by the equivalence relation between the nodes, is sufficient. If H is the arc-node incidence matrix of the spanning tree, then the following Lagrangian relaxation can be constructed by relaxing the equivalence relations:

$$z_{\text{LD}}(\Pi) = \max_\lambda L(\lambda, \Pi)$$

where

$$L(\lambda, \Pi) = \sum_{\omega \in \Omega} \min_{x(\omega), y(\omega)} (p(\omega)c(\omega)^t + (\lambda H)_\omega) x(\omega) + p(\omega) f(\omega)^t y(\omega)$$

$$\text{s.t. } \chi^L \leq Tx(\omega) \leq \chi^U$$

$$x(\omega) \in X$$

$$D(\omega) y(\omega) \geq h(\omega) + Tx(\omega)$$

$$D(\omega) y(\omega) \geq h(\omega) + \chi^L + \epsilon \cdot \mathbf{1} \qquad (6.5)$$

$$y(\omega) \in Y \cap \mathbb{Z}^{n_2}$$

where ϵ is calculated as in Algorithm EpsilonCalculate.

The following generalization of Proposition 6.3 holds:

Proposition 6.8. *If Π satisfies (A8), then*

$$z_L(\Pi) \leq z_{\text{LD}}(\Pi) \leq \inf_{x,\chi} \Big\{ c^t x + \mathrm{E}_{\omega \in \Omega}[\Psi(\chi, \omega)] \mid x \in X, \chi \in \Pi \cap \mathcal{X} \Big\}.$$

Proof. The construction of ϵ in Algorithm EpsilonCalculate guarantees that (6.5) does not chop off any solution that is valid for $\chi^L < Tx \leq \chi^U$. Therefore, the second inequality holds from the Lagrangian weak duality.

Let $\overline{x}(\omega), \overline{y}(\omega)$ be the optimal solution to the scenario subproblem corresponding to ω in $L(\lambda, P)$. Let $\tilde{x}, \tilde{\chi}$ be the optimal solution for LB with $\tilde{y}(\omega)$ as the solution to the scenario subproblem corresponding to ω in LB. Then, it follows that:

$$z_L(\Pi) = c^t \tilde{x} + \sum_{\omega \in \Omega} p(\omega) f(\omega)^t \tilde{y}(\omega)$$

$$\leq \sum_{\omega \in \Omega} p(\omega) c^t \overline{x}(\omega) + \sum_{\omega \in \Omega} p(\omega) f(\omega)^t \overline{y}(\omega)$$

$$= L(0, \Pi) \leq z_{\text{LD}}(\Pi),$$

6.3. FINITENESS ISSUES

where the first inequality follows from the fact that $\bar{x}(\omega), \bar{y}(\omega)$ is optimal to a more constrained problem than $\tilde{x}$ and $\tilde{y}(\omega)$. □

6.3.9 Extension to Mixed-Integer Recourse

In the presence of continuous variables in the second-stage, the orthogonality of discontinuities of $\Psi(\cdot, \omega)$ is lost. Therefore, the ideas presented in this section do not generalize easily to problems with mixed-integer recourse. It is indeed true that the continuous variables can be moved to the first stage to recover finiteness. However, doing so defeats the purpose of branching on the first-stage variables to reduce the dimension of the space of branching variables. In fact, it is possible to reorient the discontinuities arising from any subproblem in any arbitrary fashion with the help of the continuous variables in the second-stage problem. To support this claim, we modify Ψ in example Ex (Section 6.3.2) in a manner that the discontinuities in the χ-space in the modified problem are parallel to discontinuities in the x-space in Ex (see Figure 6.4):

$$\Psi(\chi_1, \chi_2, \omega_1, \omega_2) = \min \quad -16y_1 - 19y_2 - 23y_3 - 28y_4$$
$$\text{s.t.} \quad 2y_1 + 3y_2 + 4y_3 + 5y_4 \leq \omega_1 - y_5$$
$$6y_1 + y_2 + 3y_3 + 2y_4 \leq \omega_2 - y_6$$
$$-1.9706y_5 + 0.9706y_6 \leq -\chi_1$$
$$0.9706y_5 - 1.9706y_6 \leq -\chi_2$$
$$y_1, y_2, y_3, y_4 \in \{0, 1\}$$
$$y_5, y_6 \in [0, 5].$$

Blair & Jeroslow (1984) proved that the value function of a mixed-integer program is piece-wise polyhedral over certain cones in the space of the right-hand side vectors. At this point, it is not obvious how one can use these results to design a partitioning scheme that guarantees finiteness of the branch-and-bound algorithm.

6.3.10 Computational Results for Stochastic Programs

We demonstrate in this section that the proposed branch-and-bound scheme compares favorably to most algorithms in the literature for solving 2SSIP.

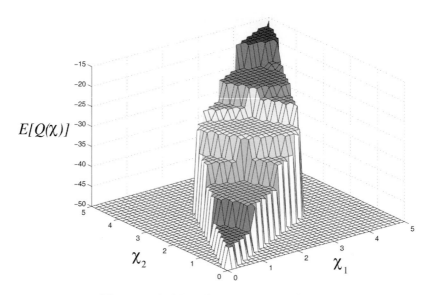

Figure 6.4: Mixed-integer second stage

For the purpose of demonstration, we use the *weaker* lower bounding problem LB instead of the Lagrangian relaxation developed in Section 6.3.8.

The test problems are generated from the following model of Carøe & Schultz (1999):

(EX1): $\quad\min\ -1.5x_1 - 4x_2 + E[Q(x_1, x_2, \omega_1, \omega_2)]$
$\quad\quad$ s.t. $\quad x_1, x_2 \in [0, 5] \cap \mathbb{Z}$,

where,

$$Q(x_1, x_2, \omega_1, \omega_2) = \min\ -16y_1 - 19y_2 - 23y_3 - 28y_4$$
$$\text{s.t.}\ \ 2y_1 + 3y_2 + 4y_3 + 5y_4 \leq \omega_1 - x_1$$
$$6y_1 + y_2 + 3y_3 + 2y_4 \leq \omega_2 - x_2$$
$$y_1, y_2, y_3, y_4 \in \{0, 1\},$$

where (ω_1, ω_2) is uniformly distributed on $\Omega \subseteq [5, 15] \times [5, 15]$. Five test problems are generated from the above instance by varying the number of scenarios by taking Ω as equidistant lattice points in $[5, 15] \times [5, 15]$ with equal probability assigned to each point. The resulting instances have 4, 9, 36, 121, and 441 scenarios.

6.3. FINITENESS ISSUES

	Carøe (1998)		Proposed	
Computer	Digital Alpha 500 MHz		IBM RS/6000 43P 332 MHz	
LINPACK	235.3		59.9	
Scenarios Obj.	CPU s.	IPs solved	CPU s.	IPs solved
4 57.00	0.2	52	0.01	48
9 59.33	0.4	189	0.01	135
36 61.22	1.4	720	0.01	540
121 62.29	4.8	2783	0.02	1815
441 61.32	25.1	9702	0.06	6615

Table 6.1: Computational results for Test Set 1

		Schultz et al. (1998)	Proposed
Problem	T Obj.	IPs solved	IPs solved
1	I 61.32	53361	12248
2	B 61.44	8379	4410

Table 6.2: Comparative performance for Test Set 2

Carøe (1998) points out that some of these problems can be solved by commercial integer programming solvers (CPLEX 5.0), thereby motivating the development of specialized approaches to stochastic programming. We compare the performance of our algorithm with the results reported by Carøe (1998):

Our second test set is a variant of EX1 modeled after Example 7.1 in Schultz et al. (1998). In particular, we drop the integrality requirements on the first-stage variables and use the following technology matrices:

$$I = \begin{pmatrix} 1 & 0 \\ 0 & 1 \end{pmatrix}, \quad B = \begin{pmatrix} 2/3 & 1/3 \\ 1/3 & 2/3 \end{pmatrix}.$$

Tables 6.2 and 6.3 provide the computational results with this test set.

Our final test set is a collection of two-stage stochastic product substitution problems described in Jorjani, Scott & Woodruff (1995) and Lokketangen & Woodruff (1996). Computational experience provided with CPLEX 5.0 in Jorjani et al. (1995) suggests that these problems are intractable for general-purpose integer programming solvers. The first- as well as second-stage problems in this test set are mixed-integer problems. Based on the

	Proposed Algorithm	
Computer	IBM RS/6000 43P 332 MHz	
	T=I	T=B
Scenarios	CPU s. IPs solved Obj.	CPU s. IPs solved Obj.
4	0.01 60 57.00	0.00 24 56.75
9	0.01 171 59.33	0.00 63 59.56
36	0.02 684 61.22	0.01 396 60.28
121	0.03 2299 62.29	0.02 1331 61.01
441	0.15 12348 61.32	0.05 4410 61.44

Table 6.3: Computational results for Test Set 2

Problem	Binary Variables	Continuous Variables	Constraints
SIZES3	40	260	142
SIZES5	60	390	186
SIZES10	110	715	341

Table 6.4: Sizes of Test Set 3

insight gained from the algorithm for 2SSIP, we branched on the second-stage variables in order to recover the finiteness while taking advantage of the problem structure. The algorithm was implemented in BARON, details of which are provided in Chapter 7. The sizes of these problem are reported in Table 6.4 and our computational experience is reported in Tables 6.5 and 6.6. To the best of our knowledge, this is the first algorithm to close the gaps for two of the problems of Table 6.5.

6.3. FINITENESS ISSUES

	Carøe (1998)			BARON		
Computer	Digital Alpha 550 MHz			IBM RS/6000 43P 332 MHz		
Problem	LB	UB	CPU s	LB	UB	CPU s
SIZES3	224.384	224.544	1,000	224.433		70.7
SIZES5	224.354	224.567	1,000	224.486		7,829.1
SIZES10	224.336	224.735	1,000	224.236	224.717	10,000.0

Table 6.5: Computational results for Test Set 3

Problem	Iterations	N^*_{mem}	$N^\dagger_{\text{sol}}$
SIZES3	1885	260	906
SIZES5	108,782	13,562	41,642
SIZES10	36,700	23,750	20,458

∗: N_{mem}: maximum nodes in memory

†: N_{sol}: node where best solution was found

Table 6.6: Nodes in branch-and-bound tree

Chapter 7
The Implementation

Synopsis

BARON (Sahinidis 1996, Ghildyal & Sahinidis 2001, Sahinidis 1999-2000, Tawarmalani & Sahinidis 1999b, Sahinidis & Tawarmalani 2000), the branch-and-reduce optimization navigator, includes an implementation of many of the global optimization algorithms described in this book. A software implementation of algorithmic work often reveals additional considerations and it is the purpose of this chapter to address some of these issues that we faced while implementing BARON. This chapter also highlights some useful enhancements to the branch-and-bound algorithm. Throughout this chapter, we assume that the problem being solved is a minimization problem.

7.1 Design Philosophy

An important aspect of a software is its design and adaptability to forthcoming changes and developments in the area. Our design of BARON relies heavily on the observation that, even though branch-and-bound provides a prototype for a global optimization algorithm, its various components including domain reduction, relaxation strategies, branching decisions, and upper bounding schemes are often largely problem-dependent. There does not seem, at the time of this writing, to be any automated procedure which provably or even empirically dominates all competing strategies in terms of time and space requirements. Therefore, BARON is conceptually classified into its core component and the modular components.

The core component of BARON is an implementation of the branch-and-bound algorithm for a general-purpose global optimization problem. In addition, it provides a library of utilities that ease the implementation of problem-specific enhancements (modules). For example, BARON's core includes an interface for various linear programming, quadratic programming, semidefinite programming and nonlinear programming solvers, and an implementation of specialized range reduction techniques, presolving methods, sparse matrix manipulation algorithms, debugging facilities, memory management functions, and various data structures and associated algorithms. The main reason behind providing a common interface for different solvers is to identify the desirable features in convex programming solvers from the viewpoint of a global optimization algorithm designer, to keep the implementation independent of any specific solver, to allow switching to new solvers, and to be able to reach a wider community of optimization researchers. The behavior of the branch-and-bound implementation is itself tuned through a series of options which include amongst others the ability to specify the module to use, various numerical tolerance values, node-selection strategies, parameters affecting the frequency or execution of different range reduction and upper bounding strategies, and the number of best solutions to locate before terminating the search.

The modular component of BARON is an implementation of problem-specific enhancements. A complete listing of BARON modules is provided in Table 7.1. As seen in this table, modules are currently available to solve various classes of concave minimization problems over linear constraints (SCQP, SCP, PES, FCP), fractional integer programs over linear constraints (FP), univariate unconstrained polynomial programs (POLY), multiplicative programs over linear constraints (LMP and GLMP), indefinite quadratic programs (IQP), mixed-integer linear and semidefinite programs (MILP and MISDP), and factorable programming problems.

All modules are capable of handling integrality requirements on some or all of the variables. The factorable programming module is the most general module at the time of this writing and solves factorable nonlinear and mixed-integer nonlinear programming problems, the subject of this book. Every module defines its own problem format so that the core of BARON is unaware of the problem format and representation. A general-purpose parser has been developed as part of the factorable programming module which defines a simple and intuitive context-free grammar for declaring variables, constraint equations, objective function, and the options affecting the performance of

7.2. PROGRAMMING LANGUAGES AND PORTABILITY

Module	What optimization problems it solves
SCQP	separable concave quadratic programming
SCP	separable concave programming
PES	concave objectives with power economies of scale
FCP	fixed-charge programming
FP	fractional programming
POLY	univariate polynomial programming
LMP	linear multiplicative programming
GLMP	general linear multiplicative programming
IQP	indefinite quadratic programming
MILP	mixed-integer linear programming
MISDP	mixed-integer semidefinite programming
NLP	factorable nonlinear programming

Table 7.1: BARON modules and their capabilities

the algorithm. A simple Web interface for BARON has also been deployed that accepts and solves problems specified in the developed grammar.

7.2 Programming Languages and Portability

The parser is written in YACC (yet another compiler compiler) and LeX (lexical analyzer). YACC is a pre-compiler that converts a Backus-Naur Form (BNF) of a LALR(1) grammar into a C code (Hopcroft & Ullman 1979, Aho, Sethi & Ullman 1986). Numerical computing components of BARON core are implemented in FORTRAN 90 and the data structures are implemented in C. The automatic differentiator and other parts of the factorable programming module are written in C. The Web interface is written in Perl. The choice of different languages was governed by considerations of execution speed, ease of usage, prior experience of the developers, and historical development reasons.

The code is mostly a mix of standard FORTRAN 90 and C. It is, therefore, portable. Thus far, it has been compiled on a number of different computing platforms, including Windows, Linux, IBM AIX, HP-UX, Sun OS, and Sun Solaris.

7.3 Supported Optimization Solvers

BARON has support for many linear programming solvers including CPLEX (ILOG 2000), OSL (IBM 1995), MINOS (Murtagh & Saunders 1995), and SNOPT (Gill, Murray & Saunders 1999). Nonlinear programs can be solved using MINOS or SNOPT. Semidefinite programming subproblems are currently solved using SDPA.

7.4 Data Storage and Associated Algorithms

The choice of structures for the storage of data depends primarily on the operations that need to be performed on the data. Therefore, the discussion of algorithms and data structures is inherently linked. In this section, we comment on the data structures we use in BARON and the rationale behind our choice.

7.4.1 Management of Work-Array

BARON core and all other modules except for the parser store all the permanent data in a work-array provided by the user. This design is primarily due to historical reasons. However, it does have its advantages. By restricting memory usage to a work-array, a tight control on the memory allocations within BARON is maintained. Also, fast memory management routines have been developed for allocating specialized structures and the dynamic memory allocation calls, which are often quite slow, are reduced significantly. The biggest disadvantage, however, is that BARON cannot dynamically increase the amount of space available for storing the tree. This problem can be circumvented by allowing the use of multiple work-arrays, which we plan to implement in the future.

The primary work-array is split up into an options array, a list of open nodes, heaps, module storage, and some smaller work-arrays. The options array contains the user-defined options. The data elements stored in the list of open nodes and the heaps are discussed in more depth in Section 7.4.2. The module storage includes the problem-specific information including problem-data and any data-space requested by the module.

7.4.2 List of Open Nodes

The main memory requirement of the branch-and-bound algorithm is for the storage of information specific to the nodes of the search tree. Information stored on each node includes bounds of the variables, a recommended branching variable and branching point, node-specific module storage, information to hot-start the solver at the node, node upper bound, and some additional information including the level of the node in the tree and generation index.

Depending on how the node-specific information is stored, the amount of space needed per node can vary from node to node. However, in order to simplify the data-management, we allocate a constant amount of space per node of the tree. This allows us to easily construct a stack of free nodes and a list of open nodes in a linear array. The requirement that the data storage per node is constant in size, however, imposes a restriction that modules can only store a constant amount of data per node in the BARON work-array. Under this paradigm, the number of open nodes that BARON can hold at any given time is pre-calculated by dividing the size of the work-array with the amount of space needed per node. Data-compression techniques are used to reduce memory requirements. As an example, the bounds on integer and binary variables are stored on bits instead of words, and the bounds on continuous variables are suitably rounded up/down to a fixed precision to save a significant amount of memory.

The order in which nodes are accessed depends on the node selection rule. The node selection rule is a mapping from nodes to their *suitability index*. The node regarded as most suitable is chosen for further exploration at every iteration of the branch-and-bound algorithm. The search tree can grow arbitrarily large while solving mixed-integer nonlinear programs. Even for integer linear programs, the bound on the size of the tree is rather large (exponential) compared to the input data size. As a result, traversing the list of open nodes to determine the most suitable node soon becomes a computationally prohibitive task. An indexed access into the list of open nodes in decreasing order of suitability is therefore essential. In BARON, we provide such an indexed access using the heap data structure. The logarithmic-time bound for the insertion, deletion, and updating steps guarantees that each of these operations is performed in polynomial time for integer programs, and fairly efficiently in practice for nonlinear programs.

When an improved upper bound is located, inferior nodes in the branch-and-bound tree need to be pruned so that the space occupied by these nodes is released for subsequent usage. This operation requires indexed access into the list of open nodes in decreasing order of the lower bound. For providing such an access capability, it suffices to maintain a heap using the negative of the lower bound as the key on which comparisons are made. It is possible to save some storage by maintaining the lower bounds in a *min-max* heap when the most suitable node corresponds to the one with lowest lower bound. Such a rule is often used, at least periodically, to guarantee convergence of the branch-and-bound algorithm. BARON does not make use of this space saving technique at this time.

7.4.3 Module Storage: Factorable Programming

In this section, we discuss the storage structures used in the factorable programming module. Factorable programming storage provides an example of how the modules store their information.

The parser reads the input file and generates binary trees that store the involved nonlinear expressions and lists that store the variable and equation declarations. Incorporating hash-table lookups for variables and equations can speed-up the parsing phase of the algorithm and is one of the improvements we plan to include in BARON.

Once the problem has been read, BARON traverses the binary trees to construct the following data structures:

1. The binary trees are recursively decomposed into linear, bilinear, logarithmic, power, and monomial expressions. The linear expressions are stored in a sparse matrix data structure while the nonlinear relations are stored in arrays linking the dependent and independent variables. Using expression unification, it is ensured that no relation is repeated in the reformulated problem. Further details on the recursive decomposition and relaxation construction are provided in Chapter 4. The reformulated problem is traversed at every node to accomplish the following tasks. First, interval arithmetic operations are used to perform range reduction on every relation (Kearfott 1996). Second, a relaxation of the nonconvex problem is constructed at every node of the branch-and-bound tree by relaxing each relation in the reformulated problem over the current partition element. For nonlinear convex relations, retained as such in the relaxation, the reformulated problem is traversed to compute the function and gradients when requested by the nonlinear

programming solver.

2. BARON converts the binary trees into evaluation procedures suitable for automatic differentiation purposes. The nonconvex problem is upper bounded in BARON using local optimization techniques. Most local solvers require function and gradient evaluation routines. Even though binary trees can be traversed to compute this information, such a procedure is extremely slow. Instead, BARON uses the automatic differentiation technique detailed in Section 7.5.

7.5 Evaluating Derivatives

Function gradients are often required for upper bounding the nonconvex program using local solvers. The technique used in BARON to evaluate derivatives is known as *automatic* or *algorithmic* differentiation. This differentiation technique is efficient and provides an accurate estimation of derivatives for functions defined by their evaluation procedures. For a comprehensive treatment of automatic differentiation, the reader is referred to Griewank (2000). The potential of the reverse method for algorithmic differentiation was realized by Ostrovskii, Volin & Borisov (1971). We briefly explain the main ideas behind the method using the example of a product function initially discussed by Speelpenning (1980):

$$f(x) = \prod_{i=1}^{n} x_i = x_1 \cdots x_n.$$

The symbolic derivative of $f(x)$ is given by:

$$\nabla f(x) = (\quad x_2 \cdots x_i x_{i+1} \cdots x_n,$$
$$\vdots$$
$$x_1 \cdots x_{i-1} x_{i+1} \cdots x_n$$
$$\vdots$$
$$x_1 \cdots x_{i-1} x_i \cdots x_{n-1} \quad).$$

which needs $O(n^2)$ computations. The numeric differencing technique can compute the gradients for the product term in $O(n)$ operations by utilizing the division operator (inverse of multiplication). However, in general, even

$$v_1 = x_1$$
$$v_2 = v_1 x_2$$
$$\vdots$$
$$v_i = v_{i-1} x_i$$
$$\vdots$$
$$v_n = v_{n-1} x_n$$

Table 7.2: Evaluation procedure for products

the numeric techniques take n function evaluations to compute the gradients. Algorithmic differentiation evaluates the derivative for the product in $O(n)$ time without resorting to division. In fact, for an arbitrary function, the gradient calculation can be done in a time proportional to the time required for performing one function evaluation. The reason algorithmic differentiation is more efficient than the other methods is that it stores intermediate quantities called "adjoints" and does not incur the cost of recomputing them.

In order to use the automatic differentiation algorithm to compute the gradients for the product function, we express the function in the form of an evaluation procedure shown in Table 7.2. We start by performing the computations listed in the evaluation procedure in forward mode (from top to bottom). Then, we set $\frac{dv_n}{dv_n} = 1$ and traverse the evaluation procedure in the reverse order (bottom to top). Note that we have already computed $\frac{dv_n}{dv_i}$ before we reach the ith step. Therefore, we can compute:

$$\frac{dv_n}{dv_{i-1}} = x_i \frac{dv_n}{dv_i},$$

and

$$\frac{dv_n}{dx_i} = v_{i-1} \frac{dv_n}{dv_i}.$$

It takes $O(n)$ steps for the forward traversal of the evaluation procedure and $O(n)$ steps to compute the gradients by traversing the evaluation procedure in the reverse order.

The ideas presented above are easily generalized to arbitrary evaluation procedures thereby providing a simultaneous function and gradient evaluator which requires less than five times the number of computations required to compute the original function. In practice, this bound has been observed to be significantly lower.

7.6 Algorithmic Enhancements

In this section, we present some of the subtle differences between BARON and other implementations of branch-and-bound methods.

7.6.1 Multiple Solutions

BARON is not only capable of identifying the global optimum for a mixed-integer nonlinear program, but can identify the K best solutions to the optimization problem at hand, where K is an option specified by the user. Note that this feature will work best when the sought-after solutions are isolated (separated by a certain distance). For most continuous problems, including linear programs, there are often infinite feasible solutions in the vicinity of the optimal solutions with approximately the same objective function value.

The ability to generate more than just the optimal solution was initially motivated by an application involving design of automotive refrigerants using global optimization techniques. This application is described in detail in Chapter 8. The need for finding multiple solutions to the problem was felt since empirical relations are used in the formulating the model. As a result, the model does not always predict the behavior of the molecules accurately. Further, the model selects molecules on the basis of thermodynamic efficiency of the associated refrigeration cycle, whereas practical usage demands that a refrigerant possess additional properties such as nontoxicity and environmental friendliness. It is certainly possible to identify the molecules one at a time by chopping off the previously known solutions. However, such a procedure is easily seen to be extremely cumbersome and time-consuming considering that 29 solutions were identified for the primary refrigerant cycle and over 3,000 solutions were found for the secondary refrigerant cycle using a similar model.

Indeed, the ability to find multiple solutions during the search requires certain changes to the standard branch-and-bound algorithm, particularly in the rules for pruning open nodes. BARON maintains for this purpose an upper bound which is provably valid for K best solutions. One such bound is available if K feasible solutions have been identified during the course of the algorithm. Another way is to derive an upper bound for the objective function value over all the feasible solutions to the problem. A bound of this sort is easily derived by lower bounding the problem with an objective function that is negative of the current objective. An upper bound for the search can also be introduced using a user-specified option. BARON then fathoms all nodes that are inferior than the computed or user-specified. In addition, nodes with extremely small volume in the space of the nonconvex variables, *i.e.*, those corresponding to a point within tolerance values, are also eliminated from the search space.

In order to store the K solutions, BARON maintains a heap of the solutions found during the search. Each time a candidate solution is found, BARON compares the solution vector with its database of solutions to ensure that the solution is unique up to a tolerance and is one of the K best solutions found. An ordered list of solutions is printed prior to termination of the algorithm.

7.6.2 Local Upper Bounds

BARON has the capability of using local upper bounds in the search tree. A standard branch-and-bound algorithm uses global upper bounds. Local upper bounds for the problem can be computed at any node of the tree by lower bounding the problem with an objective function that is negative of the original objective function. However, the upper bound so calculated is valid only for the subtree rooted at the node where the bound was derived. The local upper bound is used to perform range reduction tests and provide an expedited convergence.

7.6.3 Postponement

In branch-and-bound, a node is partitioned after the lower bounding problem is solved and a branching decision is made. However, it is not until the time that any of the generated nodes becomes the most *suitable* node for further exploration that the information in these nodes is accessed or altered.

BARON, therefore, refrains from partitioning the node right after the branching decision is made. Instead, it stores the branching decision and partitions the node only when it is chosen subsequently by the node selection rule. This decision was motivated by the fact that the memory requirements of the branch-and-bound tree can be slashed by as much as a factor of two.

An additional benefit can be reaped by altering the branch-and-bound algorithm slightly, especially when the lower bounding problem is solved by a computationally expensive but iterative algorithm like Benders decomposition, Lagrangian outer-approximation, or a dual ascent bundle method. All these methods provide a series of lower bounds before the optimum is attained and can be temporarily suspended and subsequently resumed if appropriate information about the state of the algorithm is maintained. For Lagrangian outer-approximation, this information might include the set of constraints binding at the optimal point in the current master problem. Suppose now that, before the convergence of the lower bounding algorithm, a lower bound is derived for the current node that is better than the lower bound for many other open nodes in the branch-and-bound tree. BARON can then postpone the node untouched until it becomes suitable for further exploration again at which point the lower bounding process will be resumed. A node can also be postponed if significant range reduction has been performed at the current node and the node has lost its appeal for further exploration at this time. A Benders decomposition approach to the time-dependent traveling salesman problem (Tawarmalani & Sahinidis 1996, Tawarmalani 1997, Tawarmalani & Sahinidis 2002a) gainfully employed this feature of BARON.

7.6.4 Finite Branching Schemes

Contrary to the case of combinatorial optimization problems, when applied to continuous global optimization problems, branch-and-bound is typically an infinite process. Results pertaining to finite rectangular branching schemes for continuous global optimization problems have been rather sparse in the literature (Shectman & Sahinidis 1998, Shectman 1999, Ahmed et al. 2000).

Finite branching schemes as developed in the context of separable concave programming by Shectman & Sahinidis (1998) produce beneficial results in the context of general global optimization problems and are adopted in BARON. The scheme relies on selecting a branching point based on an exhaustive partitioning scheme like bisection of the largest nonconvex interval, with the modification that the branching point is set to the incumbent whenever the latter lies in the current subdomain and is not one of the end-points of the interval of the selected branching variable. This renders the incumbent gapless and is the key to the finiteness property for separable concave programming.

For stochastic programs with pure-integer second-stage variables, a finite branch-and-bound algorithm was developed in Chapter 6. The algorithm obviates the need to branch on the second-stage integer variables thereby improving convergence. An important characteristic of the proposed algorithm is that the generated partition elements are not closed. Formally, the partition elements are expressible as Cartesian products of intervals which are open on the left end-point and closed on the right end-point. Since all the partition elements generated during the course of the algorithm share this characteristic, no special treatment is needed on the part of BARON. If, however, some module generates partition elements that are sometimes closed and sometimes not, then it is required to maintain this information in auxiliary data structures so that a distinction between the different partition types can be made.

7.7 Debugging Facilities

BARON has various built-in debugging facilities which help isolate the cause for any errors in its execution. For example, BARON can:

- dump all the information that it uses during the navigation of the search tree into a file;

- trace a point provided by the user to determine how exactly it is chopped off from the search tree;

- provide a detailed listing of the heaps, generated relaxation, and many other data structures at any instant during the progress of the algorithm.

7.8 BARON Interface

In order to test BARON on a wide variety of practically relevant problems we felt that its interface should be easy to use and comprehend. Since the current focus of BARON is on the optimization community, this requirement translated into interfacing the solver to a simple modeling language. The BARON parser provides such a language (see Figure 7.1). Even though the language currently lacks advanced features like set declarations and associated operations, we have found it fairly easy to translate a mathematical model into BARON input format in an automated fashion. However, in order to provide further flexibility, we have interfaced BARON to the GAMS modeling environment (Brook et al. 1988) as described by Sahinidis & Tawarmalani (2002) and illustrated in Chapter 11.

7.8. BARON INTERFACE

Figure 7.2 illustrates the output that is generated when the problem described in Figure 7.1 is input to BARON's parser. A typical input file is comprised of an options section, a module section, variable declarations, variable bounds, equation declarations, equation definitions and the objective specification. In the problem presented in Figure 7.1, there are no integrality requirements. However, we illustrate the syntax for imposing such restrictions in a commented line.

A password-protected online access to BARON was launched in October 1999. Through the online interface, BARON has been tested on a variety of problems and a reasonably large database of nonlinear mathematical programs has been cataloged for further development purposes. In Figure 7.3, we provide an annotated illustration of the BARON Web page (Tawarmalani & Sahinidis 1999a).

```
// Design of an insulated tank

OPTIONS{                    ← Relaxation Strategy
nlpdolin: 1;
dolocal: 0; numloc: 3;      ← Local Search Options
brstra: 7; nodesel: 0;      ← B&B options
nlpsol: 4; lpsol: 3;        ← Solver Links
pdo: 1; pxdo: 1; mdo: 1;
}
MODULE: NLP;                ← Domain Reduction Options
```

```
// INTEGER_VARIABLE y1;
POSITIVE_VARIABLES x1, x2, x4;
VARIABLE x3;

LOWER_BOUNDS{x2:14.7;   x3:-459.67;}

UPPER_BOUNDS{
x1:    15.1;   x2:    94.2;
x3:    80.0;   x4:  5371.0;
}

EQUATIONS e1, e2;
e1: x4*x1 - 144*(80-x3) >= 0;
e2: x2-exp(-3950/(x3+460)+11.86) == 0 ;
OBJ: minimize 400*x1^0.9 + 1000
    + 22*(x2-14.7)^1.2+x4;
```

Figure 7.1: Parser input for insulated tank design

7.8. BARON INTERFACE 227

```
============================================================
                    Welcome to BARON v. 5.0
              Global Optimization by BRANCH-AND-REDUCE
============================================================
                  Factorable Non-Linear Programming
============================================================
Preprocessing found feasible solution with value  .51948D+04
============================================================
         We have space for   24173   nodes in the tree
============================================================
Itn. no.  Open Nodes   Total Time  Lower Bound  Upper Bound
   1           1       000:00:00   .10000D+04   .51948D+04
   1           1       000:00:00   .30832D+04   .51948D+04
   7           0       000:00:00   .51948D+04   .51948D+04
```

```
                    *** Successful Termination ***


 Total time elapsed      :   000:00:00,   in seconds :   .18
    on preprocessing:        000:00:00,   in seconds :   .04
    on navigating    :       000:00:00,   in seconds :   .01
    on relaxed       :       000:00:00,   in seconds :   .06
    on local         :       000:00:00,   in seconds :   .00
    on tightening    :       000:00:00,   in seconds :   .01
    on marginals     :       000:00:00,   in seconds :   .00
    on probing       :       000:00:00,   in seconds :   .06


 Total no. of BaR iterations:       7
 Best solution found at node:       0
```

Figure 7.2: BARON output for problem in Figure 7.1

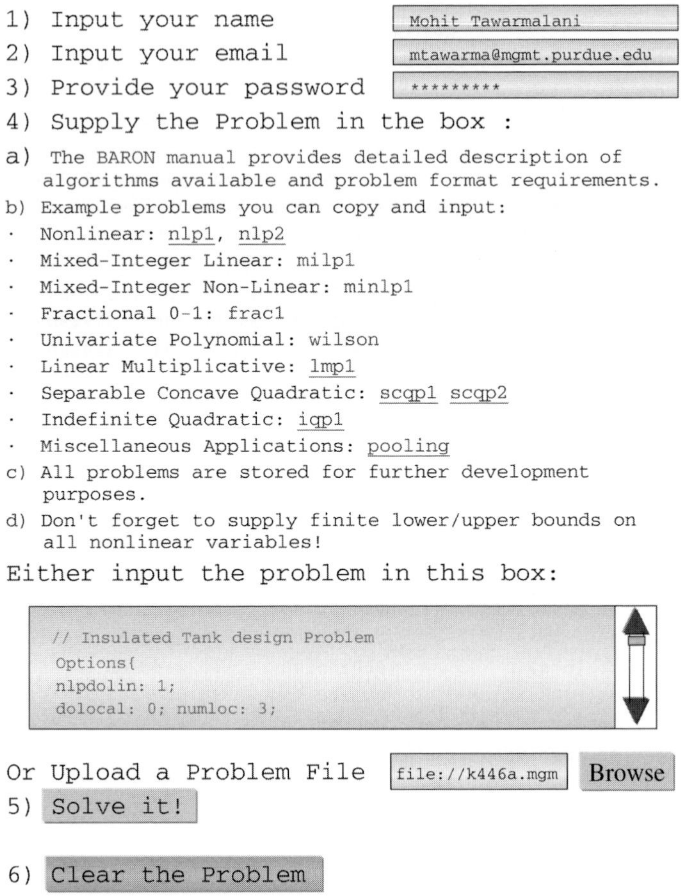

1) Input your name   [Mohit Tawarmalani]
2) Input your email   [mtawarma@mgmt.purdue.edu]
3) Provide your password   [*********]
4) Supply the Problem in the box :
a) The BARON manual provides detailed description of
   algorithms available and problem format requirements.
b) Example problems you can copy and input:
   · Nonlinear: nlp1, nlp2
   · Mixed-Integer Linear: milp1
   · Mixed-Integer Non-Linear: minlp1
   · Fractional 0-1: frac1
   · Univariate Polynomial: wilson
   · Linear Multiplicative: lmp1
   · Separable Concave Quadratic: scqp1 scqp2
   · Indefinite Quadratic: iqp1
   · Miscellaneous Applications: pooling
c) All problems are stored for further development
   purposes.
d) Don't forget to supply finite lower/upper bounds on
   all nonlinear variables!
Either input the problem in this box:

```
// Insulated Tank design Problem
Options{
nlpdolin: 1;
dolocal: 0; numloc: 3;
```

Or Upload a Problem File [file://k446a.mgm] [Browse]
5) [Solve it!]

6) [Clear the Problem]

Figure 7.3: BARON on the Web

Chapter 8

Refrigerant Design Problem

Synopsis

The enumeration of large, combinatorial search spaces presents a central conceptual difficulty in molecular design. To address this difficulty, we develop an algorithm which guarantees globally optimal solutions to a mixed-integer nonlinear programming formulation for molecular design. The formulation includes novel structural feasibility constraints while the algorithm provides all feasible solutions to this formulation through the implicit enumeration of a single branch-and-reduce tree. We use this algorithm to provide the complete solution set to the refrigerant design problem posed by Joback & Stephanopoulos (1990). In addition to rediscovering CFCs, the proposed methodology identifies a number of novel potential replacements of *Freon 12*. All of the identified alternatives are predicted to possess thermodynamic properties that would result in a more efficient refrigeration cycle than obtained via the use of *Freon 12*.

8.1 Introduction

Due to recent growing concerns regarding the depletion of the ozone layer by chlorofluorocarbons (CFCs), several actions have been taken to regulate the use of CFCs: the 1985 Vienna Convention, the 1987 Montreal Protocol, the Clean Air Act of 1990, and the 1997 Kyoto Protocol. In response, industry ended the production of CFCs in 1995 in developed countries and introduced transitional substances, such as hydrofluorocarbons (HFCs) and hydrochlorofluorocarbons (HCFCs) to replace CFCs. However, it is now clear that these are not ideal alternatives. These replacements are not as energy efficient as earlier refrigerants. Worse though, HCFCs still destroy the ozone layer, and, although HFCs may cause no damage to the ozone layer, they have strong green house effects. Today, global warming poses a problem as serious as the ozone depletion.

This chapter develops a systematic methodology for screening molecules with the aim of designing an environmentally benign and energy-efficient automotive refrigerant. Traditionally, molecular designers have followed the identify, generate, and test paradigm: candidate molecule is first identified by a group of scientists, then synthesized in the laboratory and finally tested for suitability by experimental studies. Identifying a molecule is a time-consuming process since the search space is large (Horvath 1992). Our global optimization algorithms come to aid by efficiently reducing the search space to a manageable size using property prediction techniques. As a result, we find a number of molecular designs that are novel. Some of these compounds exist in chemical databases but have never been used as refrigerants while others do not seem to have been discussed in the literature prior to the work described in this chapter.

8.2 Problem Statement

A mathematical abstraction of the problem is derived by viewing a molecule as a collection of atoms, or groups of atoms bonded with each other in a way that satisfies valence, connectivity and other chemical and physical constraints. The molecular design problem can be expressed in the form of the following optimization model:

$$\begin{aligned}
\min \quad & f(x,n) \\
\text{s.t.} \quad & g(x,n) \leq 0 \\
& x \in \mathbb{R}^m, n \in \mathbb{Z}_+^N.
\end{aligned}$$

In particular, for every $j \in \{1, \ldots, N\}$, n_j denotes the number of occurrences of atom (or group) of type j in the synthesized compound. The thermal, physical and chemical properties (heat capacity, enthalpy of vaporization, viscosity, etc.) are related to the atomic composition, n, through property prediction techniques, and are modeled through the continuous variables, x, in the optimization model above. The formulation may incorporate explicit environmental constraints, or $g(x,n) \leq 0$ can include constraints such as bounds on *ozone depletion potential* and *global warming potential* (for definitions of these terms and related discussion see Wuebbles (1981) and Wuebbles & Edmonds (1991)), if reasonably accurate models for predicting them are available as nonlinear relations.

The optimization problem described above is not easy to solve because of two reasons. First, the search space of the optimization problem is prohibitively large. A molecular design problem in which one is allowed to select up to K groups of atoms from a set of K potential groups in order to synthesize a compound, involves $\sum_{i=2}^{K} {}^{N+i-1}C_i$ potential combinations (Joback & Stephanopoulos 1995). For $N = 40$ groups, this number grows rapidly from $12,300$ when $K = 3$, to $9,366,778$ when $K = 6$, and to $2,054,455,593$ when $K = 9$. Second, the highly nonlinear nature of property prediction techniques makes development of efficient algorithms a challenging task (Reid, Prausnitz & Poling 1987, Horvath 1992). Tight structural constraints and clever optimization algorithms are needed to cope with the computational complexity of the molecular design problem.

8.3 Previous Work

The first systematic approach to designing refrigerants was pioneered by Midgley, Henne, and McNary who studied the periodic table and identified eight elements for further investigation. This drew their attention to halogens and led to the discovery of CFCs. An enormous amount of work has been done since the development of CFCs on issues regarding refrigeration and alternative refrigerants. We briefly overview the approaches that are closely related to our work.

Brignole, Bottini & Gani (1986), Gani, Nielsen & Fredenslund (1991), Gani & Fredenslund (1993), Constantinou & Gani (1994), Constantinou, Bagherpour, Gani, Klein & Wu (1996), and Pretel, Lopez, Bottini & Brignole (1994) proposed group contribution methods for molecular design using explicit enumeration to search the design space. In a mathematical programming approach, Macchietto, Odele & Omatsone (1990) and Odele & Macchietto (1993), Naser & Fournier (1991), Churi & Achenie (1996), Duvedi & Achenie (1996), Churi & Achenie (1997b), and Churi & Achenie (1997a) proposed the use of mixed-integer nonlinear programming techniques. Due to the nonconvex nature of their formulations, their approaches can at best guarantee locally optimal solutions. The approach of Joback & Stephanopoulos (1990) and Joback & Stephanopoulos (1995) avoids the multiple minima difficulty through the clever use of interval arithmetic techniques to construct bounds on the properties. However, interval lower bounds are known to be weaker than mathematical programming bounds. Joback & Stephanopoulos (1990) and Joback & Stephanopoulos (1995) eventually resort to an interactive heuristic search technique. Recognizing the limitations of standard optimization methods, some recent works have also proposed the use of guided stochastic search techniques Venkatasubramanian, Chan & Caruthers (1994), Venkatasubramanian, Chan & Caruthers (1995), Venkatasubramanian, Sundaram, Chan & Caruthers (1996), Devillers & Putavy (1996), Marcoulaki & Kokossis (1998), Ourique & Telles (1998).

8.4 Optimization Formulation

We detail our approach in the context of a case study that was initially posed by Joback & Stephanopoulos (1990) and Joback & Stephanopoulos (1995), and addressed in various forms by many others, including Gani et al. (1991), Pretel et al. (1994), Churi & Achenie (1996), Duvedi & Achenie (1996), Churi & Achenie (1997b), Churi & Achenie (1997a), Venkatasubramanian et al. (1995), Marcoulaki & Kokossis (1998), and Ourique & Telles (1998). It will be clear that our approach can be easily generalized to solve other models. In fact, variants of the model and algorithms presented in this work have been used by Nanda (2001) to screen secondary refrigerants for the retail food industry.

Our model uses the same physical property estimation relations as in Joback & Reid (1987), Joback & Stephanopoulos (1990), and Joback & Stephanopoulos (1995). We model these authors' structural constraints using integer variables and introduce new structural constraints that tighten the formulation but do not eliminate any chemically feasible molecules.

The goal of the case study is to design an alternative to dichlorodifluoromethane (CCl_2F_2), also denoted as $R12$, a refrigerant that in the past found wide use in home refrigerators and automotive air conditioners. It is assumed that the refrigerant will be used in a typical automotive refrigeration cycle (see Figure 8.1) with an evaporation temperature $T_{evp} = 30\,°F$, and an average operating temperature of $T_{avg} = 70°F$. The particular groups considered are shown in Table 8.1. In this table, "r" is used to denote bonds that are to

8.4. OPTIMIZATION FORMULATION

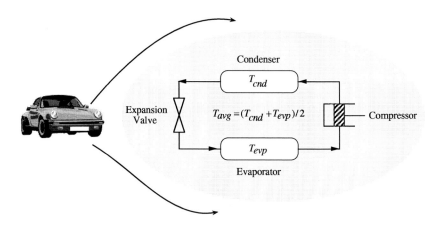

Figure 8.1: Automotive refrigeration cycle

be part of a ring only. A total of $N = 44$ groups are considered.
The optimization model takes the form:

$$\max \quad \frac{\Delta H_{\text{ve}}}{C_{\text{pla}}} \tag{8.1}$$

$$\text{s.t.} \quad \Delta H_{\text{ve}} \geq 18.4 \tag{8.2}$$

$$C_{\text{pla}} \leq 32.2 \tag{8.3}$$

$$P_{\text{vpe}} \geq 1.4 \tag{8.4}$$

$$P_{\text{vpc}} \leq 14 \tag{8.5}$$

Property Prediction Constraints

Structure Feasibility Constraints

$$n \in \mathbb{Z}_+^N$$

The case study requires that the new refrigerant's enthalpy of vaporization at T_{evp} be greater than or equal to that of $R12$ at the same temperature ($18.4 \frac{KJ}{g-mol}$). The motivation behind (8.2) is that a higher ΔH_{ve} reduces the amount of refrigerant required. Also, a smaller liquid heat capacity, C_{pla}, reduces the amount of refrigerant vapor generated in the expansion valve.

Acyclic Groups	Cyclic Groups	Halogen Groups	Oxygen Groups	Nitrogen Groups	Sulfur Groups
$-CH_3$	$^r-CH_2-^r$	$-F$	$-OH$	$-NH_2$	$-SH$
$-CH_2-$	$^r_r\!\!>\!CH-^r$	$-Cl$	$-O-$	$>\!NH$	$-S-$
$>\!CH-$	$_r\!\!>\!CH-^r$	$-Br$	$^r-O-^r$	$^r_r\!\!>\!NH$	$^r-S-^r$
$>\!C\!<$	$^r_r\!\!>\!C\!<^r_r$	$-I$	$>\!CO$	$>\!N-$	
$=CH_2$	$_r\!\!>\!C\!<^r_r$		$^r_r\!\!>\!CO$	$=N-$	
$=CH-$	$>\!C\!<^r_r$		$-CHO$	$^r=N-^r$	
$=C\!<$	$^r=CH-^r$		$-COOH$	$-CN$	
$=C=$	$^r=C\!<^r_r$		$-COO-$	$-NO_2$	
$\equiv CH$	$^r=C\!<^r_r$		$=O$		
$\equiv C-$	$\equiv C\!<^r_r$				

Table 8.1: Functional groups considered

8.4. OPTIMIZATION FORMULATION

For this reason, (8.3) enforces that the liquid heat capacity of the new refrigerant at T_{avg} be less than or equal to that of $R12$ at the same temperature ($32.2 \frac{cal}{g\text{-}mol \cdot K}$). To reduce the possibility of air and moisture leaking into the system, we require that the lowest pressure in the refrigeration cycle is greater than atmospheric pressure. In constraint (8.5), we impose a pressure ratio in the refrigeration cycle of no more than 10 to make sure that the size, weight, and cost of equipment required are reasonable. The rationale behind the objective function is that a large enthalpy of vaporization reduces the amount of refrigerant needed and a low liquid heat capacity reduces the amount of vapor generated in the expansion valve.

In Sections 8.4.1 and 8.4.2, we model the physical property relations and structure feasibility constraints, respectively.

8.4.1 Modeling Physical Properties

We include below the physical property estimation relations. The parameters used in the relations are detailed in Table 8.2.

Boiling temperature:
$$T_b = 198.2 + \sum_{i=1}^{N} n_i T_{bi}$$

Critical temperature:
$$T_c = \frac{T_b}{0.584 + 0.965 \sum_{i=1}^{N} n_i T_{ci} - (\sum_{i=1}^{N} n_i T_{ci})^2}$$

Critical pressure:
$$P_c = \frac{1}{(0.113 + 0.0032 \sum_{i=1}^{N} n_i a_i - \sum_{i=1}^{N} n_i P_{ci})^2}$$

Heat capacity at average temperature:

$$C_{p0a} = \sum_{i=1}^{N} n_i C_{p0ai} - 37.93 + \left(\sum_{i=1}^{N} n_i C_{p0bi} + 0.21\right) T_{avg}$$
$$+ \left(\sum_{i=1}^{N} n_i C_{p0ci} - 3.91 \times 10^{-4}\right) T_{avg}^2$$
$$+ \left(\sum_{i=1}^{N} n_i C_{p0di} + 2.06 \times 10^{-7}\right) T_{avg}^3$$

Reduced boiling temperature:
$$T_{br} = \frac{T_b}{T_c}$$

Reduced average temperature:
$$T_{avgr} = \frac{T_{avg}}{T_c}$$

Reduced condensing temperature:
$$T_{cndr} = \frac{T_{cnd}}{T_c}$$

Reduced evaporating temperature:
$$T_{evpr} = \frac{T_{evp}}{T_c}$$

Accentric factor:
$$\alpha = -5.97214 - \ln\left(\frac{P_c}{1.013}\right) + \frac{6.09648}{T_{br}} + 1.28862\ln(T_{br})$$
$$- 0.169347 T_{br}^6$$
$$\beta = 15.2518 - \frac{15.6875}{T_{br}} - 13.4721\ln(T_{br}) + 0.43577 T_{br}^6$$
$$\omega = \frac{\alpha}{\beta}$$

Liquid heat capacity:

8.4. OPTIMIZATION FORMULATION

$$C_{\text{pla}} = \frac{1}{4.1868}\left\{C_{\text{p0a}} + 8.314\left[1.45 + \frac{0.45}{1-T_{\text{avgr}}} + 0.25\omega\right.\right.$$
$$\left.\left.\left(17.11 + 25.2\frac{(1-T_{\text{avgr}})^{1/3}}{T_{\text{avgr}}} + \frac{1.742}{1-T_{\text{avgr}}}\right)\right]\right\}$$

Enthalpy of vaporization at boiling temperature:
$$\Delta H_{\text{vb}} = 15.3 + \sum_{i=1}^{N} n_i \Delta H_{\text{vbi}}$$

Enthalpy of vaporization at evaporating temperature:
$$\Delta H_{\text{ve}} = \Delta H_{\text{vb}}\left(\frac{1-T_{\text{evp}}/T_c}{1-T_b/T_c}\right)^{0.38}$$

Reduced vapor pressure at condensing temperature:
$$h = \frac{T_{\text{br}}\ln(P_c/1.013)}{1-T_{\text{br}}}$$

$$G = 0.4835 + 0.4605h$$

$$k = \frac{h/G - (1+T_{\text{br}})}{(3+T_{\text{br}})(1-T_{\text{br}})^2}$$

$$\ln P_{\text{vpcr}} = \frac{-G}{T_{\text{cndr}}}\left[1 - T_{\text{cndr}}^2 + k(3+T_{\text{cndr}})(1-T_{\text{cndr}})^3\right]$$

Reduced vapor pressure at evaporating temperature:
$$\ln P_{\text{vper}} = \frac{-G}{T_{\text{evpr}}}\left[1 - T_{\text{evpr}}^2 + k(3+T_{\text{evpr}})(1-T_{\text{evpr}})^3\right]$$

Vapor pressure at condensing temperature:
$$P_{\text{vpc}} = P_{\text{vpcr}}P_c$$

Vapor pressure at evaporating temperature:
$$P_{\text{vpe}} = P_{\text{vper}}P_c$$

N	total number of group types in the model (equals 44);
a_i	number of atoms of group i;
b_i	number of free bonds of group i;
ΔH_{vbi}	contribution of group i to enthalpy of vaporization at boiling temperature;
C_{p0ai}	constant contribution of group i to ideal gas heat capacity;
C_{p0bi}	first order contribution of group i to ideal gas heat capacity;
C_{p0ci}	second order contribution of group i to ideal gas heat capacity;
C_{p0di}	third order contribution of group i to ideal gas heat capacity;
P_{ci}	contribution of group i to critical pressure;
T_{bi}	contribution of group i to boiling temperature;
T_{ci}	contribution of group i to critical temperature;
T_{cnd}	condensing temperature (equals $316.48K$ or $110° F$);
T_{evp}	evaporating temperature (equals $272.04K$ or $30° F$).

Table 8.2: Parameters in property estimations

8.4.2 Modeling Structural Constraints

The purpose of the structural constraints is to eliminate combinations of groups that do not satisfy valence requirements. In this section, we model the structural constraints proposed by Joback & Stephanopoulos (1995), Odele & Macchietto (1993), Nanda (2001), and Nanda & Sahinidis (2002). Finally, we develop some new structural constraints that tighten the formulations of Joback & Stephanopoulos (1995), Odele & Macchietto (1993), and Nanda & Sahinidis (2002).

In the sequel, we associate a binary variable Y_X with a set of molecular groups X as follows:

$$Y_X = \begin{cases} 1, & \text{if } \sum_{i \in X} n_i \geq 1; \\ 0, & \text{otherwise.} \end{cases}$$

where n_i denotes the number of occurrences of group i in the final design. The relation can easily be modeled either by using big-M constraints or by relaxing:

$$(1 - Y_X) \sum_{i \in X} n_i = 0.$$

Joback-Stephanopoulos Structural Constraints

1. There must be at least two groups in a molecule:

$$\sum_{i=1}^{N} n_i \geq 2$$

2. If groups with acyclic bonds and groups with cyclic bonds are to be combined in a molecule, there must exist groups with both bond types:

$$\sum_{i \in M} n_i \geq 1 \text{ if } \sum_{i \in A} n_i \geq 1 \text{ and } \sum_{i \in C} n_i \geq 1$$

where M, A, and C, respectively, denote the union of groups having both acyclic and cyclic bonds, the union of groups having only acyclic bonds, and the union of groups having only cyclic bonds. The relation is modeled as $Y_M \geq Y_A Y_C$.

3. If there are groups with both acyclic and cyclic bonds, then there must be at least one acyclic group or one cyclic group, or both:
$$(Y_A + Y_C)Y_M \geq Y_M,$$
which is reformulated as $Y_A + Y_C \geq Y_M$.

4. Groups with cyclic bonds, if present, must be at least three in number:
$$Y_R \sum_{i \in C} n_i = 3Y_R,$$
where $R = C \cup M$, which can be rewritten in as a linear relaxation with a somewhat larger gap as:
$$3Y_R \leq \sum_{i \in C} n_i \leq N_{\max} Y_R |R|,$$
where $N_{\max}$ is an upper bound on all n_i. Similar linearization techniques can be used for all nonlinear structural constraints presented in this section.

5. The number of groups with odd bonds should be even:
$$\sum_{i \in B} n_i = 2Z_B,$$
where B is the set of groups with odd bonds and Z_B is restricted to be integral.

6. All groups must be connected:
$$\sum_{i=1}^{N} n_i b_i \geq 2 \left(\sum_{i=1}^{N} n_i - 1 \right).$$
The above constraint is easily seen to be true since any connected graph of n nodes has at least $n-1$ edges.

7. There should not be any free bonds in the molecule:
$$\sum_{i=1}^{N} n_i b_i \leq \sum_{i=1}^{N} n_i \left(\sum_{i=1}^{N} n_i - 1 \right).$$
The above constraint follows from the fact that maximum number of edges in a graph is $n(n-1)/2$.

8.4. OPTIMIZATION FORMULATION

8. For different bond types, there must exist transition groups (much in the spirit of item 2 above). For example, groups with both acyclic single and double bonds must exist where single-bonded groups are to be combined with double-bonded groups. The same is true for groups with both acyclic single and triple bonds, both acyclic and cyclic single bonds, both acyclic single and cyclic double bonds, and both acyclic double and cyclic single bonds:

$$\sum_{i \in SD} n_i \geq 1 \quad \text{iff} \quad \sum_{i \in S/D} n_i \geq 1 \text{ and } \sum_{i \in D/S} n_i \geq 1 \qquad (8.6)$$

$$\sum_{i \in ST} n_i \geq 1 \quad \text{iff} \quad \sum_{i \in S/T} n_i \geq 1 \text{ and } \sum_{i \in T/S} n_i \geq 1 \qquad (8.7)$$

$$\sum_{i \in SSR} n_i \geq 1 \quad \text{iff} \quad \sum_{i \in S/SR} n_i \geq 1 \text{ and } \sum_{i \in SR/S} n_i \geq 1 \qquad (8.8)$$

$$\sum_{i \in SDR} n_i \geq 1 \quad \text{iff} \quad \sum_{i \in S/DR} n_i \geq 1 \text{ and } \sum_{i \in DR/S} n_i \geq 1 \qquad (8.9)$$

$$\sum_{i \in DSR} n_i \geq 1 \quad \text{iff} \quad \sum_{i \in D/SR} n_i \geq 1 \text{ and } \sum_{i \in SR/D} n_i \geq 1, \qquad (8.10)$$

where the index set notation is explained in Table 8.3.

All the above constraints are of the form: *"Constraint x must be enforced if and only if constraints y and z are met."* To model (8.6), we begin by defining binary variables Y_{SDx}, Y_{SDy}, and Y_{SDz}, where:

$$Y_{SDx} = \begin{cases} 1, & \text{if } \sum_{i \in SD} n_i \geq 1 \\ 0, & \text{otherwise.} \end{cases}$$

$$Y_{SDy} = \begin{cases} 1, & \text{if } \sum_{i \in S/D} n_i \geq 1 \\ 0, & \text{otherwise} \end{cases}$$

$$Y_{SDz} = \begin{cases} 1, & \text{if } \sum_{i \in D/S} n_i \geq 1 \\ 0, & \text{otherwise} \end{cases}$$

Then, we introduce the constraints:

$$Y_{SDx} \leq \sum_{i \in SD} n_i \leq N_{max} Y_{SDx} \|SD\| \qquad (8.11)$$

$$Y_{SDy} \leq \sum_{i \in S/D} n_i \leq N_{max} Y_{SDy} \|S/D\| \qquad (8.12)$$

D/S	union of groups having acyclic double bonds and no acyclic single bonds;
D/SR	union of groups having acyclic double bonds and no cyclic single bonds;
DR/S	union of groups having cyclic double bonds and no acyclic single bonds;
DSR	union of groups having acyclic double bonds and cyclic single bonds;
S/D	union of groups having acyclic single bonds and no double bonds;
S/DR	union of groups having acyclic single bonds and no cyclic double bonds;
S/SR	union of groups having acyclic single bonds and no cyclic single bonds;
S/T	union of groups having acyclic single bonds and no triple bonds;
SD	union of groups having acyclic single bonds and acyclic double bonds;
SDR	union of groups having acyclic single bonds and cyclic double bonds;
SR/D	union of groups having cyclic single bonds and no acyclic double bonds;
SR/S	union of groups having cyclic single bonds and no acyclic single bonds;
SSR	union of groups having both acyclic and cyclic single bonds;
ST	union of groups having acyclic single bonds and acyclic triple bonds;
T/S	union of groups having acyclic triple bonds and no single bonds.

Table 8.3: Index notation for structural groups

8.4. OPTIMIZATION FORMULATION

$$Y_{SDz} \leq \sum_{i \in D/S} n_i \leq N_{max} Y_{SDz} \|D/S\| \tag{8.13}$$

$$Y_{SDy} + Y_{SDz} - 1 \leq Y_{SDx} \leq Y_{SDy} + Y_{SDz} \tag{8.14}$$

where $\|SD\|$, $\|S/D\|$, and $\|D/S\|$ are the cardinalities of SD, S/D, and D/S, respectively. Constraints (8.11)-(8.13) force variables Y_{SDx}, Y_{SDy}, and Y_{SDz} to a value of one when the corresponding condition is met. Constraint (8.14) enforces the logical requirement that $Y_{SDx} = 1$ if $Y_{SDy} = Y_{SDz} = 1$, while $Y_{SDx} = 0$ if $Y_{SDy} = Y_{SDz} = 0$. The modeling of (8.7) to (8.10), similarly, requires the introduction of triplets of binary variables $(Y_{STx}, Y_{STy}, Y_{STz})$, $(Y_{SSRx}, Y_{SSRy}, Y_{SSRz})$, $(Y_{SDRx}, Y_{SDRy}, Y_{SDRz})$, and $(Y_{DSRx}, Y_{DSRy}, Y_{DSRz})$, along with constraint sets analogous to (8.11)-(8.14) above.

In implementing the model, the above constraints were disaggregated in order to tighten their continuous relaxation. For example, the right side of constraint (8.11) was disaggregated as:

$$n_i \leq N_{max} Y_{SDx} \qquad i \in SD.$$

9. If there are mixed groups, then the number of acyclic groups is no more than the number of free bonds in other groups:

$$Y_M \sum_{\substack{i \in A \\ b_i = 1}} n_i \leq Y_M \sum_{i \in M} (b_i - 2) n_i + Y_M \sum_{\substack{i \in A \\ b_i \geq 3}} (b_i - 2) n_i.$$

The constraint is verified by considering a forest of $\sum_{i \in M}(b_i - 2)n_i$ rooted trees each with one dangling bond. If the degree of node i in a tree is $\deg(i)$, then $\sum_{i \in V}(\deg(i) - 2) = -1$, where V is the vertex set of the graph.

10. When there are no groups with cyclic bonds, i.e., $Y_R = 0$, there is a single tree with no dangling bond. Therefore:

$$(1 - Y_R) \sum_{i \in A} (b_i - 2) n_i = 2(1 - Y_R)$$

11. Clearly, each bond type, single, double or triple, has to be an even number:

$$\sum_{i \in S} n_i = 2 Z_S$$

$$\sum_{i \in D} n_i = 2Z_D$$

$$\sum_{i \in T} n_i = 2Z_T,$$

where Z_S, Z_D, and Z_T are integral.

Odele-Macchietto Structural Constraints

The above constraints are necessary but not sufficient to guarantee structural feasibility. Odele & Macchietto (1993) presented two constraints in a different approach to structural feasibility. If there are L cycles in a compound, then the first constraint is:

$$\sum_{i=1}^{N} n_i(b_i - 2) = 2(L - 1).$$

The second constraint is:

$$\sum_{i=1}^{N} n_i \geq n_j(b_j - 2) + 2 \quad \forall j \in A.$$

The compounds of type j might be bonded together. However, if $j \in A$, then at most $n_j - 1$ bonds can be shared between groups of type j. This leaves us with $n_j b_j - 2n_j + 2$ bonds which must connect to groups of other types.

Nanda-Sahinidis Constraints

The groups $> \mathrm{CO}$, ${}^r_r> \mathrm{CO}$, $-\mathrm{CHO}$, $-\mathrm{COOH}$ and $-\mathrm{COO}-$ are available in the set of potential groups and should not be formed using other groups for the prediction to be accurate. Nanda (2001) and Nanda & Sahinidis (2002) develop constraints which enforce that:

1. Groups $> \mathrm{C} =$ and $= \mathrm{O}$ can not be linked to form group $> \mathrm{CO}$.

2. Groups ${}^r_r > \mathrm{C} =$ and $= \mathrm{O}$ can not be linked to form group ${}^r_r > \mathrm{CO}$.

3. Groups $-\mathrm{CH} =$ and $= \mathrm{O}$ can not be linked to form group $-\mathrm{CHO}$.

4. Groups $> \mathrm{CO}$ and $-\mathrm{OH}$ can not be linked to form group $-\mathrm{COOH}$.

8.4. OPTIMIZATION FORMULATION

5. Groups $-CO-$ and $-O-$ can not be linked to form group $-COO-$.

We include the constraints described in Nanda (2001) and Nanda & Sahinidis (2002) in our formulation.

New Structural Constraints

Acyclic Constraints: We now present new structural constraints which chop off additional collections of groups that cannot bond to form a molecule. Assume that f^0 is a type of acyclic bond. If the forest formed by bonds of type f^0 has F_{f^0} components and f_i^0 denotes the number of bonds of type f^0 in the group i, then:

$$\sum_{i=1}^{N} n_i(f_i^0 - 2) = -2F_{f^0}. \tag{8.15}$$

We develop valid bounds on F_{f^0}. Let O_{f^0} denote groups with only bonds of type f^0, S_{f^0} the set of groups that have at least one bond of type f^0, and U the set of all groups. If some groups in the molecule belong to O_{f^0} but there are no groups in S_{f^0}, then the groups in O_{f^0} form a component by themselves. Otherwise, there exists in each component at least one group that bonds with a group outside the component and therefore:

$$F_{f^0} \leq \sum_{i \in U \setminus O_{f^0}} n_i. \tag{8.16}$$

Combining the two cases:

$$F_{f^0} \leq 2Y_{O_{f^0}} \left(1 - Y_{S_{f^0} \setminus O_{f^0}}\right) + \sum_{i \in U \setminus O_{f^0}} n_i. \tag{8.17}$$

From (8.15) with (8.17), we get:

$$\sum_{i \in O_{f^0}} n_i(2 - f_i^0) \leq \sum_{i \in U \setminus O_{f^0}} n_i f_i^0 + 2(1 - Y_{U \setminus O_{f^0}}). \tag{8.18}$$

If the bond types s^a, d^a, and t^a denote, respectively, the acyclic single, double, and triple bonds, then:

$$\sum_{i \in O_{s^a}} n_i(2 - s_i^a) \leq \sum_{i \in U \setminus O_{s^a}} n_i s_i^a + 2Y_{O_{s^a}} \left(1 - Y_{S_{s^a} \setminus O_{s^a}}\right)$$

$$\sum_{i \in O_{d^a}} n_i(2 - d_i^a) \leq \sum_{i \in U \setminus O_{d^a}} n_i d_i^a + 2Y_{O_{d^a}}\left(1 - Y_{S_{d^a} \setminus O_{d^a}}\right)$$

$$\sum_{i \in O_{t^a}} n_i(2 - t_i^a) \leq \sum_{i \in U \setminus O_{t^a}} n_i t_i^a + 2Y_{O_{t^a}}\left(1 - Y_{S_{t^a} \setminus O_{t^a}}\right).$$

There are seven such constraints depending on which classes (single, double, or triple) of acyclic bonds are included and which are left out. For example, when all acyclic bonds are included, (8.18) reduces to:

$$\sum_{i \in A} n_i(2 - b_i) \leq \sum_{i \in U \setminus A} n_i(s_i^a + d_i^a + t_i^a) + 2Y_A(1 - Y_M).$$

Clearly, $s_i^a + d_i^a + t_i^a > 0$ for $i \in U \setminus A$ only if $i \in M$. Therefore,

$$\sum_{i \in A} n_i(2 - b_i) \leq \sum_{i \in M} n_i(s_i^a + d_i^a + t_i^a) + 2Y_A(1 - Y_M).$$

In addition, the transition constraint $Y_{S_{f^0}} \geq Y_{O_{f^0}} Y_{U \setminus S_{f^0}}$ should be imposed.

A lower bound on F_{f^0} can also be developed easily. If S_{f^0} denotes the set of groups that have at least one bond of type f^0, then:

$$F_{f^0} \geq \sum_{i \in U \setminus S_{f^0}} n_i + Y_{S_{f^0}}. \qquad (8.19)$$

Simplifying (8.15), we get:

$$\sum_{i \in S_{f^0}} n_i f_i^0 \leq 2 \sum_{i \in S_{f^0}} n_i - 2Y_{S_{f^0}}.$$

Equation (8.15) can also be used in a hierarchical way if f^0 does not include all the acyclic bonds. Notice that, after contracting every component containing edges of type f^0 to a single node, the resulting graph, G^0, is still a simple tree with no loops or multiple edges. If we now analyze G for bonds of type f^1, then:

$$2(F_{f^0} - F_{f^{01}}) = \sum_{i=1}^{N} n_i f_i^1,$$

where $F_{f^{01}}$ is the number of components of the graph obtained after deleting all bonds that are not of type f^1 from G^0. Clearly, bounds on $F_{f^{01}}$ can be

8.4. OPTIMIZATION FORMULATION

used to develop valid constraints. We consider for example the bond type f^1 that matches all acyclic bond types except f^0. It follows from (8.19) that $F_{f^{01}} \geq \sum_{i \in C} n_i + Y_{U \backslash C}$. Performing Fourier-Motzkin elimination:

$$F_{f^0} \geq \frac{1}{2} \sum_{i=1}^{N} n_i f_i^1 + \sum_{i \in C} n_i + Y_{U \backslash C}.$$

Using the resulting lower bound, (8.15) yields:

$$\sum_{i=1}^{N} n_i (f_i^0 - 2) \leq -2 \sum_{i=1}^{N} n_i f_i^1 - 2 \sum_{i \in C} n_i - 2 Y_{U \backslash C}.$$

If we denote by a_i the number of acyclic bonds in group i, then the above constraint simplifies to:

$$\sum_{i \in U \backslash C} n_i (a_i - 2) + 2 Y_{U \backslash C} \leq 0.$$

Observe that the acyclic constraints developed here eliminate the group combination: 1 (> CH−), 3 (= O), and 1 (= N−), which is feasible to all the structural constraints of Joback & Stephanopoulos (1995) and Odele & Macchietto (1993).

Clique Constraints for Rings: Let r denote a bond type that can participate in rings. Consider the subgraph of the molecular structure obtained after deleting all edges except those corresponding to r. Clearly, each component of the induced subgraph has fewer edges than in a clique. If the number of components of the graph is F_r, then:

$$\sum_{i=1}^{N} n_i r_i \leq E\left(K_{\sum_{i=1}^{N} n_i - F_r + 1}\right), \tag{8.20}$$

where $E(\cdot)$ is the number of edges in $\cdot$, and K_n denotes the complete graph of n vertices. The relation can be verified by observing that each component has at least one vertex. If not, the edges are only increased by moving all vertices except one to one of the components. Let R_r denote the set of groups that have at least one bond of type r. Then, $F_r \geq Y_{R_r} + \sum_{i \in U \backslash R} n_i$. Therefore:

$$\sum_{i=1}^{N} n_i - F_r + 1 \leq \sum_{i \in R_r} n_i - Y_{R_r} + 1. \tag{8.21}$$

Using (8.20) and (8.21), we get:

$$\sum_{i \in R_r} n_i r_i \leq (n_{R_r} - Y_{R_r} + 1)(n_{R_r} - Y_{R_r}),$$

where $n_{R_r} = \sum_{i \in R_r} n_i$. The above relation is simplified to:

$$\sum_{i \in R_r} n_i r_i \leq n_{R_r}(n_{R_r} - 1), \quad (8.22)$$

by noting that $Y_{R_r} = 1$ whenever $n_{R_r} \geq 1$. The right-hand-side in the above constraint can be tightened to $3n_{R_r} - 6$ if it is assumed that the induced subgraph is planar (West 1996, Theorem 7.1.18).

We assume for the remainder of this section that r matches all ring bonds. If R_a is the number of acyclic bonds between groups in R, then clearly:

$$0 \leq 2R_a \leq n_R(n_R - 1) - \sum_{i \in R} n_i r_i. \quad (8.23)$$

If we denote the number of acyclic bonds of type f^0 between groups of R by R_{f^0}, then $0 \leq R_{f^0} \leq R_a$ and $R_{f^0} \leq \sum_{i \in S_{f^0} \cap M} n_i - 1$. Let G^R be the simple tree resulting after contracting the groups in R to one node that has R_{f^0} free bonds of type f^0. Then, from (8.15):

$$\sum_{i \in M} n_i f_i^0 - 2R_{f^0} - 2Y_R + \sum_{i \in A} n_i(f_i^0 - 2) = -2F_{fR0}, \quad (8.24)$$

where F_{fR0} is the number of components of G^R that remain after deleting all the bonds of type other than f^0. Clearly, using (8.19):

$$F_{fR0} \geq \sum_{i \in A \setminus S_{f^0}} n_i + Y_{S_{f^0}}.$$

Then, simplifying (8.24), we get:

$$\sum_{i \in M} n_i f_i^0 \leq 2R_{f^0} + \sum_{i \in A \cap S_{f^0}} n_i(2 - f_i^0) + 2(Y_R - Y_{S_{f^0}}). \quad (8.25)$$

Consider the collection of groups: 2 ($_r$> C <$_r^r$), 1 (= C <$_r^r$), and 1 (= O), which was feasible to the constraints of Joback & Stephanopoulos (1995) and Odele & Macchietto (1993) and is eliminated by constraint (8.23). The group combination: 3 ($_r$> CH– ^{r}) and 1(−F) is eliminated by (8.25).

8.5 Multiple Solutions

In formulating a mathematical program for refrigerant design, we made use of property prediction techniques based on group contribution methods. These prediction techniques are empirical approximations and do not, in general, provide extremely accurate property estimates. Therefore, it is imperative that many good solutions to the refrigerant design problem be identified by the algorithm. As explained in Section 7.6.1, our implementation is capable of isolating not just the best but K best solutions during the search procedure. This feature works well for the refrigerant design problem since the solution set is discrete. It is certainly possible to eliminate a collection of groups, say $\bar{n}$, after it is found in the search tree by introducing the reverse convex constraint:

$$\|n - \bar{n}\|_p \geq 1,$$

where $\|\cdot\|_p$ denotes the l_p norm. However, such a procedure is cumbersome, time-consuming, and requires manual intervention and dynamic model changes that alter the constraint matrix data structures significantly. For this reason, the procedure described in Section 7.6.1 identifies all the solutions in a much shorter time since the search is combined in one tree.

8.6 Computational Results

Using BARON, we identified all the molecules that are feasible to the optimization model in about half an hour of CPU time on an IBM/RS 6000 43P workstation with a 332 MHz processor and a LINPACK rating of 59.9.

The complete set of solutions identified for the above model is shown in Tables 8.4 and 8.5. Among these 29 compounds, one notices CH_3Cl, a well-known former refrigerant. One also observes that, in the identified compounds, there are 41, 38, 5, 11, and 8 groups containing C, F, Cl, O, and N, respectively. It is rather remarkable that the model clearly points towards CFCs as the ideal refrigerants.

Among the solutions of Tables 8.4 and 8.5, the following ten compounds were not identified in earlier works on this problem:

1. F – N = O

2. Cl – O – F

3. $CH_2 = C = C = O$

4. $CH_3 - CH = C = O$

5. $F - O - CH = C = O$

6. $CH_2 = C = CH_2$

7. $(F-)(CH3-) > C = C = O$

8. $CH \equiv C - O - F$

9. $CH_2 = CH - O - F$

10. $CH2 = C < (-O - F)(F-)$

Except for the first compound on this list (nitrosyl fluoride), which is available commercially, minimal information currently exists in the literature about the other compounds. While many or all of these compounds may turn out to be unstable, toxic, or otherwise not fit to be used as refrigerants, their identification demonstrates that the proposed methodology has the potential of discovering novel compounds.

8.6. COMPUTATIONAL RESULTS

	Molecular Structure	$\frac{\Delta H_{ve}}{C_{pla}}$
FNO	F – N = O	1.2880
FSH	F – SH	1.1697
CH_3Cl	CH_3 – Cl	1.1219
ClFO	(Cl–)(–O–)(–F)	0.9822
C_2HClO_2	O = C < (–CH = O)(–Cl)	1.1207
C_3H_4O	CH_3 – CH = C = O	0.9619
C_3H_4	CH_3 – C ≡ CH	0.9278
C_2F_2	F – C ≡ C – F	0.9229
CH_2ClF	F – CH2 – Cl	0.9202
C_2HO_2F	F – O – CH = C = O	0.8705
C_3H_4	CH_2 = C = CH_2	0.8656
C_2H_6	CH_3 – CH_3	0.8632
C_3H_3FO	(F–)(CH3–) > C = C = O	0.8531
NHF_2	F – NH – F	0.8468
C_2HOF	CH ≡ C – O – F	0.8263

Table 8.4: Set of solutions to refrigerant design problem (Part I)

	Molecular Structure	$\frac{\Delta H_{ve}}{C_{pla}}$
C_3H_3F	CH ≡ C − CH$_2$ − F	0.7802
CHF_2Cl	(F−)(F−) > CH − Cl	0.7770
C_2H_3OF	CH$_2$ = CH − O − F	0.7685
NF_2Cl	(F−)(F−) > N − Cl	0.7658
C_2H_6NF	(CH3−)(CH3−) > N − F	0.6817
N_2HF_3	(F−)(F−) > N − NH − F	0.6711
$C_2H_2OF_2$	CH2 = C < (−O − F)(−F)	0.6705
$C_3H_2F_2$	(F−)(F−) > CH − C ≡ CH	0.6686
C_2HNF_2	CH ≡ C − N < (−F)(−F)	0.6587
$C_3H_4F_2$	(F−)(F − CH$_2$−) > C = CH$_2$	0.6377
$C_3H_4F_2$	(F−)(F−) > CH − CH = CH$_2$	0.6263
$C_2H_3NF_2$	CH2 = CH − N < (−F)(−F)	0.6176
CH_3NOF_2	(F−)(CH$_3$−) > N − O − F	0.6139
$C_3H_3F_3$	($_r$> CH− ^{r})$_3$(−F)$_3$	0.5977

Table 8.5: Set of solutions to refrigerant design problem (Part II)

Chapter 9

The Pooling Problem

Synopsis

Convexification techniques based on disjunctive programming have been extensively employed to solve many hard combinatorial optimization problems. In this chapter, we demonstrate the potential of convexification techniques in the context of continuous nonlinear programming by applying them to the pooling problem. The pooling problem is a nonlinear network flow problem that models the operation of the supply chain of petrochemicals where crude oils are mixed to produce intermediate quality streams that are blended at retail locations to produce the final products. The main challenge in finding optimal solutions to pooling problems is that the nonlinearities result in many local optima. We address the question of finding a tight polyhedral outer-approximation of the convex hull of solutions of the pooling problem. In particular, we consider a formulation of the pooling problem that is derived after the addition of one class of nonlinear equalities to the classical q-formulation of the problem. These constraints are obtained using the reformulation-linearization technique (Sherali & Adams 1990). Using convexification and disjunctive programming arguments, we prove that the linear programming relaxation of this formulation dominates the linear programming and Lagrangian relaxations of the classical p- and q-formulations of the problem. We finally present extensive computational results demonstrating that the use of this tight formulation in the context of a spatial branch-and-bound algorithm makes it trivial to solve to global optimality all pooling problems from the open literature.

9.1 Introduction

Motivated by the seminal works of Rockafellar (1970) and Balas (1998), convexification techniques based on disjunctive programming have been studied extensively in the integer programming literature (Balas 1985, Jeroslow 1989, Sherali & Adams 1990, Lovàsz & Schrijver 1991, Balas, Ceria & Cornuèjols 1993) and have been gainfully employed to solve many hard combinatorial optimization problems. In contrast, development of convexification techniques in continuous nonlinear programming has been somewhat limited (Rikun 1997, Sherali & Adams 1999, Tawarmalani & Sahinidis 2002a). In this chapter, we demonstrate the potential that convexification techniques hold in developing and analyzing relaxations for nonconvex nonlinear programs. We do so by applying these techniques in the context of the pooling problem.

The pooling problem arises in the petrochemical and agriculture industries, when streams such as different crude oils or various types of animal feeds are mixed together to produce end-products that must satisfy certain requirements on the qualities of attributes like chemical composition and nutrient content. As shown in Figure 9.1, the pooling problem may be defined as a network flow problem over three sets of nodes: supply, transshipment, and demand nodes. Supply nodes represent the raw materials that flow to final product destinations (demand nodes) either directly or indirectly through pools (transshipment nodes). The unit costs as well as attributes, such as component concentrations, of raw materials and final products are given. The problem then calls for finding the optimal flows in the network so as to maximize net profit. Nonlinearities arise in attribute balances around pools since the pool attribute qualities as well as the inflows and outflows are all variables.

Numerical algorithms for solving pooling problems have included sensitivity and feasible analysis and local optimization techniques. We refer the reader to Greenberg (1995) for a review of these techniques and to Rigby, Lasdon & Waren (1995) for a discussion of successful industrial applications. More recently, global optimization algorithms have also been proposed (Foulds, Haugland & Jornsten 1992, Ben-Tal, Eiger & Gershovitz 1994, Quesada & Grossmann 1995b, Visweswaran & Floudas 1996, Adhya, Tawarmalani & Sahinidis 1999, Audet, Brimberg, Hansen & Mladenovic 2000). Even though pooling problems have been demonstrated to exhibit multiple local minima (Haverly 1978, Adhya, Tawarmalani & Sahinidis 1999), the application of global optimization algorithms to pooling continues to be a

9.1. INTRODUCTION

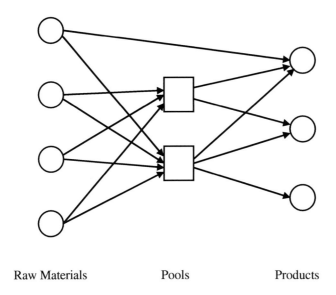

Raw Materials   Pools   Products

Figure 9.1: The pooling problem

challenge due to the slow convergence of the proposed methods, apparently due to lower bounding schemes that provide weak approximations of the convex hull of solutions.

From the modeling point of view, two distinct, but equivalent, bilinear programming formulations have been proposed for the pooling problem. Following the terminology of Ben-Tal, Eiger & Gershovitz (1994), we refer to these formulations as the "p-formulation" and the "q-formulation." The p-formulation, which first appeared in the works of Haverly (1978) and Haverly (1979), models the pooling problem using flows and pool attribute qualities as the variables. Instead of using variables for the pool input flows, the more recent q-formulation proposed by Ben-Tal et al. (1994) models the problem using variables for the relative proportions of pool input flows.

Some computational comparisons have been made with different pooling problem relaxations (Adhya, Tawarmalani & Sahinidis 1999, Audet, Brimberg, Hansen & Mladenovic 2000). However, the relaxation quality of formulations for the pooling problem has not been analytically investigated heretofore.

In this chapter, we develop an efficient global optimization algorithm for the pooling problem. Towards this end, we present a comprehensive analysis of the quality of relaxations for the problem and use convexification and disjunctive programming techniques in order to prove dominance results for a variant of the q-formulation.

The remainder of this chapter is structured as follows. Section 9.2 reviews the p- and q-formulations of the pooling problem as well as their corresponding linear programming relaxations. Starting from the q-formulation, in Section 9.3, we add a few valid constraints to derive a third formulation for the pooling problem. The constraints are derived using the reformulation-linearization technique (Sherali & Adams 1990). We refer to the resulting model as the "pq-formulation." Then, we use our recent theory of convex extensions (Tawarmalani & Sahinidis 2002a) in order to analyze the relaxation quality for each of the formulations and prove that the linear programming relaxation of the pq-formulation using standard bilinear envelopes (Al-Khayyal & Falk 1983) provides tighter bounds than the bounds obtained through the relaxations of the p-formulation and the q-formulation using either the bilinear envelopes or the Lagrangian relaxation. We further prove that even the Lagrangian relaxation of the pq-formulation is no tighter than the linear programming relaxation obtained using bilinear envelopes. Finally, in Section 9.4, we develop a branch-and-bound algorithm to solve the pq-formulation to global optimality and present extensive comparative computational results with earlier approaches.

9.2 The p- and q-Formulations of the Pooling Problem

9.2.1 The p-Formulation

The bilinear programming formulation for the pooling problem first proposed by Haverly (1978) is often referred to as the p-formulation. Using the variables, index sets and parameters described in Table 9.1, the p-formulation of the pooling problem is as follows:

$$(\text{P}) \quad \min \quad \sum_{i=1}^{I} c_i \sum_{l=1}^{L} x_{il} - \sum_{j=1}^{J} d_j \sum_{l=1}^{L} y_{lj} - \sum_{i=1}^{I} \sum_{j=1}^{J} (d_j - c_i) z_{ij}$$

9.2. THE P- AND Q-FORMULATIONS

Indices	i	raw materials, $i = 1, \ldots, I$
	j	products, $j = 1, \ldots, J$
	k	qualities, $k = 1, \ldots, K$
	l	pools, $l = 1, \ldots, L$
Variables	p_{lk}	k^{th} quality of pool l from pooling of raw materials
	x_{il}	flow of ith raw material into pool l
	y_{jk}	total flow from pool j to product k
	z_{ij}	direct flow of raw material i to product j
Parameters	c_i	unit cost of the ith raw material
	d_j	price of jth product
	A_i	availability of ith raw material
	C_{ik}	k^{th} quality of raw material i
	D_j	demand of jth product
	P^U_{jk}	upper bound on k^{th} quality of jth product
	S_l	l^{th} pool capacity

Table 9.1: Indices, variables, and parameters in the p-formulation

$$\text{s.t.} \sum_{l=1}^{L} x_{il} + \sum_{j=1}^{J} z_{ij} \leq A_i \qquad i = 1, \ldots, I \qquad (9.1)$$

$$\sum_{i=1}^{I} x_{il} - \sum_{j=1}^{J} y_{lj} = 0 \qquad l = 1, \ldots, L \qquad (9.2)$$

$$\sum_{i=1}^{I} x_{il} \leq S_l \qquad l = 1, \ldots, L \qquad (9.3)$$

$$\sum_{l=1}^{L} y_{lj} + \sum_{i=1}^{I} z_{ij} \leq D_j \qquad j = 1, \ldots, J \qquad (9.4)$$

$$\sum_{l=1}^{L} (p_{lk} - P_{jk}^U) y_{lj} + \sum_{i=1}^{I} (C_{ik} - P_{jk}^U) z_{ij} \leq 0$$
$$j = 1, \ldots, J; \; k = 1, \ldots, K \qquad (9.5)$$

$$\sum_{i=1}^{I} C_{ik} x_{il} - p_{lk} \sum_{j=1}^{J} y_{lj} = 0 \quad l = 1, \ldots, L; \; k = 1, \ldots, K \qquad (9.6)$$

$$x_{il} \geq 0, \forall (i,l); \quad y_{lj} \geq 0, \forall (l,j); \quad z_{ij} \geq 0, \forall (i,j); \quad p_{lk} \geq 0, \forall (l,k)$$

In P, the objective function is the difference between the cost of purchasing the input raw materials and the returns from selling the end-products. Constraints (9.1), (9.2), (9.3), (9.4), (9.5), and (9.6), model raw material availabilities, mass balances around pools, pool capacities, product demands, product quality requirements, and quality mass balances around pools, respectively. Depending on the structure of the underlying network, appropriate x_{il}, y_{lj}, and z_{ij} variables are forced to zero. We show later that, in the absence of such restrictions, the pooling problem is trivial to solve.

An example of the p-formulation is illustrated in Figure 9.2 for a classical problem taken from Haverly (1978). This problem involves three raw materials, two of which are pooled together before blending produces the two final products. Note that the quality requirements can be reformulated into linear constraints.

Model P may be relaxed by replacing bilinearities by their convex and concave envelopes as follows:

$$(\text{P}_\text{R}) \qquad \min \sum_{i=1}^{I} c_i \sum_{l=1}^{L} x_{il} - \sum_{j=1}^{J} d_j \sum_{l=1}^{L} y_{lj} - \sum_{i=1}^{I} \sum_{j=1}^{J} (d_j - c_i) z_{ij}$$

9.2. THE P- AND Q-FORMULATIONS

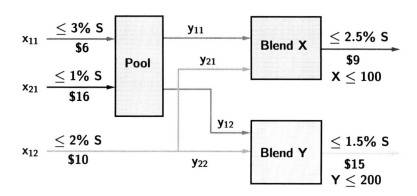

$$\min \quad \overbrace{6x_{11} + 16x_{21} + 10x_{12}}^{\text{cost}} - \overbrace{9(y_{11} + y_{21})}^{X\text{-revenue}} - \overbrace{15(y_{12} + y_{22})}^{Y\text{-revenue}}$$

s.t. $\quad q = \dfrac{3x_{11} + x_{21}}{y_{11} + y_{12}} \quad$ Sulfur mass balance

$$\begin{aligned} x_{11} + x_{21} &= y_{11} + y_{12} \\ x_{12} &= y_{21} + y_{22} \end{aligned} \quad \text{Total mass balances}$$

$$\begin{aligned} \dfrac{qy_{11} + 2y_{21}}{y_{11} + y_{21}} &\leq 2.5 \\ \dfrac{qy_{12} + 2y_{22}}{y_{12} + y_{22}} &\leq 1.5 \end{aligned} \quad \text{Quality requirements}$$

$$\begin{aligned} y_{11} + y_{21} &\leq 100 \\ y_{12} + y_{22} &\leq 200 \end{aligned} \quad \text{Demands}$$

Figure 9.2: An example of the p-formulation

$$\text{s.t.} \sum_{l=1}^{L} x_{il} + \sum_{j=1}^{J} z_{ij} \leq A_i \qquad i=1,\ldots,I \qquad (9.7)$$

$$\sum_{i=1}^{I} x_{il} - \sum_{j=1}^{J} y_{lj} = 0 \qquad l=1,\ldots,L \qquad (9.8)$$

$$\sum_{i=1}^{I} x_{il} \leq S_l \qquad l=1,\ldots,L \qquad (9.9)$$

$$\sum_{l=1}^{L} y_{lj} + \sum_{i=1}^{I} z_{ij} \leq D_j \qquad j=1,\ldots,J \qquad (9.10)$$

$$\sum_{l=1}^{L} u_{lkj} - \sum_{l=1}^{L} P_{jk}^{U} y_{lj} + \sum_{i=1}^{I}(C_{ik} - P_{jk}^{U})z_{ij} \leq 0$$
$$j=1,\ldots,J;\ k=1,\ldots,K \qquad (9.11)$$

$$\sum_{i=1}^{I} C_{ik}x_{il} - \sum_{j=1}^{J} u_{lkj} = 0 \qquad l=1,\ldots,L;\ k=1,\ldots,K \qquad (9.12)$$

$$u_{lkj} \geq \operatorname*{convenv}_{[p_{lk}^L, p_{lk}^U] \times [y_{lj}^L, y_{lj}^U]}(p_{lk}y_{lj}) \quad l=1,\ldots,L;\ j=1,\ldots,J;$$
$$k=1,\ldots,K \qquad (9.13)$$

$$u_{lkj} \leq \operatorname*{concenv}_{[p_{lk}^L, p_{lk}^U] \times [y_{lj}^L, y_{lj}^U]}(p_{lk}y_{lj}) \quad l=1,\ldots,L;\ j=1,\ldots,J;$$
$$k=1,\ldots,K \qquad (9.14)$$

$$x_{il} \geq 0, \forall(i,l); \quad y_{lj} \geq 0, \forall(l,j); \quad z_{ij} \geq 0, \forall(i,j)$$

$$p_{lk} \geq 0, \forall(l,k); \quad u_{lkj} \geq 0, \forall(j,k,l)$$

where p_{lk}^L and p_{lk}^U are lower and upper bounds on p_{lk} and y_{lj}^L and y_{lj}^U are lower and upper bounds on y_{lj}. The bilinear envelopes are (McCormick 1976, Al-Khayyal & Falk 1983):

$$u_{lkj} \geq y_{lj}^L p_{lk} + p_{lk}^L y_{lj} - y_{lj}^L p_{lk}^L$$
$$u_{lkj} \geq y_{lj}^U p_{lk} + p_{lk}^U y_{lj} - y_{lj}^U p_{lk}^U$$
$$u_{lkj} \leq y_{lj}^L p_{lk} + p_{lk}^U y_{lj} - y_{lj}^L p_{lk}^U$$
$$u_{lkj} \leq y_{lj}^U p_{lk} + p_{lk}^L y_{lj} - y_{lj}^U p_{lk}^L.$$

9.2. THE P- AND Q-FORMULATIONS

We assume in the sequel that:

$$p_{lk}^L = \min_i C_{ik}, \quad p_{lk}^U = \max_i C_{ik}. \quad (9.15)$$

This assumption can be relaxed so that the minimum and maximum are only taken over the raw materials that flow into the pool. Our arguments in this chapter can be suitably modified to account for this relaxed assumption. However, for the sake of clarity, we work with (9.15).

9.2.2 The q-Formulation

Ben-Tal et al. (1994) derived the q-formulation of the pooling problem introducing new variables q_{il} ($i = 1,\ldots,I$ and $l = 1,\ldots,L$) satisfying the relationship:

$$x_{il} = q_{il} \sum_{j=1}^{J} y_{lj}. \quad (9.16)$$

Using the q variables, the q-formulation is as follows:

$$(Q) \quad \min \sum_{j=1}^{J} \left(\sum_{l=1}^{L} y_{lj} \sum_{i=1}^{I} c_i q_{il} - d_j \sum_{l=1}^{L} y_{lj} + \sum_{i=1}^{I} c_i z_{ij} - \sum_{i=1}^{I} d_j z_{ij} \right)$$

$$\text{s.t.} \sum_{l=1}^{L} \sum_{j=1}^{J} q_{il} y_{lj} + \sum_{j=1}^{J} z_{ij} \leq A_i \quad i = 1,\ldots,I$$

$$\sum_{j=1}^{J} y_{lj} \leq S_l \quad l = 1,\ldots,L$$

$$\sum_{l=1}^{L} y_{lj} + \sum_{i=1}^{I} z_{ij} \leq D_j \quad j = 1,\ldots,J$$

$$\sum_{l=1}^{L} \left(\sum_{i=1}^{I} C_{ik} q_{il} - P_{jk}^U \right) y_{lj} + \sum_{i=1}^{I} (C_{ik} - P_{jk}^U) z_{ij} \leq 0$$
$$\quad k = 1,\ldots,K;\, l = 1,\ldots,L$$

$$\sum_{i=1}^{I} q_{il} = 1 \quad l = 1,\ldots,L$$

$$q_{il} \geq 0, \forall (i,l); \quad y_{lj} \geq 0, \forall (l,j); \quad z_{ij} \geq 0, \forall (i,j).$$

An example of the q-formulation is illustrated in Figure 9.3 for the same problem illustrated in Figure 9.2. As the example illustrates, the nature of nonlinearities in both models is the same (bilinearities).

Using (9.16), it can easily be shown that the p- and q-formulations are equivalent. As noted earlier, some of the variables may be set to zero depending on the structure of the underlying network. The main advantage of the q-formulation is that, in most applications, the number of extreme points of the simplex containing the variables q_{il} is much smaller than the number of extreme points of the hypercube containing p_{lk}. This is exploited algorithmically by Ben-Tal et al. (1994).

Model Q can be relaxed in a manner similar to the construction of P_R from P as below:

$$(Q_R) \quad \min \sum_{j=1}^{J} \left(\sum_{l=1}^{L} \sum_{i=1}^{I} c_i v_{ilj} - d_j \sum_{l=1}^{L} y_{lj} + \sum_{i=1}^{I} c_i z_{ij} - \sum_{i=1}^{I} d_j z_{ij} \right)$$

$$\text{s.t.} \sum_{l=1}^{L} \sum_{j=1}^{J} v_{ilj} + \sum_{j=1}^{J} z_{ij} \leq A_i \quad i = 1, \ldots, I \qquad (9.17)$$

$$\sum_{j=1}^{J} y_{lj} \leq S_l \qquad l = 1, \ldots, L \qquad (9.18)$$

$$\sum_{l=1}^{L} y_{lj} + \sum_{i=1}^{I} z_{ij} \leq D_j \qquad j = 1, \ldots, J \qquad (9.19)$$

$$\sum_{l=1}^{L} \sum_{i=1}^{I} C_{ik} v_{ilj} - \sum_{l=1}^{L} P_{jk}^U y_{lj} + \sum_{i=1}^{I} (C_{ik} - P_{jk}^U) z_{ij} \leq 0$$

$$k = 1, \ldots, K; \; l = 1, \ldots, L \quad (9.20)$$

$$v_{ilj} \geq \underset{[0,1]\times[y_{lj}^L, y_{lj}^U]}{\text{conveny}} (q_{il} y_{lj}) \qquad i = 1, \ldots, I; \; l = 1, \ldots, L;$$

$$j = 1, \ldots, J \qquad (9.21)$$

$$v_{ilj} \leq \underset{[0,1]\times[y_{lj}^L, y_{lj}^U]}{\text{conceny}} (q_{il} y_{lj}) \qquad i = 1, \ldots, I; \; l = 1, \ldots, L;$$

$$j = 1, \ldots, J \qquad (9.22)$$

$$\sum_{i=1}^{I} q_{il} = 1 \qquad l = 1, \ldots, L$$

9.2. THE P- AND Q-FORMULATIONS

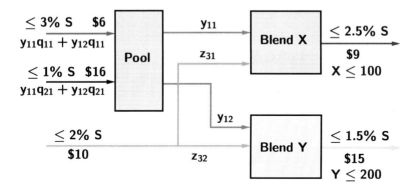

$$\min \; \overbrace{6(y_{11}q_{11} + y_{12}q_{11}) + 16(y_{11}q_{21} + y_{12}q_{21}) + 10(z_{31} + z_{32})}^{\text{cost}}$$

$$- \overbrace{9(y_{11} + y_{21})}^{X\text{-revenue}} - \overbrace{15(x_{12} + x_{22})}^{Y\text{-revenue}}$$

s.t. $\quad q_{11} + q_{21} = 1 \qquad\qquad\qquad$ Mass balances

$\qquad\begin{aligned} -0.5z_{31} + 3y_{11}q_{11} + y_{11}q_{21} &\le 2.5y_{11} \\ 0.5z_{32} + 3y_{12}q_{11} + y_{12}q_{21} &\le 1.5y_{12} \end{aligned}\quad$ Quality requirements

$\qquad\begin{aligned} y_{11} + z_{31} &\le 100 \\ y_{12} + z_{32} &\le 200 \end{aligned} \qquad\qquad$ Demands

Figure 9.3: An example of the q-formulation

$$q_{il} \geq 0, \forall (i,l); \quad y_{lj} \geq 0, \forall (l,j)$$

$$v_{ilj} \geq 0, \forall (i,l,j); \quad z_{ij} \geq 0, \forall (i,j).$$

9.3 The pq-Formulation

We construct the pq-formulation by adding the following constraints to the q-formulation:

$$\sum_{i=1}^{I} q_{il} y_{lj} = y_{lj} \quad l = 1, \ldots, L;\ j = 1 \ldots, J. \tag{9.23}$$

In Figure 9.4, we illustrate the pq-formulation for the same problem for which the p- and q-formulations were presented in Figures 9.2 and 9.3, respectively. As the example illustrates, the newly added constraints are nonlinear constraints that are redundant as far as the feasible region of the q-formulation is concerned. Thus, the addition of these constraints is against standard practice and understanding of the state-of-the-art in nonlinear programming.

Constraints (9.23) can be relaxed as:

$$\sum_{i=1}^{I} v_{ilj} = y_{lj} \quad l = 1, \ldots, L;\ j = 1 \ldots, J. \tag{9.24}$$

Clearly, constraints (9.23) are valid as they express the mass balances across the pools. The motivation for adding these constraints is as follows. Starting with the solution to Q_R, we can define x_{il} using relation (9.16). However, such a definition does not guarantee that constraint (9.2) is satisfied. We show later that including (9.24) in the relaxation allows us to define the input flows starting from the relaxation solution in a manner consistent with the mass balances across the pools.

Constraints (9.23) were independently derived by Quesada & Grossmann (1995b). These authors used the reformulation-linearization technique (RLT) (Sherali & Adams 1999) by multiplying the constraint $\sum_{i=1}^{I} q_{ij} = 1$ with y_{lj} in the context of global optimization of bilinear process networks. Since pooling and blending problems can be considered as a special case of process networks, the pq-formulation follows from the work of Quesada & Grossmann (1995b). Further, since the pq-formulation is a special case of RLT,

9.3. THE PQ-FORMULATION

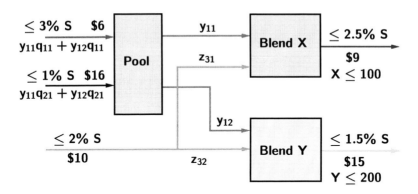

$$\min \quad \overbrace{6\left(y_{11}q_{11}+y_{12}q_{11}\right)+16\left(y_{11}q_{21}+y_{12}q_{21}\right)+10\left(z_{31}+z_{32}\right)}^{\text{cost}}$$
$$-\overbrace{9(y_{11}+y_{21})}^{X\text{-revenue}}-\overbrace{15(x_{12}+x_{22})}^{Y\text{-revenue}}$$

s.t. $\quad q_{11}+q_{21}=1 \qquad\qquad$ Mass balances

$\quad\;\; -0.5z_{31}+3y_{11}q_{11}+y_{11}q_{21} \leq 2.5y_{11}$
$\quad\;\; 0.5z_{32}+3y_{12}q_{11}+y_{12}q_{21} \leq 1.5y_{12}$ $\quad$ Quality requirements

$\quad\;\; y_{11}+z_{31} \leq 100$
$\quad\;\; y_{12}+z_{32} \leq 200 \qquad\qquad$ Demands

$\quad\;\; q_{11}y_{11}+q_{21}y_{11}=y_{11}$ $\qquad$ Convexification
$\quad\;\; q_{11}y_{12}+q_{21}y_{12}=y_{12}$ $\qquad$ constraints

Figure 9.4: An example of the pq-formulation

it follows trivially from Sherali & Adams (1999) that the pq-formulation is tighter than the q-formulation. However, the main question of interest is the relationship between the pq-formulation and the p-formulation as well as the relative tightness of the linear and Lagrangian relaxations of the three formulations. These are the questions addressed in the remainder of this section. In particular, several nice structural properties and dominance relations associated with the pq-formulation will be derived next by using our theory of convex extensions (Tawarmalani & Sahinidis 2002a).

In addition to constraints (9.23), the SS-RLT scheme (Sherali, Adams & Driscoll 1999) also advocates the introduction of the following constraints:

$$\sum_{j=1}^{J} q_{il} y_{lj} \leq q_{il} S_l \quad i = 1, \ldots, I; l = 1, \ldots, L. \tag{9.25}$$

For most benchmark problems in the literature, the pool capacity is a redundant constraint that can be derived using the bounds on y_{lj}. We show in Proposition 9.4 below that constraints (9.25) are redundant when this is indeed the case. However, for problems where the pool capacity is not derivable using bounds on y_{lj}, these constraints have the potential of further tightening the lower bound if they are relaxed as follows:

$$\sum_{j=1}^{J} v_{ilj} \leq q_{il} S_l \quad i = 1, \ldots, I; l = 1, \ldots, L. \tag{9.26}$$

It may be noted that, in the presence of (9.23) and (9.25), constraints (9.18) are redundant. We do not use (9.25) in the computations reported in Section 9.4.2, but will make use of them in Section 11.9.

In the remainder of this section, we say that formulation F1 is tighter than formulation of F2 if it can be shown that every solution feasible to F1 is also feasible to F2 and produces the same objective function value. In all cases to be considered, strict tightness follows from the numerical examples of Section 9.4 which demonstrate points that are feasible in F2 but not in F1.

9.3.1 Properties of the pq-Formulation

Proposition 9.1. *The pq-formulation when relaxed using bilinear envelopes produces a tighter lower bound than P_R and Q_R.*

9.3. THE PQ-FORMULATION

Proof. Clearly, the pq-formulation is tighter than the q-formulation, since we have only added the constraints in (9.23).

In the relaxation of the pq-formulation, we introduce a variable v_{ilj} for the product $q_{il}y_{ij}$. Let $(\bar{q}, \bar{y}, \bar{z}, \bar{v})$ be a feasible solution to the relaxation of the pq-formulation with an objective function value of $f(\bar{q}, \bar{y}, \bar{z}, \bar{v})$. Define:

$$\bar{x}_{il} = \sum_{j=1}^{J} \bar{v}_{ilj} \qquad i=1,\ldots,I;\; l=1,\ldots,L$$

$$\bar{u}_{lkj} = \sum_{i=1}^{I} C_{ik}\bar{v}_{ilj} \qquad l=1,\ldots,L;\; k=1,\ldots,K;\; j=1\ldots,J$$

$$\bar{p}_{lk} = \sum_{i=1}^{I} C_{ik}\bar{q}_{il} \qquad l=1,\ldots,L;\; k=1,\ldots,K$$

Then, we claim that $(\bar{x}, \bar{y}, \bar{z}, \bar{u}, \bar{p})$ is feasible to P_R with the objective function value of $f(\bar{q}, \bar{y}, \bar{z}, \bar{v})$. Clearly:

$$\sum_{i=1}^{I} c_i \sum_{l=1}^{L} \bar{x}_{il} - \sum_{j=1}^{J} d_j \sum_{l=1}^{L} \bar{y}_{lj} - \sum_{i=1}^{I}\sum_{j=1}^{J}(d_j - c_i)\bar{z}_{ij}$$

$$= \sum_{i=1}^{I} c_i \sum_{l=1}^{L}\sum_{j=1}^{J} \bar{v}_{ilj} - \sum_{j=1}^{J} d_j \sum_{l=1}^{L} \bar{y}_{lj} - \sum_{i=1}^{I}\sum_{j=1}^{J}(d_j - c_i)\bar{z}_{ij}$$

$$= \sum_{j=1}^{J} \left(\sum_{l=1}^{L}\sum_{i=1}^{I} c_i \bar{v}_{ilj} - d_j \sum_{l=1}^{L} \bar{y}_{lj} + \sum_{i=1}^{I} c_i \bar{z}_{ij} - \sum_{i=1}^{I} d_j \bar{z}_{ij} \right).$$

It is trivial to verify that (9.7), (9.9), (9.10), (9.11), and (9.12) are satisfied by $(\bar{x}, \bar{y}, \bar{z}, \bar{u}, \bar{p})$. (9.2) can be verified as follows:

$$\sum_{i=1}^{I} \bar{x}_{il} - \sum_{j=1}^{J} \bar{y}_{lj}$$

$$= \sum_{i=1}^{I}\sum_{j=1}^{J} \bar{v}_{ilj} - \sum_{j=1}^{J} \bar{y}_{lj}$$

$$= \sum_{i=1}^{I} \bar{y}_{lj} - \sum_{j=1}^{J} \bar{y}_{lj}$$

$$= 0,$$

where the second equality follows from (9.24).

The main challenge is in proving that $\bar{u}_{jkl}$, $\bar{y}_{lj}$ and $\bar{p}_{lk}$ as defined satisfy (9.13) and (9.14). We show that, for each $l \in \{1, \ldots, L\}$ and $j \in 1, \ldots, J$, the relaxation of the pq-formulation includes the convex and concave envelopes of $\sum_{i=1}^{I} C_{ik} q_{il} y_{lj}$ over

$$W = \left\{ (q_{1l}, \ldots, q_{Il}, y_{lj}) \;\middle|\; q_{il} \in [0,1] \; \forall i, \; \sum_{i=1}^{I} q_{il} = 1, \; y_{lj} \in [y_{lj}^L, y_{lj}^U] \right\}.$$

When y_{lj} is fixed, the function $\sum_{i=1}^{I} C_{ik} q_{il} y_{lj}$ is linear. Therefore, it follows from Theorem 2.17 that the points in the generating set of the convex (and concave) envelope of $\sum_{i=1}^{I} C_{ik} q_{il} y_{lj}$ over W are such that $(q_{1l}, \ldots, q_{Il})$ is an extreme point of the simplex:

$$\Delta = \left\{ (q_{1l}, \ldots, q_{Il}) \;\middle|\; q_{il} \in [0,1], \; \sum_{i=1}^{I} q_{il} = 1 \right\},$$

i.e., $(q_{1l}, \ldots, q_{Il})$ is a column of the identity matrix. Define the set B_i as:

$$B_i = \left\{ (q_{1l}, \ldots, q_{Il}, y_{lj}) \;\middle|\; (q_{1l}, \ldots, q_{Il}) = e_i, \; y_{lj} \in [y_{lj}^L, y_{lj}^U] \right\},$$

where e_i is the ith column of the identity matrix of dimension I. Then, treating q_{il} as the convex multiplier of B_i, the convex envelope of $\sum_{i=1}^{I} C_{ik} q_{il} y_{lj}$ can be expressed as:

$$\min \sum_{i=1}^{I} C_{ik} v_{ilj}$$

$$\text{s.t.} \;\; y_{lj} = \sum_{i=1}^{I} v_{ilj} \tag{9.27}$$

$$q_{il} y_{lj}^L \leq v_{ilj} \leq q_{il} y_{lj}^U \quad i = 1, \ldots, I \tag{9.28}$$

$$\sum_{i=1}^{I} q_{il} = 1, \tag{9.29}$$

where $v_{ilj} = q_{il} y_{lj}^{B_i}$ for some $(e_i, y_{lj}^{B_i}) \in B_i$. The above constraints are present in the relaxation of the pq-formulation. (9.28) is present as a result of the bilinear envelope and (9.27) relaxes (9.23). A similar argument proves the containment of the concave envelope.

9.3. THE PQ-FORMULATION

Now, consider the set:

$$(D_{lkj}) \qquad u_{lkj} = \sum_{i=1}^{I} C_{ik} q_{il} y_{lj}$$
$$(q_{1l}, \ldots, q_{Il}) \in \Delta$$
$$y_{lj} \in [y_{lj}^L, y_{lj}^U],$$

and reformulate it as:

$$(R_{lkj}) \qquad u_{lkj} = p_{lk} y_{lj}$$
$$p_{lk} = \sum_{i=1}^{I} C_{ik} q_{il}$$
$$(q_{1l}, \ldots, q_{Il}) \in \Delta$$
$$y_{lj} \in [y_{lj}^L, y_{lj}^U].$$

Let X_{lkj} be any convex relaxation of R_{lkj}. Construct the convex relaxation C_{lkj} such that:

$$C_{lkj} = X_{lkj} \cap \left\{ (p_{lk}, q_{1l}, \ldots, q_{Il}) \,\middle|\, p_{lk} = \sum_{i=1}^{I} C_{ik} q_{il} \right\}.$$

It is clear that, after projecting out the p_{lk} variables from C_{lkj}, we are left with a convex outer-approximation of the convex hull of D_{lkj}. We argued earlier that setting $\bar{u}_{lkj} = \sum_{i=1}^{I} C_{ik} \bar{v}_{ilj}$ guarantees that $(\bar{u}_{lkj}, \bar{q}_{il}, \bar{y}_{lj})$ lies in the convex hull of the solutions of D_{lkj}. Therefore, the point $U = (\bar{u}_{lkj}, \bar{p}_{lk}, \bar{q}_{il}, \bar{y}_{lj})$ lies in C_{lkj}. Since the bounds on p_{lk} assumed in (9.15) are implicit in R_{lkj}, it is apparent that (9.13) and (9.14) form a convex outer-approximation of R_{lkj} guaranteeing the containment of U.

Hence, we have shown that the relaxation of the p-formulation is a relaxation of the relaxation of the pq-formulation. □

The relaxation of the pq-formulation can be simplified without affecting the quality of the bounds. We present a few results in this direction. The first result is motivated by the fact that the characterization of the convex envelope of $\sum_{i=1}^{I} C_{ik} q_{il} y_{lj}$ in the proof of Proposition 9.1 did not require two out of the four inequalities used in the bilinear envelope. As a result, the two inequalities are redundant in the relaxation of the pq-formulation. We provide a direct algebraic proof next.

Proposition 9.2. *The relations $v_{ilj} \geq y_{lj}^U q_{il} + y_{lj} - y_{lj}^U$ and $v_{ilj} \leq y_{lj}^L q_{il} + y_{lj} - y_{lj}^L$ are redundant in the relaxation of the pq-formulation.*

Proof. Consider an arbitrary raw-material i':

$$v_{i'lj} \stackrel{(9.27)}{=} y_{lj} - \sum_{\substack{i=1 \\ i \neq i'}}^{I} v_{ilj} \stackrel{(9.28)}{\geq} y_{lj} - \sum_{\substack{i=1 \\ i \neq i'}}^{I} q_{il} y_{lj}^U \stackrel{(9.29)}{=} y_{lj} - (1 - q_{i'l}) y_{lj}^U = y_{lj}^U q_{i'l} + y_{lj} - y_{lj}^U.$$

The relations used in making the arguments for each step are provided above $\geq$ and $=$ signs. Similarly:

$$v_{i'lj} \stackrel{(9.27)}{=} y_{lj} - \sum_{\substack{i=1 \\ i \neq i'}}^{I} v_{ilj} \stackrel{(9.28)}{\leq} y_{lj} - \sum_{\substack{i=1 \\ i \neq i'}}^{I} q_{il} y_{lj}^L \stackrel{(9.29)}{=} y_{lj} - (1 - q_{i'l}) y_{lj}^L = y_{lj}^L q_{i'l} + y_{lj} - y_{lj}^L.$$

□

Including (9.26) may further improve the quality of the relaxation as we show in the next proposition.

Proposition 9.3. *When constraints (9.26) are added to the relaxation of the pooling problem, the convex hull of $\sum_{i,l,j} \alpha_{ilj} q_{il} y_{lj}$ over:*

$$\sum_{j=1}^{J} y_{lj} \leq S_l \qquad l = 1, \ldots, L$$

$$y_{lj}^L \leq y_{lj} \leq y_{lj}^U \qquad l = 1, \ldots, L; j = 1, \ldots, J$$

$$\sum_{i=1}^{I} q_{il} = 1 \qquad l = 1, \ldots, L$$

$$q_{il} \geq 0 \qquad i = 1, \ldots, I; l = 1, \ldots, L$$

is embedded in the relaxation.

Proof. The proposition can be easily proven by first invoking Lemma 9.10 to decompose $\sum_{i,l,j} \alpha_{ilj} q_{il} y_{lj}$ over l and then using arguments similar to those presented in Proposition 9.1. □

The constraint on the pool capacity is often redundant since S_l equals $\sum_{j=1}^{J} y_{lj}^U$. We show next that, in this case, constraints (9.26) do not improve the quality of the relaxation.

9.3. THE PQ-FORMULATION

Proposition 9.4. *Constraints (9.26) are redundant if $S_l \geq \sum_{j=1}^{J} y_{lj}^U$.*

Proof. The relaxation already includes constraints $v_{ilj} \leq q_{il} y_{lj}^U$. Therefore, $\sum_{j=1}^{J} v_{ilj} \leq \sum_{j=1}^{J} y_{lj}^U \leq S_l$. □

During the course of branch and bound algorithm, branching on y_{lj} variables reduces the feasible ranges of these variables further diminishing the need for these constraints. We show next by optimality arguments that certain variables in the pq-formulation can be deleted when certain conditions are met.

Proposition 9.5. *Consider a pool $\hat{l}$. Let $I_{\hat{l}}$ denote the set of raw materials that flow into pool $\hat{l}$ and let $J_{\hat{l}}$ denote the set of products that receive flow from pool $\hat{l}$. If for some $\hat{j} \in J_{\hat{l}}$ and all $i \in I_{\hat{l}}$, $z_{i\hat{j}}$ is unrestricted, then $y_{\hat{l}\hat{j}}$ can be set to zero.*

Proof. Let $(\bar{q}, \bar{y}, \bar{z})$ be an optimal solution to Q. Then, construct $(\tilde{q}, \tilde{y}, \tilde{z})$ using $(\bar{q}, \bar{y}, \bar{z})$ such that the two solutions agree on all coordinates except $z_{i\hat{j}}$ where $i \in I_{\hat{l}}$ and $y_{\hat{l}\hat{j}}$. Set $\tilde{y}_{\hat{l}\hat{j}} = 0$ and $\tilde{z}_{i\hat{j}} = \bar{z}_{i\hat{j}} + \bar{q}_{i\hat{l}} \bar{y}_{\hat{l}\hat{j}}$. Then, $(\tilde{q}, \tilde{y}, \tilde{z})$ can be verified to be optimal to Q since it attains the same objective function value as $(\bar{q}, \bar{y}, \bar{z})$ and is feasible. □

Proposition 9.6. *Consider a pool $\hat{l}$. Let $I_{\hat{l}}$ denote the set of raw materials that flow into pool $\hat{l}$ and let $J_{\hat{l}}$ denote the set of products that receive flow from pool $\hat{l}$. If for some $\hat{i} \in I_{\hat{l}}$ and all $j \in J_{\hat{l}}$, $z_{\hat{i}j}$ is unrestricted, then $q_{\hat{i}\hat{l}}$ can be set to 1 if $I_{\hat{l}} = \{\hat{i}\}$ and to 0 otherwise.*

Proof. Let $(\bar{q}, \bar{y}, \bar{z})$ be an optimal solution to Q. If $\bar{q}_{\hat{i}\hat{l}} = 1$, we apply Proposition 9.5 for all $j \in J_{\hat{l}}$. Otherwise, construct $(\tilde{q}, \tilde{y}, \tilde{z})$ using $(\bar{q}, \bar{y}, \bar{z})$ such that the two solutions agree on all coordinates except $(y_{\hat{l}j}, z_{ij})$ where $j \in J_{\hat{l}}$ and $q_{i\hat{l}}$ for $i \in I_{\hat{l}}$. For all $i \neq \hat{i}$, set $\tilde{q}_{i\hat{l}}$ as follows:

$$\tilde{q}_{i\hat{l}} = \frac{\bar{q}_{il}}{\sum_{i \neq \hat{i}} \bar{q}_{il}}.$$

Set $\tilde{q}_{\hat{i}\hat{l}} = 0$, $\tilde{z}_{\hat{i}j} = \bar{q}_{\hat{i}\hat{l}} \bar{y}_{\hat{l}j} + \bar{z}_{\hat{i}j}$, and $\tilde{y}_{\hat{l}j} = \bar{y}_{\hat{l}j} - \bar{q}_{\hat{i}\hat{l}} \bar{y}_{\hat{l}j}$. Note that the construction guarantees that for all $i \neq \hat{i}$ and $j \in J_{\hat{l}}$:

$$\tilde{q}_{i\hat{l}} \tilde{y}_{\hat{l}j} = \tilde{q}_{i\hat{l}} (\bar{y}_{\hat{l}j} - \bar{q}_{\hat{i}\hat{l}} \bar{y}_{\hat{l}j})$$

$$= \tilde{q}_{i\hat{l}} \left(\sum_{\bar{i}=1}^{I} \bar{q}_{\bar{i}\hat{l}} \bar{y}_{\hat{l}j} - \bar{q}_{\hat{i}\hat{l}} \bar{y}_{\hat{l}j} \right)$$

$$= \tilde{q}_{i\hat{l}} \sum_{\bar{i}\neq\hat{i}} \bar{q}_{\bar{i}\hat{l}} \bar{y}_{\hat{l}j}$$

$$= \bar{q}_{il} \bar{y}_{\hat{l}j}.$$

Then, $(\tilde{q}, \tilde{y}, \tilde{z})$ is optimal to Q since it attains the same objective function value as $(\bar{q}, \bar{y}, \bar{z})$ and is feasible. $\square$

Remark 9.7. *It follows from Proposition 9.5 (also from Proposition 9.6) that, when z_{ij} is unrestricted for all i and j, the y_{lj} variables can be set to zero. This reduces Q to a linear program. As the next proposition shows, even without modifications proposed in Proposition 9.5 and Proposition 9.6, the relaxation of the pq-formulation attains the same value.*

Proposition 9.8. *Assuming $y_{lj}^L = 0$, the relaxation of the pq-formulation attains the same lower bound irrespective of whether or not the variables are fixed using Propositions 9.5 and 9.6.*

Proof. Let us assume that the conditions of Proposition 9.5 are satisfied for some pool $\hat{l}$ and some product $\hat{j}$. If $(\bar{z}, \bar{q}, \bar{y}, \bar{v})$ is the optimal solution to the relaxation of the pq-formulation, then define a solution $(\tilde{z}, \tilde{q}, \tilde{y}, \tilde{v})$ that agrees with the $(\bar{z}, \bar{q}, \bar{y}, \bar{v})$ in all coordinates except $(z_{i\hat{j}}, v_{i\hat{l}\hat{j}})$ for all $i \in I_{\hat{l}}$. Define $\tilde{z}_{i\hat{j}} = \bar{z}_{i\hat{j}} + \bar{v}_{i\hat{l}\hat{j}}$, $\tilde{y}_{\hat{l}\hat{j}} = \bar{y}_{\hat{l}\hat{j}} - \sum_{i \in I_{\hat{l}}} \bar{v}_{i\hat{l}\hat{j}}$ and $\tilde{v}_{i\hat{l}\hat{j}} = 0$. Then, $(\tilde{z}, \tilde{q}, \tilde{y}, \tilde{v})$ is feasible to relaxation of the modified pq-formulation with the same objective function value and hence optimal.

Now, let us assume that the conditions of Proposition 9.6 are satisfied for some pool $\hat{l}$ and some raw material $\hat{i}$. If $(\bar{z}, \bar{q}, \bar{y}, \bar{v})$ is the optimal solution to the relaxation of the pq-formulation, then define a solution $(\tilde{z}, \tilde{q}, \tilde{y}, \tilde{v})$ that agrees with the $(\bar{z}, \bar{q}, \bar{y}, \bar{v})$ in all coordinates except $(z_{\hat{i}j}, v_{\hat{i}\hat{l}j}, y_{\hat{l}j})$ for all $j \in J_{\hat{l}}$ and $q_{i\hat{l}}$ where $i \in I_{\hat{l}}$. Define:

$$\tilde{z}_{\hat{i}j} = \bar{z}_{\hat{i}j} + \bar{v}_{\hat{i}\hat{l}j}$$
$$\tilde{v}_{\hat{i}\hat{l}j} = 0$$
$$\tilde{y}_{\hat{l}j} = \bar{y}_{\hat{l}j} - \bar{v}_{\hat{i}\hat{l}j}$$
$$\tilde{q}_{\hat{i}\hat{l}} = 0$$
$$\tilde{q}_{i\hat{l}} = \frac{\bar{q}_{il}}{\sum_{\bar{i}\neq\hat{i}} \bar{q}_{\bar{i}l}}.$$

9.3. THE PQ-FORMULATION

$(\tilde{z}, \tilde{q}, \tilde{y}, \tilde{v})$ is clearly feasible to (9.17), (9.18), (9.19), (9.20) and has the same objective function value as $(\bar{z}, \bar{q}, \bar{y}, \bar{v})$. It remains to verify that $(\tilde{z}, \tilde{q}, \tilde{y}, \tilde{v})$ is feasible to (9.21), (9.22) and (9.24). Define B_i as follows:

$$B_i = \left\{ (q_{1l}, \ldots, q_{Il}, y_{lj}) \mid (q_{1l}, \ldots, q_{Il}) = e_i, y_{lj} \in [y_{lj}^L, y_{lj}^U] \right\}.$$

It is easy to show (using arguments similar to those used in the proof of Proposition 9.1, also see special structured RLT (Sherali et al. 1999)) that the convex hull of the union of the sets B_i, $i = 1, \ldots, I$ is given by:

$$y_{lj} = \sum_{i=1}^{I} v_{ilj}$$

$$q_{il} y_{lj}^L \leq v_{ilj} \leq q_{il} y_{lj}^U \quad i = 1, \ldots, I$$

$$\sum_{i=1}^{I} q_{il} = 1$$

$$q_{il} \geq 0.$$

In other words, there exists a $\bar{y}_{lj}^{B_i} \in [y_{lj}^L, y_{lj}^U]$ such that $\bar{v}_{ilj} = \bar{q}_{il} \bar{y}_{lj}^{B_i}$. If $\tilde{q}_{i\hat{l}} = 0$ for some $i \in I_{\hat{l}} \backslash \{\hat{i}\}$, define $\tilde{y}_{\hat{l}j}^{B_i} = \bar{y}_{\hat{l}j}^{B_i}$. Otherwise, define $\tilde{y}_{\hat{l}j}^{B_i} = \frac{\bar{q}_{il}}{\tilde{q}_{il}} \bar{y}_{\hat{l}j}^{B_i}$. It is easy to verify that $\tilde{y}_{\hat{l}j}^{B_i} \in [y_{\hat{l}j}^L, y_{\hat{l}j}^U]$ because $\bar{q}_{i\hat{l}} \leq \tilde{q}_{i\hat{l}}$, and we assumed that $y_{\hat{l}j}^L = 0$. Furthermore, $\tilde{v}_{i\hat{l}j} = \bar{v}_{i\hat{l}j} = \tilde{q}_{i\hat{l}} \tilde{y}_{\hat{l}j}^{B_i}$ for all $i \in I_{\hat{l}} \backslash \{\hat{i}\}$. Since $\tilde{v}_{\hat{i}\hat{l}j} = 0 = \tilde{q}_{\hat{i}\hat{l}} \tilde{y}_{\hat{l}j}$ and $\sum_{i=1}^{I} \tilde{v}_{i\hat{l}j} = \tilde{y}_{\hat{l}j}$, we have shown that $(\tilde{z}, \tilde{q}, \tilde{y}, \tilde{v})$ is feasible and therefore optimal to the relaxation of pq-formulation. □

9.3.2 Lagrangian Relaxations

A Lagrangian relaxation for the pooling problem was constructed in Adhya, Tawarmalani & Sahinidis (1999) by dualizing all constraints except the bounds on y_{lj} and p_{lk} variables. The authors showed that the resulting relaxation provides a tighter lower bound than the one obtained after solving the P_R. We show next that the pq-formulation provides a better bound than the relaxation of Adhya et al. (1999).

Proposition 9.9. *The relaxation of the pq-formulation provides a tighter lower bound than the Lagrangian relaxation of Adhya et al. (1999).*

Proof. It was shown in the proof of Proposition 9.1 that the relaxation of the pq-formulation dominates any convex relaxation of the p-formulation where the bilinear terms are relaxed over the hypercube formed by the constraints $p_{lk}^L \leq p_{lk} \leq p_{lk}^U$ and $y_{lj}^L \leq y_{lj} \leq y_{lj}^U$. By Theorem 5.2 in Adhya et al. (1999), the Lagrangian relaxation formed by dualizing all constraints except the bounds is one such relaxation. $\square$

The results of Adhya et al. (1999) suggest the possibility of deriving an improved bound by constructing a similar Lagrangian relaxation of the pq-formulation. However, we show in the next result that the Lagrangian relaxation is not tighter than the linear programming relaxation studied in Section 9.3. We need the following lemma from Rikun (1997).

Lemma 9.10 (Rikun 1997). *Let P be a Cartesian product of polytopes, $P = P_0 \times P_1 \times \ldots P_K$, $P_k \in \mathbb{R}^{n_k}$, and let $g_k(x_0, x_k)$ be a continuous function defined on $P_0 \times P_k$, $k = 1, \ldots, K$. If each $g_k(x_0, x_k)$ is a concave function of x_0 when x_k is fixed and P_0 is a simplex, then*

$$\operatorname*{conv}_P \left(\sum_{k=1}^K g_k(x_0, x_k) \right) = \sum_{k=1}^K \operatorname*{conv}_{P_0 \times P_k} g_k(x_0, x_k)$$

where conv_P refers to the convex envelope over the region P.

Proposition 9.11. *Consider the Lagrangian relaxation of the pq-formulation obtained when all constraints are dualized except for the set:*

$$\left\{ y_{lj}^L \leq y_{lj} \leq y_{lj}^U, \, l = 1, \ldots, L; \, j = 1, \ldots, J; \, \sum_{i=1}^I q_{il} = 1, \, l = 1, \ldots, L \right.$$

$$\left. q_{il} \geq 0, \, i = 1, \ldots, I; \, l = 1, \ldots, L \right\}.$$

The above dual provides the same lower bound as the relaxation of the pq-formulation using bilinear envelopes.

Proof. The Lagrangian relaxation that dualizes all the constraints of the relaxation of the pq-formulation except

$$(C_S) \qquad y_{lj} = \sum_{i=1}^I v_{ilj} \qquad\qquad l = 1, \ldots, L; \, j = 1, \ldots, J$$

9.3. THE PQ-FORMULATION

$$q_{il} y_{lj}^L \leq v_{ilj} \leq q_{il} y_{lj}^U \qquad i = 1, \ldots, I$$

$$\sum_{i=1}^{I} q_{il} = 1 \qquad l = 1, \ldots, L$$

$$q_{il} \geq 0 \qquad i = 1, \ldots, I; \, l = 1, \ldots, L,$$

attains the same optimal value as the relaxation of the pq-formulation (strong duality in linear programming). Define F_S as the set of solutions feasible to S. It follows from the arguments in the proof of Proposition 9.1 and Lemma 9.10 that optimizing a bilinear function of the form $\sum_{i,l,j} \alpha_{ijl} q_{il} y_{jl}$ over F_S is equivalent to optimizing $\sum_{i,l,j} \alpha_{ijl} v_{ijl}$ over C_S. This implies that, for each dual solution, the Lagrangian subproblem returns the same solution value. The Lagrangian bound obtained as maximum over all dual solutions is then also identical. □

Proposition 9.12. *Consider the Lagrangian relaxation of the q-formulation obtained when all constraints are dualized except for the set:*

$$\left\{ y_{lj}^L \leq y_{lj} \leq y_{lj}^U, \, l = 1, \ldots, L; \, j = 1, \ldots, J; \, \sum_{i=1}^{I} q_{il} = 1, \, l = 1, \ldots, L; \right.$$

$$\left. q_{il} \geq 0, \, i = 1, \ldots, I; \, l = 1, \ldots, L \right\}.$$

The above dual provides the same lower bound as the relaxation of the pq-formulation using bilinear envelopes.

Proof. The constraints (9.23) are redundant in the presence of the nondualized constraints above. Therefore, they always satisfy the complementary slackness conditions and do not penalize the objective in the Lagrangian subproblem. As a result, the Lagrangian relaxation of the pq-formulation and the q-formulation achieve the same lower bound. The remainder of the result follows from Proposition 9.11. □

Ben-Tal et al. (1994) proposed a different relaxation of the q-formulation, where they dualized all constraints except for

$$\left\{ \sum_{i=1}^{I} q_{il} = 1, \, l = 1, \ldots, L; \, q_{il} \geq 0, \, i = 1, \ldots, I; \, l = 1, \ldots, L \right\}.$$

Since the bound constraints were dualized, the bilinear envelopes were not implied in their relaxation. As a result, the following result follows easily from Proposition 9.12.

Corollary 9.13. *The relaxation of the pq-formulation using bilinear envelopes is tighter than the Lagrangian relaxation of the q-formulation proposed by Ben-Tal et al. (1994).*

9.4 Global Optimization of the Pooling Problem

In this section, we use the pq-formulation to develop a branch-and-bound algorithm for solving the pooling problem. The notation we use is as follows:

Notation:

k	Iteration number
p	number of probing variables
i	Current node of the branch-and-bound tree ($i = 0$ for the root node)
$\mathcal{L}$	List of active nodes
PQ_i	pq-formulation of the pooling problem restricted to the feasible domain of ith node
P_i	p-formulation of the pooling problem restricted to the feasible domain of ith node
R_i	relaxation of PQ_i at the ith node
$L(PQ_i)$	Lower bound on the solution of PQ_i
U^k	Upper bound on the pooling problem at the kth iteration
L^k	Lower bound on the pooling problem at the kth iteration

BRANCH-AND-BOUND ALGORITHM:
Initialization:

- Fix variables according to Propositions 9.5 and 9.6. Set $k = 0$, $L^k = -\infty$, and $L(PQ_0) = -\infty$.

- Include the root node in the list: $\mathcal{L} = \{0\}$.

9.4. GLOBAL OPTIMIZATION OF THE POOLING PROBLEM

Main Step(Iteration k)
Step 1. Termination:

- Set the lower bound $L^k := \min_{i \in \mathcal{L}}\{L(\text{PQ}_i)\}$.

- For all i such that $L(\text{PQ}_i) \geq U^k$ set $\mathcal{L} = \mathcal{L}\setminus\{i\}$.

- If $\mathcal{L} = \emptyset$, stop; the current best solution is optimal. Else, set $k = k+1$, $L^k = L^{k-1}$ and $U^k = U^{k-1}$.

Step 2. Node Selection:

- Select node i from $\mathcal{L}$ according to some node selection rule that is bound improving (for example, select the node corresponding to the best lower bound).

- Set $\mathcal{L} = \mathcal{L}\setminus\{i\}$.

Step 3. Node Tightening:

- Perform interval arithmetic to improve bounds on the bilinear terms.

- Construct PQ_i using the new bounds.

Step 4. Lower Bounding:

- Form R_i by relaxing the bilinear terms in PQ_i using their convex and concave envelopes.

- Solve R_i to obtain $L(\text{PQ}_i)$. If $L(\text{PQ}_i) \geq U^k$, go to Step 7.

Step 5. Marginals and Probing:

- Perform marginals-based range reduction.

- Use the branching strategy of Section 9.4.1 to identify the first n variables for branching and probe on these variables to improve their bounds.

- If significant reduction in bounds occurred, go to Step 3.

Step 6. Upper Bounding:

- Locally optimize P_i to find an upper bound and update U^k. Record the best known solution.

- If $L(\text{PQ}_i) \geq U^k$, go to Step 1.

Step 7. Branching:

- Use the branching strategy in Section 9.4.1 to obtain two new problems PQ_{i_1} and PQ_{i_2} from PQ_i.

- Update the node list $\mathcal{L} = \mathcal{L} \cup \{i_1, i_2\}$.

- Go to Step 1.

The algorithm uses various domain reduction strategies including those based on marginals and probing. Using the branching strategy of Section 9.4.1, the first n variables are chosen and probing linear programs are solved to determine improved bounds on these variables. Dual solutions from the probing linear programs and the relaxation at the current node enable us to improve the bounds on the remaining variables. A more elaborate discussion of these reduction strategies can be found in Chapter 5.

9.4.1 Branching Strategy

Contrary to integer linear programming, branching at the current relaxation solution does not exhaustively partition the feasible space and is therefore not guaranteed to converge to the optimal solution. The branching variables for the pooling problem are the proportions of the input flows q_{il}, $i = 1, \ldots, I$, $l = 1, \ldots, L$ and the outgoing flows from the pools y_{lj}, $l = 1, \ldots, L$, $j = 1, \ldots, J$, since the only nonconvexities in the model arise from the bilinear terms of the form $q_{il}y_{lj}$. In theory, branching exclusively on only one of these two sets of variables suffices for convergence (Sherali & Alameddine 1992). Nonetheless, we consider branching on both q and y variables as this provides more flexibility to the algorithm.

The branching variable is selected as follows. For every term of the form $v_{ilj} = q_{il}y_{lj}$, we construct the interval $[\min\{q_{il}^R y_{lj}^R, v_{ilj}^R\}, \max\{q_{il}^R y_{lj}^R, v_{ilj}^R\}]$, where q_{il}^R, y_{il}^R and v_{il}^R are the optimal values for the relaxation at the current node. If $v_{ilj}^R > q_{il}^R y_{lj}^R$, we obtain δ as the minimum norm solution of $(q_{il}^R + \delta)(y_{il}^R + \delta) = v_{il}^R$, otherwise we compute δ as the minimum norm solution of $(q_{il}^R + \delta)(y_{il}^R - \delta) = v_{ilj}^R$. We construct the smallest interval, q_{il}^I, that contains q_{il}^R and $q_{il}^R + \delta$. Similarly, we construct the smallest interval y_{lj}^I containing y_{lj}^R and either $y_{lj}^R + \delta$ or $y_{lj}^R - \delta$ depending on the case above. Then, we intersect q_{il}^I and y_{lj}^I with the bounds on q_{il} and y_{lj} in the current partition element. The length of the resulting intervals is defined as the violation of variables q_{il} and y_{lj}, respectively. Let $M_{\cdot q_{il}}$ denote the column in the linear relaxation corresponding to q_{il} and π^R be the optimal dual solution. Define the branching priority of q_{il} to be the L_1 norm of the termwise product of $M_{rq_{il}}$ and π_r^R. Similarly, the branching priority for y_{lj} is computed. Then, the variable with the largest product of the branching priority and the violation is chosen as the branching variable. The idea behind this branching scheme is presented in greater detail in Chapter 6. Once the branching variable is selected, the branching point is chosen as a convex combination of the point that bisects the current feasible interval into equal halves (bisection rule) and the current relaxation optimal value (ω−branching). Finally, the variables with the n largest priorities are considered candidates for probing.

9.4.2 Computational Experience

Pooling problems with up to twelve raw materials, five products, and five pools are constructed in Androulakis, Visweswaran & Floudas (1996) over graphs that include all possible arcs from the raw material nodes to the final products nodes. Per Remark 9.7, these problems can be formulated as linear programs after eliminating all the pools from the pooling model. The relaxation of the pq-formulation is also exact in this case and solves these problems as shown in Proposition 9.6. Hence, the proposed algorithm solves all these problems at the root node.

In Table 9.2, we present characteristics of all other pooling problems from the open literature. These problems involve up to eleven raw materials, sixteen products, five qualities, and eight pools. Problem Haverly 1 in this table is taken from Haverly (1978). Problems Haverly 2 and 3 are from Haverly (1979). The Foulds problems are from Foulds, Haugland & Jornsten (1992). The Ben-Tal problems are taken from Ben-Tal et al. (1994). Problem RT 2 is from Audet, Brimberg, Hansen & Mladenovic (2000). Finally, the Adhya problems are from Adhya et al. (1999).

In Table 9.3, the following lower bounds are presented for this collection of problems:

P_R linear programming relaxation of the p-formulation
L(P)   Lagrangian relaxation of Adhya et al. (1999) for p-formulation (values taken from Adhya et al. (1999))
Q_R linear programming relaxation of the q-formulation
L(Q)   Lagrangian relaxation of Ben-Tal et al. (1994) for q-formulation
PQ_R linear programming relaxation of the pq-formulation as developed in Section 9.3

These lower bounds demonstrate that the proposed PQ_R strictly dominates all other relaxations. For any other relaxation, there is at least one benchmark pooling problem for which the proposed PQ_R provides a strictly higher lower bound. In addition to lower bounds, the global solutions obtained with the proposed algorithm are also provided. For the four Foulds and the two Ben-Tal problems, PQ_R is exact.

In Table 9.4, we present computational results and comparisons with prior global optimization approaches for the problems of Table 9.2. The algorithms considered in this table are denoted as follows:

Foulds '92	Foulds et al. (1992)
Ben-Tal '94	Ben-Tal et al. (1994)
GOP '96	Visweswaran & Floudas (1996)
BARON '99	Adhya et al. (1999)
Audet '00	Audet et al. (2000)
BARON '02	the proposed algorithm

For each problem/algorithm combination, Table 9.4 presents the total number of algorithm iterations (N_{tot}). With the exception of the decomposition algorithm of Visweswaran & Floudas (1996), N_{tot} corresponds to nodes of the branch-and-bound tree. Also shown in the table are the CPU seconds (T_{tot}) required by each algorithm, the computer type used and its LINPACK score in order to facilitate comparisons. As seen, the proposed algorithm solves ten out of these 14 problems at the root node of the search tree. This includes the three Haverly problems and Adhya 4 for which the proposed relaxation provided a nonzero gap as shown in Table 9.3. In these four cases, the domain reduction strategies of our algorithm are effective in closing the gap at the root node itself. Finally, we note that even the four Adhya problems that were previously thought to be hard are now solved within seventeen nodes. These computational results show that all pooling problems from the open literature are trivially solvable by the proposed algorithm.

9.4. GLOBAL OPTIMIZATION OF THE POOLING PROBLEM

Problem	Raw Materials (I)	Products (J)	Qualities (K)	Pools (L)
Haverly 1	3	2	1	1
Haverly 2	3	2	1	1
Haverly 3	3	2	1	1
Foulds 2	6	4	1	2
Foulds 3	11	16	1	8
Foulds 4	11	16	1	4
Foulds 5	11	16	1	4
Ben-Tal 4	4	2	1	1
Ben-Tal 5	4	5	2	3
RT 2	3	3	4	2
Adhya 1	5	4	4	2
Adhya 2	5	4	6	2
Adhya 3	8	4	6	3
Adhya 4	8	5	4	2

Table 9.2: Characteristics of benchmark pooling problems

Problem	P_R	$L(P)$	Q_R	$L(Q)$	PQ_R	Global
Haverly 1	-500	-500	-1687.5	-500	-500	-400
Haverly 2	-1000	-1000	-2625.0	-1000	-1000	-600
Haverly 3	-800	-800	-1687.5	-875	-800	-750
Foulds 2	-1100	-1100	-5100	-1100	-1100	-1100
Foulds 3	-∞	-∞	-100	-∞	-∞	-∞
Foulds 4	-∞	-∞	-100	-∞	-∞	-∞
Foulds 5	-∞	-∞	-100	-∞	-∞	-∞
Ben-Tal 4	-550	-550	-1864.29	-550	-550	-450
Ben-Tal 5	-3500	-3500	-6100	-3500	-3500	-3500
RT 2	-6331.73		-19260.9	-6034.87	-6034.87	-4391.83
Adhya 1	-999.31	-939.29	-1185.0	-856.25	-840.27	-549.80
Adhya 2	-854.10	-825.59	-1185.0	-574.78	-574.78	-549.80
Adhya 3	-882.84	-864.81	-1185.0	-574.78	-574.78	-561.05
Adhya 4	-1012.50	-988.50	-1090.0	-976.39	-961.93	-877.65

Table 9.3: Comparison of bounds for benchmark pooling problems

9.4. GLOBAL OPTIMIZATION OF THE POOLING PROBLEM

Algorithm	Foulds '92		Ben-Tal '94		GOP '96		BARON '99		Audet '00		BARON '02	
Computer*	CDC 4340				HP9000/730		IBM RS/6000 43P		Sun Ultra-1/167		Dell 530, P4 1.7GHz	
LINPACK	> 3.5				49		59.9		70		363	
Tolerance*					**		10^{-6}				10^{-6}	
Problem	N_{tot}	T_{tot}	N_{tot}	T_{tot}	N_{tot}	T_{tot}	N_{tot}	T_{tot}	N_{tot}	T_{tot}	N_{tot}	T_{tot}
Haverly 1	5	0.7	3		12	0.22	3	0.09	13	0.59	1	0.02
Haverly 2			3		12	0.21	9	0.09	13	0.48	1	0.03
Haverly 3			3		14	0.26	5	0.13	5	0.49	1	0.02
Foulds 2	9	3.0					1	0.10			1	0.03
Foulds 3	1	10.5					1	2.33			1	0.13
Foulds 4	25	125.0					1	2.59			1	0.14
Foulds 5	125	163.6					1	0.86			1	0.07
Ben-Tal 4			25		7	0.95	3	0.11	47	1.42	1	0.03
Ben-Tal 5			283		41	5.80	1	1.12	49	118.13	1	0.10
RT 2									185	20.04	19	0.89
Adhya 1							6174	425			17	1.75
Adhya 2							10743	1115			15	1.49
Adhya 3							79944	19314			13	0.79
Adhya 4							1980	182			1	0.26

*: Blank entries in this table indicate that data was not provided or problems were not solved by prior approaches

**: Tolerances used by Visweswaran & Floudas (1996) were 0.05% for Haverly 1, 2, 3; 0.5% for Ben-Tal 4; and 1% for Ben-Tal 5

Table 9.4: Comparative computational results for pooling problems

Chapter 10

Miscellaneous Problems

Synopsis

In this chapter, we provide computational experience with BARON on a variety of factorable nonlinear programs, including separable concave quadratic programs, indefinite quadratic programs, linear multiplicative programs, univariate polynomial programs, and benchmark problems from diverse application areas. All problems were solved to global optimality with an absolute tolerance of 10^{-6} unless otherwise specified. Our computational experience demonstrates that it is now possible to solve (mixed-integer) nonlinear optimization problems to global optimality in an automated fashion with reasonable computational resources.

10.1 Separable Concave Quadratic Programs

The separable concave quadratic programs take the following form

$$(\text{SCQP}): \quad \min \ f(x,y) = \sum_{i=1}^{r}(c_i y_i + q_i y_i^2) + \sum_{i=1}^{n}(c_{r+i} x_i + q_{r+i} x_i^2)$$
$$\text{s.t.} \quad a \leq Ax + By \leq b$$
$$l_y \leq y \leq u_y$$
$$l_x \leq x \leq u_x,$$

where $y \in \mathbb{Z}^r$, $x \in \mathbb{R}^n$, $c \in \mathbb{R}^{r+n}$, $q \in \mathbb{R}^{r+n}_-$, $a \in \mathbb{R}^m$, $b \in \mathbb{R}^m$, $l_y \in \mathbb{R}^r$, $u_y \in \mathbb{R}^r$, $l_x \in \mathbb{R}^n$, $u_x \in \mathbb{R}^n$, $B \in \mathbb{R}^{m \times r}$ and $A \in \mathbb{R}^{m \times n}$.

Separable concave quadratic programs have been addressed by numerous researchers in the literature (Tuy, Thieu & Thai 1985, Rosen & Pardalos 1986, Kalantari & Rosen 1987, Phillips & Rosen 1988, Tuy & Horst 1988, Phillips & Rosen 1992, Sherali & Alameddine 1992, Visweswaran & Floudas 1993, Sherali & Tuncbilek 1995, Visweswaran & Floudas 1996, Liu, Sahinidis & Shectman 1996, Shectman & Sahinidis 1998). Test problems of this kind have been proposed by Rosen & van Vliet (1987), Phillips & Rosen (1988), and Floudas & Pardalos (1990) and used to compare the efficiency of most of the above algorithms.

Even verifying local optimality of a separable concave quadratic program with $r = 0$ is $\mathcal{NP}$-hard (Murty & Kabadi 1987). In Tables 10.1 and 10.2, we present computational results for various standard separable concave quadratic programs from Shectman & Sahinidis (1998) and Shectman (1999). We use the BARON SCQP module without probing and the NLP module with and without probing to solve these problems. The time T_{tot} reported in each case is the CPU time required to solve the problem. Clearly, our general purpose factorable NLP module takes approximately the same number of branch-and-bound iterations to solve the problems as the specialized SCQP module.

The problems tackled in Tables 10.1 and 10.2 are small-sized. We also provide computational results on larger problems that are generated using the test problem generator of Phillips & Rosen (1988). The problems generated are of the form:

$$\min \quad \frac{1}{2}\theta_1 \sum_{j=1}^{n} \lambda_j (x_j - \bar{\omega}_j)^2 + \theta_2 \sum_{j=1}^{k} d_j y_j$$

$$\text{s.t.} \quad A_1 x + A_2 y \leq b$$

$$x \geq 0, y \geq 0$$

where

$$x, \lambda, \bar{\omega} \in \mathbb{R}^n,$$

$$y, d \in \mathbb{R}^k,$$

$$b \in \mathbb{R}^m,$$

10.1. SEPARABLE CONCAVE QUADRATIC PROGRAMS

Problem	Obj.	BARON SCQP no probing			BARON NLP probing			BARON NLP no probing		
		T_{tot}	N_{tot}	N_{mem}	T_{tot}	N_{tot}	N_{mem}	T_{tot}	N_{tot}	N_{mem}
FP1	-17.00	0.02	7	2	0.07	7	2	0.03	11	3
FP2	-213.00	0.01	1	1	0.03	1	1	0.02	1	1
FP3	-15.00	0.01	1	1	0.06	1	1	0.04	1	1
FP4	-11.00	0.01	1	1	0.03	1	1	0.02	1	1
FP5	-268.02	0.04	3	2	0.05	1	1	0.08	3	2
FP6	-39.00	0.03	9	3	0.16	11	3	0.11	12	6
FP7a	-354.75	0.17	77	5	0.32	7	2	0.34	41	6
FP7b	-634.75	0.19	93	7	0.33	9	2	0.39	39	6
FP7c	-8695.01	0.19	81	7	0.33	7	2	0.38	37	7
FP7d	-114.75	0.19	81	4	0.25	7	2	0.42	43	9
FP7e	-3730.41	0.38	197	16	1.35	39	9	2.14	309	22
FP8	15639.00	0.07	9	2	0.18	3	2	0.2	9	2
Z	0.00	0.02	7	3	0.06	3	2	0.02	7	3
BSJ2	1.00	0.02	7	3	0.07	3	2	0.05	19	4
BSJ3	-86768.60	0.00	1	1	0.00	1	1	0	1	1
BSJ4	-70262.10	0.02	1	1	0.05	1	1	0.03	1	1
RV1	-59.94	0.04	35	3	0.09	3	2	0.09	25	5
RV2	-64.48	0.16	103	7	0.41	13	4	0.4	75	12
RV3	-35.76	0.40	233	15	0.79	39	8	0.84	145	21
RV7	-138.19	0.32	155	6	0.92	27	5	1.25	149	17
RV8	-132.66	0.47	187	13	1.40	25	5	1.62	173	19
RV9	-120.15	1.25	526	38	4.44	81	14	5.17	527	64

Table 10.1: Computational results for small SCQPs (Part I)

| | | BARON SCQP ||| BARON NLP ||| BARON NLP |||
| | | no probing ||| probing ||| no probing |||
Problem	Obj.	T_{tot}	N_{tot}	N_{mem}	T_{tot}	N_{tot}	N_{mem}	T_{tot}	N_{tot}	N_{mem}
KR	-85.00	0.01	3	2	0.02	1	1	0.01	3	2
M1	-461357.00	0.36	189	3	0.45	5	2	0.61	79	5
M2	-856649.00	0.93	291	3	0.98	7	2	1.31	117	7
Ph1	-230.12	0.01	1	1	0.04	1	1	0.02	1	1
Ph2	-1028.12	0.01	1	1	0.02	1	1	0.02	1	1
Ph3	-420.24	0.02	1	1	0.02	1	1	0.01	1	1
Ph10	-9.00	0.02	3	2	0.03	3	2	0.01	3	2
Ph11	-11.28	0.00	1	1	0.03	1	1	0.05	3	2
Ph12	-22.63	0.02	1	1	0.04	1	1	0.03	3	2
Ph13	-11.28	0.00	1	1	0.04	1	1	0.05	3	2
Ph14	-229.72	0.00	1	1	0.03	1	1	0.02	1	1
Ph15	-392.70	0.02	1	1	0.04	1	1	0.02	1	1
Ph20	-158.00	0.00	1	1	0.04	1	1	0.04	1	1
PhEx	-85.00	0.02	3	2	0.04	1	1	0.03	3	2
ht	-1.60	0.01	1	1	0.02	1	1	0.05	5	2
pan1	-5.28	0.02	1	1	0.03	3	2	0.02	5	2
pan2	-17.00	0.01	7	2	0.08	7	2	0.07	11	3

Table 10.2: Computational results for small SCQPs (Part II)

$$A_1 \in \mathbb{R}^{m\times n},$$
$$A_2 \in R^{m\times k},$$
$$\theta_1, \theta_2 \in \mathbb{R}.$$

The values of θ_1 and θ_2 were chosen to be -0.001 and 0.1, respectively. Problems of this kind have been studied by Phillips & Rosen (1988), Visweswaran & Floudas (1993), and Visweswaran & Floudas (1996). The results of Table 10.3 were obtained by solving 5 randomly generated instances for every problem size. The algorithms compared in Table 10.3 are denoted as follows:

BARON SCQP	the SCQP model of BARON
GOP96	Visweswaran & Floudas (1996)
GOP/MILP	Visweswaran & Floudas (1996)

All algorithms used a relative termination tolerance of $\epsilon_r = 0.1$. Our algorithm is up to an order of magnitude faster than that of Visweswaran & Floudas (1996) for large problems.

10.2 Indefinite Quadratic Programs

The indefinite quadratic programs take the following form:

$$(\text{IQP}): \quad \min \ f(x,y) = \sum_{i=1}^{r}\sum_{j=1}^{i} q_{ij} y_i y_j + \sum_{i=1}^{r}\sum_{j=1}^{n} q_{i\,r+j} y_i x_j +$$
$$\sum_{i=1}^{n}\sum_{j=1}^{i} q_{r+i\,r+j} x_i x_j + \sum_{i=1}^{r} d_i y_i + \sum_{i=1}^{n} c_i x_i$$
$$\text{s.t.} \quad a \leq Ax + By \leq b$$
$$l_y \leq y \leq u_y$$
$$l_x \leq x \leq u_x,$$

where $y \in \mathbb{Z}^r$, $x \in \mathbb{R}^n$, $c \in \mathbb{R}^{r+n}$, $q \in \mathbb{R}^{(r+n)\times(r+n)}$, $a \in \mathbb{R}^m$, $b \in \mathbb{R}^m$, $l_y \in \mathbb{R}^r$, $u_y \in \mathbb{R}^r$, $l_x \in \mathbb{R}^n$, $u_x \in \mathbb{R}^n$, $B \in \mathbb{R}^{m\times r}$ and $A \in \mathbb{R}^{m\times n}$.

Computational results with a collection of indefinite QPs from Shectman (1999) are provided in Tables 10.4 and 10.5. The results demonstrate that the use of probing typically reduces the number of branch-and-bound nodes and, occasionally, increases the CPU requirements.

Algorithm	GOP96	GOP/MILP	BARON SCQP							
Computer	HP9000/730	HP9000/730	IBM RS/6000 43P							
	T_{tot}	T_{tot}	T_{tot}			N_{tot}			N_{mem}	
m n k	avg	avg	min	avg	max	min	avg	max	min avg max
50 50 50	0.12	0.12	0.10	0.10	0.10	1	2	3	1 1 2
50 50 100	0.15	0.14	0.10	0.22	0.30	1	13	24	1 5 9
50 50 200	6.05	1.57	0.30	0.70	1.60	9	51	170	5 17 45
50 50 500	—	14.13	1.00	1.86	2.80	21	68	108	9 16 30
50 100 100	0.22	1.37	0.30	0.72	1.10	20	62	127	6 23 48
50 100 200	0.36	11.98	0.50	1.48	3.30	17	98	246	4 20 39
100 100 100	0.31	0.31	0.40	0.46	0.50	3	7	14	2 4 8
100 100 200	0.38	0.36	0.70	0.78	0.90	14	26	37	5 8 11
100 100 500	—	80.03	1.61	4.40	15.2	43.7	740	2314	44 161 425
100 150 400	—	180.2	1.20	18.8	82.6	8	927	4311	3 163 721

Table 10.3: Comparative computational results for large SCQPs

10.2. INDEFINITE QUADRATIC PROGRAMS

		BARON NLP no probing			BARON NLP probing		
Problem	Obj.	T_{tot}	N_{tot}	N_{mem}	T_{tot}	N_{tot}	N_{mem}
bpAF1a	-45.38	0.06	7	2	0.04	1	1
bpAF1b	-42.96	0.02	1	1	0.04	1	1
bpK1	-13.00	0.02	1	1	0.03	1	1
bpK2	-13.00	0.03	1	1	0.03	1	1
bpV1	10.00	0.01	1	1	0.01	1	1
bpV2	-8.00	0.03	1	1	0.04	1	1
cqpF	-2.75	0.02	1	1	0.02	1	1
cqpJK1	-12.44	0.34	158	7	0.37	25	4
cqpJK2	-12.50	0.33	117	9	0.33	46	10
iqpBK1	-621.49	0.38	38	16	0.94	13	5
iqpBK2	-1195.23	0.34	34	15	0.75	17	6
jcbpAF2	-794.86	0.08	7	3	0.31	20	7
jcbpAFEx	-1.08	0.06	34	5	0.1	19	3
miqp1	281.00	0.03	5	2	0.03	3	2
miqp2	2.00	0.02	13	3	0.06	5	2
miqp3	-6.00	0.04	5	2	0.01	3	2
miqp4	-4574.00	0.02	3	2	0.02	1	1
miqp5	-333.89	0.18	55	4	0.16	11	2

Table 10.4: Computational results for indefinite quadratic programs (Part I)

		BARON NLP no probing			BARON NLP probing		
Problem	Obj.	T_{tot}	N_{tot}	N_{mem}	T_{tot}	N_{tot}	N_{mem}
qpC-M0	-5.00	0.01	3	2	0.03	1	1
qpC-M1	-473.78	0.06	7	2	0.1	1	1
qpC-M3a	-382.70	1.25	53	6	1.93	25	5
qpC-M3b	0.00	0.38	25	4	0.5	3	2
qpC-M3c	0.00	0.02	1	1	0.02	1	1
qpC-M4	0.00	0.01	1	1	0	1	1
qpK1	-3.00	0.02	7	3	0.04	3	2
qpK2	-12.25	0.12	21	7	0.29	17	5
qpK3	-36.00	1.02	133	24	2.36	63	13
test1	0.00	0.03	7	3	0.04	7	3
test2	-9.25	0.04	3	2	0.04	3	2
test3	-7.00	0.05	6	3	0.03	5	2
test4	-7.00	0.03	3	2	0.02	1	1
test5	-110.00	0.04	17	5	0.1	13	3
test6	471.00	0.06	11	3	0.09	11	3
test8	-29605.00	0.13	5	2	0.19	1	1
testGR1	-12.81	0.84	156	25	1.59	100	20
testGR3	-20.59	1.12	122	31	1.41	72	17
testPh14	-80.50	0.03	5	2	0.06	5	2

Table 10.5: Computational results for indefinite quadratic programs (Part II)

10.3 Linear Multiplicative Programs

The linear multiplicative programs take the following form:

$$\text{(LMP):} \quad \min \quad f(x,y) = \prod_{i=1}^{p} f_i(x,y) = \prod_{i=1}^{p}(d_i y + c_i x + f_i)$$

$$\text{s.t.} \quad l_r \leq Ax + By \leq u_r$$

$$l_y \leq y \leq u_y$$

$$l_x \leq x \leq u_x,$$

where $y \in \mathbb{Z}^r$, $x \in \mathbb{R}^n$, $d_i \in \mathbb{R}^r$, $c_i \in \mathbb{R}^n$ and $f_i \in \mathbb{R}$ for $i \in \{1,\ldots,p\}$, $l_r \in \mathbb{R}^m$, $l_u \in \mathbb{R}^m$, $B \in \mathbb{R}^{m \times r}$, $A \in \mathbb{R}^{m \times n}$.

These problems are known to be $\mathcal{NP}$-hard (Matsui 1996) with $r = 0$ justifying the use of branch-and-bound algorithms for their solution. Various researchers have worked on problems related to LMP (Glover & Woolsey 1973, Glover & Woolsey 1974, Konno & Kuno 1989, Sherali & Adams 1990, Konno, Yajima & Matsui 1991, Thoai 1991, Konno & Kuno 1992, Kuno, Yajima & Konno 1993, Hansen et al. 1993, Konno & Kuno 1995, Benson & Boger 1997, Liu, Umegaki & Yamamoto 1999, Ryoo 1999).

In Table 10.6, we present computational results on a set of linear multiplicative programs from Ryoo (1999) using the LMP module of BARON without probing and NLP module with and without probing. Each row corresponds to results for a single instance of a randomly generated problem with m rows, n variables, and p products. By "type" in this table, we denote the problem type, for a total of three different types that are solved by Ryoo (1999). As was the case for SCQPs, we conclude that our general purpose NLP module is competitive to the specialized LMP implementation. Similarly, probing helps reduce the number of iterations and memory requirements.

To test the approach on larger problems, we generated a set of large LMPs based on three different generation strategies used for constructing test problems by Thoai (1991), Konno & Kuno (1992), Kuno, Yajima & Konno (1993), Benson & Boger (1997), and Ryoo (1999). In particular, problems of the following types were generated:

$$\text{(LMP1):} \quad \min \quad \prod_{i=1}^{p} c_i x$$

Problem				BARON LMP no probing			BARON NLP no probing			BARON NLP probing		
type	m	n	p	T_{tot}	N_{tot}	N_{mem}	T_{tot}	N_{tot}	N_{mem}	T_{tot}	N_{tot}	N_{mem}
1	20	30	2	0.3	17	4	0.14	17	3	0.14	5	2
1	20	30	3	0.8	38	9	0.31	42	8	0.30	11	4
1	20	30	4	2.1	117	17	0.78	108	14	0.93	21	4
1	120	120	3	42.8	131	15	10.43	93	14	18.69	31	6
1	100	100	4	51.9	243	36	16.50	217	34	31.35	63	10
1	200	200	5	567.0	575	77	176.69	521	65	452.40	131	17
2	10	20	2	0.1	5	2	0.07	9	2	0.05	1	1
3	20	30	2	0.2	9	2	0.16	19	3	0.20	3	2

Table 10.6: Computational results for LMPs

$$\text{s.t.} \quad b \leq Ax$$
$$0 \leq x,$$

where the elements of c, A, and b are generated randomly in the range [0,100] as was done by Konno & Kuno (1992) and Kuno et al. (1993);

$$\text{(LMP2):} \quad \min \quad \prod_{i=1}^{2}(c_i x + f_i)$$
$$\text{s.t.} \quad b \leq Ax$$
$$0 \leq x,$$

where c_i and f_i ($i = 1, 2$) are random numbers in [0,1], the constraint matrix elements a_{ij} are generated in $[-1, 1]$ via $a_{ij} := 2\alpha - 1$ where α are random numbers in [0,1], and the left-hand-side values are generated via $b_i := \sum_j a_{ij} + 2\beta$, where β are random numbers in [0,1]. This is the way random LMPs are generated by Thoai (1991); and

$$\text{(LMP3):} \quad \min \quad \prod_{i=1}^{p} c_i x$$
$$\text{s.t.} \quad b \leq Ax$$
$$1 \leq x \leq u,$$

10.3. LINEAR MULTIPLICATIVE PROGRAMS

Algorithm	KK92/KYK93		BARON NLP								
Machine	SUN4/75		IBM RS/6000 43P								
	T_{tot}		T_{tot}			N_{tot}			N_{mem}		
m n p	avg	std	max	avg	min	max	avg	min	max	avg	min
20 30 2	0.46	0.05	0.19	0.16	0.13	19	15	17	4	3	2
120 100 2	34.4	n/a	2.97	2.25	1.85	29	20	23	6	3	2
220 200 2	263	n/a	14.63	11.23	8.49	35	26	17	6	5	4
20 30 3	1.27	0.25	0.60	0.36	0.25	76	47	27	13	8	3
120 120 3	179	43.1	12.87	9.60	6.57	103	82	85	15	12	9
200 180 3	914	130	41.40	29.09	18.98	132	93	61	16	12	7
20 30 4	14.2	10.5	0.89	0.70	0.33	127	95	99	14	11	5
100 100 4	524	210	21.72	14.67	8.18	286	186	112	38	26	12
200 200 4	n/a	n/a	162.02	103.30	76.30	462	298	281	52	40	33
200 200 5	n/a	n/a	361.09	275.09	177.96	1098	806	603	143	100	65

Table 10.7: Computational results for LMPs of Type 1

where the elements of c and A are chosen randomly from the set of integers $\{1, 2, \ldots, 10\}$, $b_i = \sum_j a_{ij}^2$ for $i = 1, \ldots, m$, and $u = \max_{i=1,\ldots,m} \{b_i\}$, as in Benson & Boger (1997) and Liu, Umegaki & Yamamoto (1999).

Results with the three different problem types, denoted as Type 1, 2, and 3, are provided in Tables 10.7, 10.8, and 10.9, respectively. Five different, randomly generated problem instances were generated for each problem size and the results were averaged for presentation.

The algorithms compared in these tables are denoted as follows:

BARON NLP	the NLP module of BARON
KK92/KYK93	Konno & Kuno (1992), Kuno et al. (1993)
T91	Thoai (1991)

From Tables 10.7 and 10.8, it is clear that our BARON NLP implementation is faster than the competing specialized implementations in the literature on machines that are comparable in speed. Further, we are able to solve much larger problems than reported earlier in the literature.

Algorithm	T91		BARON NLP								
Machine	PS2		IBM RS/6000 43P								
	T_{tot}		T_{tot}			N_{tot}			N_{mem}		
m n p	avg	std	max	avg	min	max	avg	min	max	avg	min
10 20 2	1.85	9	0.10	0.08	0.06	13	9	7	2	2	2
20 30 2	5.04	16	0.16	0.14	0.11	15	13	13	4	3	2
60 100 2	290	30	1.89	1.78	1.59	21	20	19	5	5	4
100 100 2	635	32	4.39	4.16	3.83	23	15	15	5	3	2
200 200 2	n/a	n/a	46.19	42.53	37.89	27	21	17	5	4	3

Table 10.8: Computational results for LMPs of Type 2

Algorithm	BARON NLP								
	T_{tot}			N_{tot}			N_{mem}		
m n p	max	avg	min	max	avg	min	max	avg	min
20 30 2	0.22	0.17	0.13	29	19	13	6	4	2
120 100 2	3.98	2.84	1.88	32	21	31	5	4	3
220 200 2	27.84	17.96	10.40	37	31	29	6	5	4
20 30 3	0.60	0.44	0.23	83	59	29	12	9	5
120 120 3	27.35	20.54	14.65	179	132	121	24	17	12
200 180 3	222.85	134.43	87.44	280	192	168	39	26	19
20 30 4	1.52	0.94	0.52	209	125	120	31	17	11
100 100 4	69.95	39.50	26.70	692	399	459	79	50	35
200 200 4	875.39	668.96	496.66	893	776	754	115	100	83

Table 10.9: Computational results for LMPs of Type 3

10.4. GENERALIZED LINEAR MULTIPLICATIVE PROGRAMS

Problem	Source	Obj.	BARON NLP no probing			BARON NLP probing		
			T_{tot}	N_{tot}	N_{mem}	T_{tot}	N_{tot}	N_{mem}
glmp_fp1	Falk & Polocsay (1994)	10.00	0.01	1	1	0.04	1	1
glmp_fp2	Falk & Polocsay (1994)	7.34	0.01	3	2	0.03	3	2
glmp_fp3	Falk & Polocsay (1994)	-12.00	0.04	1	1	0.03	1	1
glmp_kk90	Konno & Kuno (1990)	3.00	0.04	1	1	0.02	1	1
glmp_kk92	Konno & Kuno (1992)	-12.00	0.03	3	2	0.03	3	2
glmp_kky	Konno, Kuno & Yajima (1994)	-2.50	0.05	5	2	0.04	3	2
glmp_ss1	Schaible & Sodini (1995)	-24.57	0.11	33	4	0.1	11	2
glmp_ss2	Schaible & Sodini (1995)	3.00	0.01	1	1	0.03	1	1

Table 10.10: Computational results for GLMPs

10.4 Generalized Linear Multiplicative Programs

The generalized linear multiplicative programs take the following form:

$$(\text{GLMP}): \quad \min \quad f(x,y) = \sum_{i=1}^{T} \prod_{j=1}^{p_i} (d_{ij} y + c_{ij} x + f_{ij})$$

$$\text{s.t.} \quad a \leq Ax + By \leq b$$

$$l_y \leq y \leq u_y$$

$$l_x \leq x \leq u_x,$$

where $y \in \mathbb{Z}^r$, $x \in \mathbb{R}^n$, $d_{ij} \in \mathbb{R}^r$, $c_{ij} \in \mathbb{R}^n$ and $f_{ij} \in \mathbb{R}$ for $i \in \{1, \ldots, p\}$, $a \in \mathbb{R}^m$, $b \in \mathbb{R}^m$, $B \in \mathbb{R}^{m \times r}$, $A \in \mathbb{R}^{m \times n}$.

Results for a collection of generalized multiplicative problems from Ryoo (1999) are provided in Table 10.10. Once again, the results demonstrate that extensive range reduction (probing) reduces the number of branch-and-bound nodes.

Algorithm	Nested form		Centered form		BARON POLY		
Machine	SUN 3/50-12				IBM RS/6000 43P		
Problem Obj.	T_{tot}	N_{tot}	T_{tot}	N_{tot}	T_{tot}	N_{tot}	N_{mem}
p1 −0.30	0.38	21	1.12	20	0.01	43	4
p2 −0.66	6.30	44	72.96	34	0.01	19	4
p3 −0.44	0.30	19	0.78	18	0.00	29	4
p4 0.00	0.40	32	1.16	31	0.01	31	4
p5 0.00	0.34	21	1.22	23	0.01	3	2
p6 0.70	0.70	37	2.04	37	0.01	45	6
p7 −0.75	0.22	16	0.50	16	0.01	11	3
p8 −0.61					0.01	3	2
p9 −0.16					0.00	3	2

Table 10.11: Computational results on univariate polynomial programs

10.5 Univariate Polynomial Programs

$$\text{(POLY):} \quad \min \quad f(x,y) = \sum_{i=0}^{k} c_i x^i$$
$$\text{s.t.} \quad l \leq x \leq u,$$

where either $x \in \mathbb{R}$ or $x \in \mathbb{Z}$, $c \in \mathbb{R}^{k+1}$, $l \in \mathbb{R}$, $u \in \mathbb{R}$.

Our test problems are taken from Hansen, Jaumard & Xiong (3), in which the authors used an interval method to solve these problems. We present the computational results using BARON's POLY module for this problem in Table 10.11. The CPU times are negligible for all these problems. Algorithms compared in this table are:

> BARON NLP the POLY module of BARON
> Nested form   Hansen et al. (3)
> Centered form Hansen et al. (3)

10.6 Miscellaneous Benchmark Problems

Miscellaneous engineering design and global optimization test problems typically involve factorable functions and can therefore be handled by BARON's

10.6. MISCELLANEOUS BENCHMARK PROBLEMS

Problem	Constraints	Variables	Description
e01	1	2	bilinear program
e02	3	3	design of water pump
e03	7	10	alkylation process design
e04	2	4	design of insulated tank
e05	3	5	heat exchanger network design
e06	3	3	chemical equilibrium
e07	7	10	pooling problem
e08	2	2	bilinear and quadratic constraints
e09	1	2	bilinear constraints and objective
e10	1	2	nonlinear equality constraint
e11	2	3	bilinearities, economies of scale
e12	3	4	design of two-stage process systems
e13	2	2	MINLP, process synthesis
e14	9	7	MINLP, process synthesis
e15	5	5	MINLP, process synthesis
e16	9	12	heat exchanger network synthesis
e17	2	1	design of a reinforced concrete beam
e18	4	2	quadratically constrained LP
e19	2	2	quadratically constrained QP
e20	7	5	reactor network design
e21	6	6	design of three-stage process system

Table 10.12: Brief description of miscellaneous benchmark problems (Part I)

NLP module. We present results for a collection of 42 such problems that have been addressed in global/local optimization literature. A brief description of these problems can be found in Tables 10.12 and 10.13, while the source of the problems is provided in Tables 10.14 and 10.15. Problems e1-e27 were solved in an earlier work by Ryoo & Sahinidis (1996). However, a specialized branch-and-bound implementation was coded by Ryoo & Sahinidis (1996) for each different example. Here, we solve all these problems using our BARON NLP module and present the computational results in Tables 10.16 and 10.17.

Problem	Constraints	Variables	Description
e22	5	2	linearly constrained concave QP
e23	2	2	biconvex program
e24	4	2	linearly constrained QP
e25	8	4	linear multiplicative program
e26	4	2	linearly constrained concave QP
e27	6	4	nonlinear fixed-charge problem
e28	3	5	quadratically constrained QP
e29	7	11	reliability problem
e30	15	14	fixture design
e31	133	61	fixture design
e32	18	35	molecular design
e33	6	9	bilinearly constrained problem
e34	4	6	bilinearly constrained problem
e35	39	32	heat exchanger network synthesis
e36	2	2	MINLP
e37	1	4	parameter estimation (exponentials)
e38	3	4	pressure vessel design
e39	0	2	Shekel function
e40	9	4	truss design
e41	1	4	reliability problem
e42	2	5	design of experiments

Table 10.13: Brief description of miscellaneous benchmark problems (Part II)

10.6. MISCELLANEOUS BENCHMARK PROBLEMS

Problem	Source
e01	Sahinidis & Grossmann (1991)
e02	Stoecker (1971)
e03	Bracken & McCormick (1968)
e04	Stoecker (1971)
e05	Liebman, Lasdon, Schrage & Waren (1986)
e06	Liebman et al. (1986)
e07	Haverly (1978)
e08	Swaney (1990)
e09	Swaney (1990)
e10	Soland (1971)
e11	Westerberg & Shah (1978)
e12	Stephanopoulos & Westerberg (1975)
e13	Kocis & Grossmann (1988)
e14	Yuan, Zhang, Pibouleau & Domenech (1988)
e15	Kocis & Grossmann (1988)
e16	Floudas & Ciric (1989)
e17	Liebman et al. (1986)
e18	Visweswaran & Floudas (1990)
e19	Manousiouthakis & Sourlas (1992)
e20	Manousiouthakis & Sourlas (1992)
e21	Stephanopoulos & Westerberg (1975)

Table 10.14: Sources of miscellaneous benchmark problems (Part I)

Problem	Source
e22	Kalantari & Rosen (1987)
e23	Al-Khayyal & Falk (1983)
e24	Konno & Kuno (1989)
e25	Thoai (1991)
e26	Thakur (1990)
e27	Falk & Soland (1969)
e28	Colville (1968)
e29	Ashrafi & Berman (1992)
e30	Marin (1998)
e31	Marin (1998)
e32	Sahinidis et al. (2002)
e33	Xia (1996)
e34	Hock & Schittkowski (1981)
e35	Biegler, Grossmann & Westerberg (1997)
e36	Li & Chou (1994)
e37	Beck & Arnold (1977)
e38	Sandgren (1990)
e39	Dixon & Szegoe (1975)
e40	Shin, Gurdal & Griffin (1990)
e41	Kim (1998)
e42	Tawarmalani & Ahmed (1997)

Table 10.15: Sources of miscellaneous benchmark problems (Part II)

		BARON NLP without probing			BARON NLP with probing		
Problem	Obj.	T_{tot}	N_{tot}	N_{mem}	T_{tot}	N_{tot}	N_{mem}
e01	-6.67	0.02	1	1	0.02	1	1
e02	201.16	0.02	1	1	0.02	1	1
e03	-1161.34	1.48	275	17	1.41	61	5
e04	5194.87	0.08	17	5	0.18	7	2
e05	7049.25	0.64	268	14	1.23	120	10
e06	0.00	0.01	1	1	0.01	1	1
e07	-400.00	0.08	15	4	0.2	9	2
e08	0.74	0.03	11	5	0.06	7	2
e09	-0.50	0.06	19	3	0.07	9	2
e10	-16.74	0.03	5	2	0.03	3	2
e11	189.31	0.03	1	1	0.01	1	1
e12	-4.51	0.03	1	1	0.04	1	1
e13	2.00	0.01	1	1	0.03	1	1
e14	4.58	0.09	13	6	0.13	7	3
e15	7.67	0.02	1	1	0.03	1	1
e16	12292.50	0.13	17	3	0.36	7	2
e17	0.00	0.01	1	1	0.01	1	1
e18	-2.83	0.03	3	2	0.04	3	2
e19	-118.71	0.21	83	9	0.37	45	6
e20	-0.39	0.33	115	11	0.45	39	8
e21	-13.40	0.03	1	1	0.04	1	1

Table 10.16: Computational results on miscellaneous benchmarks (Part I)

		BARON NLP without probing			BARON NLP with probing		
Problem	Obj.	T_{tot}	N_{tot}	N_{mem}	T_{tot}	N_{tot}	N_{mem}
e22	-85.00	0.02	3	2	0.02	1	1
e23	-1.08	0.06	19	2	0.1	13	2
e24	3.00	0.04	3	2	0.04	1	1
e25	0.89	0.04	17	3	0.03	1	1
e26	-185.78	0.01	1	1	0.02	1	1
e27	2.00	0.02	3	2	0.03	1	1
e28	-30665.50	0.02	2	2	0.08	1	1
e29	-0.94	0.18	47	11	0.23	20	8
e30	-1.58	0.37	33	7	0.51	19	5
e31	-2.00	3.75	351	56	6.31	169	22
e32	-1.43	13.7	906	146	18.5	227	31
e33	-400.00	0.08	15	4	0.19	9	2
e34	0.02	0.02	1	1	0.01	1	1
e35	64868.10	16.4	465	57	81.2	580	54
e36	-246.00	2.59	768	72	4.31	229	23
e37	0.00	1.3	81	8	1.87	35	7
e38	7197.73	0.38	5	2	0.35	5	2
e39	-10.09	0.05	5	2	0.1	5	2
e40	30.41	0.15	24	6	0.31	15	4
e41	641.82	0.07	5	2	0.12	3	2
e42	18.78	0.04	1	1	0.01	1	1

Table 10.17: Computational results on miscellaneous benchmarks (Part II)

10.7 Selected Mixed-Integer Nonlinear Programs

Mixed-integer nonlinear programs were solved in this book in Chapters 3 and 9. In this section, we present solutions of additional MINLPs. The principal aim of this section is to demonstrate that global optimization reveals new solutions to problems that have been addressed by local optimization methods in the past.

10.7.1 Design of Just-in-Time Flowshops

In this section, we consider a production flowshop problem addressed earlier by Gunasekaran, Goyal, Martikainen & Yli-Olli (1993), Gutierrez & Sahinidis (1996), and Sahinidis & Tawarmalani (2000). The problem involves P products to be processed in S stages. Each stage contains identical equipment performing the same type of operation on different products. An operation for a particular product-stage combination can be performed on any of the machines of that stage. Thus, scheduling issues are nonexistent. The demand rate for each product is assumed to be known. The objective is to determine the number of machines for each stage so as to minimize the total equipment-related cost. The latter consists of costs due to product processing, imbalance in production rates between successive stages, and investment in buying the equipment.

We use i to denote the products ($i = 1, \ldots, P$) and j the stages ($j = 1, \ldots, S$) of the production system. The following are the known parameters for this problem:

- α_{ij} cost of processing product i at stage j per unit time;
- β_{ij} cost for unit imbalance in production rates of product i between stages j and $j+1$;
- δ_i demand for product i per unit time;
- κ_{ij} rate of decrease in processing time of product i and stage j for a unit increase in machine investment;
- μ_j cost per machine of stage j;
- M maximum amount of money available for investing in machines;

π_{ij} priority weight given in processing of product i at stage j ($\sum_{i=1}^{P} \pi_{ij} = 1$ for $j = 1, \ldots, S$).

σ_{ij} number of production cycles of product i at stage j per unit time: $\sigma_{ij} = \delta_i/q_{ij}$, where q_{ij} is the batch size of product i at stage j;

$\bar{\tau}_{ij}$ processing time per unit of product i at stage j when investment in the equipment is zero;

τ_{ij} processing time for a batch of product i on one machine of stage j: $\tau_{ij} = q_{ij}(\bar{\tau}_{ij} - \kappa_{ij}\mu_j)$;

ω_j resource requirements for a machine at stage j;

Ω maximum resource level available for all machines in the plant.

The following are the variables in the model:

n_j number of machines at stage j;

p_{ij} rate of production for product i at stage j (number of batches per unit time, considering all machines in the stage).

The problem of determining the number of machines required for each stage of processing while minimizing the total equipment-related costs can then be modeled as an MINLP:

$$(\text{JIT}): \quad \min \quad \sum_{i=1}^{P}\sum_{j=1}^{S} \frac{\alpha_{ij}\sigma_{ij}}{p_{ij}} + \sum_{i=1}^{P}\sum_{j=1}^{S-1} \beta_{ij}|p_{ij} - p_{ij+1}| + \sum_{j=1}^{S} \mu_j n_j$$

$$\text{s.t.} \quad p_{ij} = \frac{\pi_{ij}}{\tau_{ij}} n_j \qquad i = 1, \ldots, P; \quad j = 1, \ldots, S$$

$$\sum_{j=1}^{S} \mu_j n_j \leq M$$

$$\sum_{j=1}^{S} \omega_j n_j \leq \Omega$$

$$n_j \geq 1, \text{ integer} \qquad j = 1, \ldots, S.$$

The objective function of JIT consists of three terms corresponding to the costs due to: product processing, production imbalance between successive stages, and equipment investment. The unit production cost, α_{ij}, in the first term of the objective accounts for inventory costs, transportation costs, setup costs, labor costs, and other machine processing costs. The inverse of the production rate, p_{ij}, appears in the first term of the objective of JIT as it provides the average processing time for a batch of product i in stage j. The second term in the objective function is in line with the just-in-time philosophy. As just-in-time systems aim to maintain a balanced flow of products between successive stages, the model penalizes the absolute value of any imbalance in production rates by multiplying imbalance in production rates between successive stages by an imbalance cost (β_{ij}). The penalty cost coefficient, β_{ij}, accounts for work-in-process waiting costs, shortage costs, and costs due to idleness of the facility. The model does not differentiate among imbalance costs generated due to shortage and surplus; both cases are considered equally disrupting. An asymmetric imbalance cost function could be easily handled, though. The third term in the objective of JIT simply accounts for cost for buying equipment. For simplicity, this cost is assumed to be proportional to the number of machines in each stage.

The first constraint of JIT expresses the production rate for each product-stage combination as a function of the number of machines and batch processing times. The second constraint represents a budget constraint whereas the third constraint expresses a resource constraint.

Problem JIT involves a convex objective function and a linear constraint set. The only source of nonconvexities is the integrality requirement for the n_j variables. The model is equivalent to that first provided by Gunasekaran et al. (1993). Using a standard transformation described in introductory linear programming textbooks, the nondifferentiability in the objective can be easily eliminated while still maintaining a linearly constrained set (Gutierrez & Sahinidis 1996). In particular, one would need to replace $|p_{ij} - p_{ij+1}|$ by new variables, say Δp_{ij} and add the constraints:

$$\Delta p_{ij} \geq p_{ij} - p_{ij+1} \quad i = 1, \ldots, P; j = 1, \ldots, S - 1$$
$$\Delta p_{ij} \geq p_{ij+1} - p_{ij} \quad i = 1, \ldots, P; j = 1, \ldots, S - 1.$$

Because this is a minimization problem, Δp_{ij} will be driven towards their smallest possible values and one of these two constraints will be binding for each Δp_{ij}, effectively implying that $\Delta p_{ij} = |p_{ij} - p_{ij+1}|$.

Although continuous variables appear in the formulation, due to the first constraint of JIT, there are no degrees of freedom once the values of n_j ($j = 1, \ldots, S$) are fixed. This makes natural the application of a direct pattern search (Hooke and Jeeves with discrete steps, cf. Bazaraa et al.

1993) for the solution of the model. As this method does not guarantee global solutions, we wish to contrast the solutions that it provides to those obtained through global optimization. The problems solved are taken from Gunasekaran et al. (1993) who applied direct pattern search. The global optimization results were obtained using an early implementation of BARON (Gutierrez & Sahinidis 1996).

All problems solved involve the production of $P = 3$ products in $S = 4$ stages. Thus, there are only four degrees of freedom in this problem ($n_j, j = 1, \ldots, 4$). The product demands are $\delta = (3000, 2000, 4000)$. The machine costs and resource requirements are $\mu = (5000, 5500, 4000, 6000)$ and $\omega = (60, 50, 80, 40)$, respectively. The corresponding budget and resource limits are 6×10^6 and 3000. The remaining problem parameters for each product-stage combination are given in Table 10.18. This problem instance will be referred to as the "base case" and designated as Case 1. Eight additional cases will be considered by varying the machine costs ($\mu_j, j = 1, \ldots, S$) and lot sizes ($q_{ij}, i = 1, \ldots, P; j = 1, \ldots, S$) of the base case as follows. Cases 2, 3, 4, and 5 are obtained from the base case by multiplying all machine costs by 0.5, 1.5, 2, and 2.5, respectively. Cases 6, 7, 8, and 9 are similarly obtained from the base case by multiplying all lot sizes by 0.5, 1.5, 2, and 2.5, respectively.

	$i=1$				$i=2$				$i=3$			
j	1	2	3	4	1	2	3	4	1	2	3	4
α_{ij}	2	1.5	3	2	3	2.5	1	2	2	2	2	1
$\beta_{ij} \times 10^{-5}$	60	90	60	80	90	80	80	70	80	100	80	90
$\kappa_{ij} \times 10^5$	0.2	0.3	0.4	0.3	0.1	0.4	0.3	0.2	0.2	0.3	0.2	0.2
π_{ij}	0.2	0.4	0.5	0.5	0.6	0.3	0.3	0.2	0.2	0.3	0.2	0.3
$\bar{\tau}_{ij}$	1	1	1	1	1	1	1	1	1	1	1	1
q_{ij}	800	800	800	800	700	700	700	700	900	900	900	900

Table 10.18: JIT system design problem data for base case

The results of Gunasekaran et al. (1993) for the nine cases using a direct pattern search method (DPSM) are summarized in Table 10.19. Results for the same problems with BARON (Gutierrez & Sahinidis 1996) are shown in Table 10.20.

We first compare the results of Tables 10.19 and 10.20 for the base case (Case 1). The best solution obtained by the DPSM for this case (Table 10.19)

10.7. SELECTED MIXED-INTEGER NONLINEAR PROGRAMS

	Number of machines				Costs, $			
Case	n_1	n_2	n_3	n_4	Processing	Imbalance	Investment	Total
1	4	2	2	2	100960	55993	51000	211883
2	6	2	2	2	94990	59462	30500	184952
3	3	2	2	2	106860	60401	69000	236261
4	3	2	2	2	106160	60882	92000	259042
5	3	2	2	1	127340	52340	100000	279680
6	4	1	1	1	182100	59928	35500	277528
7	4	2	2	2	123320	34064	45000	202384
8	4	2	3	2	90290	35677	55000	180967
9	4	2	3	2	90290	28542	55000	173832

Table 10.19: JIT designs using DPSM

	Number of machines				Costs, $			
Case	n_1	n_2	n_3	n_4	Processing	Imbalance	Investment	Total
1	3	3	3	3	80512	31971	61500	173983
2	3	3	3	3	81020	31755	30750	143525
3	3	2	2	2	106861	25906	69000	201767
4	2	2	2	2	119245	21610	82000	222856
5	2	2	2	2	118484	21762	102500	242746
6	2	2	2	2	120769	42628	41000	204396
7	3	3	3	3	80512	21314	61500	163326
8	3	3	3	3	80512	15985	61500	157998
9	3	3	3	3	80512	12788	61500	154801

Table 10.20: Globally optimal results for JIT system design

yields an assignment of (4, 2, 2, 2) machines to the four stages with a total equipment related cost of $211,883. On the other hand, the global solution for the same problem (Table 10.20) yields an assignment of (3, 3, 3, 3) machines to four stages and results into an optimal cost of $173,983. This represents an improvement of the heuristic solution by $37,900 or 18%. Comparing the heuristic solution with the optimal solution, we observe that the optimal solution requires two additional machines to be installed and thus increases the investment cost. However, the additional machines allow for a significant reduction of the processing and imbalance cost, thus reducing the total cost.

Case	Reduction (increase) in machines				Cost reduction (increase), $			Savings	
	$j=1$	$j=2$	$j=3$	$j=4$	Processing	Imbalance	Investment	$	%
1	1	(1)	(1)	(1)	20448	24022	(10500)	37900	18
2	3	(1)	(1)	(1)	13970	27707	(250)	41427	22
3	0	0	0	0	(1)	34495	0	34494	15
4	1	0	0	0	(13085)	39272	10000	36187	14
5	1	0	0	(1)	8856	30577	(2500)	36937	13
6	2	(1)	(1)	(1)	61332	17300	(5500)	73131	26
7	1	(1)	(1)	(1)	42808	12750	(16500)	39058	19
8	1	(1)	0	(1)	9778	19692	(6500)	22969	13
9	1	(1)	0	(1)	9778	15753	(6500)	19031	11
Avg.	1.2	(0.7)	(0.5)	(0.8)	17433	24559	(4875)	37903	17

Table 10.21: Comparison of heuristic and exact solutions for JIT system design

Similar comparisons can be made for all other cases considered in Tables 10.19 and 10.20. In all cases, the global solutions yield lower costs as compared to the best heuristic solutions. Table 10.21 summarizes the differences between the heuristic and global search approaches. Clearly, there are significant improvements in the solutions by solving the equipment selection problem to global optimality. The percentage savings vary from 11% to 26%, with an overall average of 17%. Recall that the problem under study is very small with only 4 degrees of freedom. Larger savings can be expected for larger problems.

10.7.2 The Gupta-Ravindran Benchmarks

In this section, we address the solution of benchmark mixed-integer nonlinear programs. In particular, we focus our attention on a collection of MINLPs from Gupta & Ravindran (1985). In solving these problems, we required that the integer variables take values between 0 and 200, thereby constructing a bounded starting box. For problem 21, BARON found the upper bound on the first variable too high to guarantee reliable computations. This upper bound had to be reduced to approximately 144 to guarantee globality of the solution.

The computational results are presented in Table 10.22 where, for each problem, we report the globally optimal objective function value, the CPU seconds taken by BARON, the total number of nodes in the branch-and-bound tree, the node where the optimal solution was found, and the maximum number of nodes in memory. These runs were done on a Dell Precision 530 workstation with a 1.7 GHz Pentium IV Xeon processor and 1 GB RDRAM. Only 16 MB of memory were made available to BARON.

The results of Table 10.22 were obtained by using default search strategies for all but problems 5, 2, and 24. For the latter, the default search options require a significant amount of time. For these three problems, the results presented in Table 10.22 were obtained by restricting branching to the original problem variables only and selecting the longest edge for bisection at each node. Observe that the solutions presented here for problems 2, 14, 20, and 21 correspond to better objective function values than those reported in Gupta & Ravindran (1985).

The results presented in this as well as earlier chapters demonstrate that it is possible to solve mixed-integer nonlinear programs from a variety of disciplines within reasonable computing times through an entirely automated procedure.

Problem		Obj.	T_{tot}	N_{tot}	N_{mem}
1		12.47	0.05	29	5
2	*	5.96	0.04	13	6
3		16.00	0.01	3	2
4		0.72	0.02	1	1
5		5.47	1.91	241	52
6		1.77	0.03	11	5
7		4.00	0.02	3	2
8		23.45	0.39	21	5
9		-43.13	0.16	39	7
10		-310.80	0.04	12	4
11		-431.00	0.07	34	8
12		-481.20	0.11	70	13
13		-585.20	0.40	208	30
14	*	-40358.20	0.03	8	5
15		1.00	0.02	11	3
16		0.70	0.03	23	12
17		-1100.40	12.00	3497	388
18		-778.40	2.75	1037	122
19		-1098.40	43.45	9440	1022
20	*	230.92	1.64	156	19
21	*	-5.68	0.05	31	3
22		6.06	0.08	14	6
23		-1125.20	168.23	25926	3296
24		-1033.20	1095.87	130487	15967

*: Solution found for these problems are better than those earlier reported by Gupta & Ravindran (1985).

Table 10.22: Selected MINLP problems from Gupta and Ravindran (1985)

Chapter 11

GAMS/BARON: A Tutorial and Empirical Performance Analysis

Synopsis

The purpose of this chapter is threefold. First, to provide modelers, students, and practitioners with easily accessible models and reproducible computational results. Second, to demonstrate that the algorithms proposed in this book can be used to solve global optimization models with minimal user intervention. Third, to provide a tutorial for the recently developed version of BARON integrated with the GAMS modeling environment. To achieve these goals, we provide in GAMS format many of the models used throughout the book, including the molecular design problem of Chapter 8, the pooling problems of Chapter 9, as well as multiplicative programs, benchmark factorable problems from the engineering and optimization literature, just-in-time flow-shop design problems, and the Gupta-Ravindran collection of MINLPs of Chapter 10. Finally, we demonstrate that the techniques proposed in this book can be easily used to find all feasible solutions to systems of nonlinear equations as well as hard combinatorial optimization problems with no user intervention.

To ensure easy access and reproducibility of the results presented in this chapter, all models solved here have been placed on the book Web site and contributed to appropriate model collection libraries on the Internet (`globallib` and `minlplib`).

11.1 Introduction

A number of high level programming languages have been developed in recent years and provide a natural and simple way to interface with optimization solvers. These languages include GAMS (Brook et al. 1988), AMPL (Fourer, Gay & Kernighan 1993), LINGO (Schrage 1999), and OPL (Van Hentenryck 1999). Modelers as well as algorithm developers rely on these systems to experiment with new models and to develop new algorithmic concepts in a solver-independent framework.

While we plan to provide BARON under additional modeling languages in the future, we chose to initially interface BARON with GAMS for a number of reasons. First, GAMS is one of the oldest modeling systems that has been widely used in academia, industry, and government for over three decades. A large user community is familiar with the GAMS simple and intuitive modeling language, especially in the economics, agriculture, and engineering fields. Thus, by interfacing BARON with GAMS, we are able to reach a very large community of academics, practitioners, and policy makers. Second, many recent GAMS initiatives are complementary to our work (GAMS Development Corporation 2002). These initiatives include the establishment of the `globallib` library of test problems for global optimization models (Meeraus 2002), the creation of the `minlplib` library of test problems for mixed-integer nonlinear programming models (Bussieck 2002), and the development of `Performance World`, an automated system for empirical analysis of algorithms (Mittelmann 2002). Finally, the GAMS development team has committed a significant amount of effort to ensure that our implementation is stable and robust for widespread use. In particular, our development of BARON benefitted greatly through the extensive testing done by Alex Meeraus, Michael Bussieck, Steven Dirske, and Armin Pruessner.

A detailed manual of the GAMS/BARON software is available elsewhere (Sahinidis & Tawarmalani 2002). This chapter does not attempt to supplant the manual but to supplement it by providing a tutorial describing the use of the solver. In the remainder of this chapter, we first provide a brief description of the GAMS/BARON solver. Then, we illustrate its use by solving a variety of problems. The main goal is to demonstrate how BARON options can be employed gainfully in order to control CPU time and memory requirements of the software, and experiment systematically with different modeling and algorithmic constructs. We also outline some of BARON's unique features such as the ability to provide guaranteed global optima as well as to identify all or the K best feasible solutions through a single tree search. All computations in this chapter were done on a Dell Precision 530 workstation with a 1.7 GHz Pentium IV Xeon processor and 1 GB RDRAM. Only 16 MB of memory were made available to BARON unless otherwise noted.

11.2 Types of Problems GAMS/BARON Can Solve

While BARON is still available in the form of a callable library that facilitates computational experimentation with novel global optimization algorithms, GAMS/BARON provides a fully automated way for solving NLPs and MINLPs to global optimality.

11.2.1 Factorable Nonlinear Programming: MIP, NLP, and MINLP

The most general class of problems addressed by the software is as follows:

$$\begin{aligned}
\min \quad & f(x) \\
\text{s.t.} \quad & l_r \leq g(x) \leq u_r \\
& l_c \leq x \leq u_c \\
& x_i \in \mathbb{Z} \text{ for } i = 1, \ldots, k \\
& x_i \in \mathbb{R} \text{ for } i = k+1, \ldots, n
\end{aligned}$$

where, $x \in \mathbb{Z}^k \times \mathbb{R}^{n-k}$, $f : \mathbb{R}^n \to \mathbb{R}$, $g : \mathbb{R}^n \to \mathbb{R}^m$, $l_r \in \mathbb{R}^m$, $u_r \in \mathbb{R}^m$, $l_c \in \mathbb{R}^n$, $u_c \in \mathbb{R}^n$, and f, g are factorable functions, *i.e.*, recursive compositions of sums and products of functions of single variables. Most functions of several variables used in nonlinear optimization are factorable and can be easily brought into separable form (McCormick 1983). Some examples of factorable functions are:

- $f(x, y) = xy$.
- $f(x, y) = x/y$.

- $f(x,y,z,w) = \sqrt{\exp(xy + z \ln w)z^3}$.
- $f(x,y,z,w) = \frac{x^2 y^{0.3} z}{w^2} + \exp\left(\frac{x^2 w}{y}\right) - xy$.
- $f(x) = \sum_{i=1}^{n} \ln_i(x_i)$, where $x \in \mathbb{R}^n$.
- $f(x) = \sum_{i=1}^{T} \prod_{j=1}^{p_i} \left(c_{ij}^0 + c_{ij}^t x\right)$, where $x \in \mathbb{R}^n$, $c_{ij} \in \mathbb{R}^n$, and $c_{ij}^0 \in \mathbb{R}$ $(i = 1, \ldots, T; j = 1, \ldots, p_i)$.

These functions may appear in the objective and/or constraints of the mathematical program to be solved by BARON. In addition, integrality requirements may be explicitly placed on a subset of the problem variables.

What we refer to here as factorable NLP subsumes the GAMS model types LP, MIP, NLP, and MINLP. We next mention a number of special cases of the above model that the user may be interested in solving. BARON will automatically recognize some of these structures and exploit problem-structure in solving the model.

11.2.2 Special Cases of BARON's Factorable Nonlinear Programming Solver

Nonlinear Integer Programming

$$\begin{aligned}
\min \quad & f(x) \\
\text{s.t.} \quad & l_r \leq g(x) \leq u_r \\
& l_c \leq x \leq u_c \\
& x_i \in \mathbb{Z} \text{ for } i = 1, \ldots, n
\end{aligned}$$

where, $x \in \mathbb{Z}^n$, $f : \mathbb{R}^n \to \mathbb{R}$, $g : \mathbb{R}^n \to \mathbb{R}^m$, $l_r \in \mathbb{R}^m$, $u_r \in \mathbb{R}^m$, $l_c \in \mathbb{R}^n$, $u_c \in \mathbb{R}^n$, and f, g are factorable functions.

Continuous Nonlinear Programming

$$\min \quad f(x)$$

11.2. TYPES OF PROBLEMS GAMS/BARON CAN SOLVE

$$\text{s.t.} \quad l_r \leq g(x) \leq u_r$$
$$l_c \leq x \leq u_c$$
$$x_i \in \mathbb{R} \text{ for } i = 1, \ldots, n$$

where, $x \in \mathbb{R}^n$, $f : \mathbb{R}^n \to \mathbb{R}$, $g : \mathbb{R}^n \to \mathbb{R}^m$, $l_r \in \mathbb{R}^m$, $u_r \in \mathbb{R}^m$, $l_c \in \mathbb{R}^n$, $u_c \in \mathbb{R}^n$, and f, g are factorable functions.

Separable Concave Quadratic Programming

$$\min \ f(x) = \sum_{i=1}^{n}(c_i x_i + q_i x_i^2)$$
$$\text{s.t.} \quad l_r \leq Ax \leq u_r$$
$$l_c \leq x \leq u_c$$
$$x_i \in \mathbb{Z} \text{ for } i = 1, \ldots, k$$
$$x_i \in \mathbb{R} \text{ for } i = k+1, \ldots, n$$

where $x \in \mathbb{Z}^k \times \mathbb{R}^{n-k}$, $c \in \mathbb{R}^n$, $q \in \mathbb{R}^n_-$, $l_r \in \mathbb{R}^m$, $u_r \in \mathbb{R}^m$, $l_c \in \mathbb{R}^n$, $u_c \in \mathbb{R}^n$, and $A \in \mathbb{R}^{m \times n}$.

Problems with Power Economies of Scale (Cobb-Douglas functions)

$$\min \ f(x) = \sum_{i=1}^{n} c_i x_i^{q_i}$$
$$\text{s.t.} \quad l_r \leq Ax \leq u_r$$
$$l_c \leq x \leq u_c$$
$$x_i \in \mathbb{Z} \text{ for } i = 1, \ldots, k$$
$$x_i \in \mathbb{R} \text{ for } i = k+1, \ldots, n$$

where $x \in \mathbb{Z}^k \times \mathbb{R}^{n-k}$, $c \in \mathbb{R}^n_+$, $q \in (0,1]^n$, $l_r \in \mathbb{R}^m$, $u_r \in \mathbb{R}^m$, $l_c \in \mathbb{R}^n_+$, $u_c \in \mathbb{R}^n_+$, and $A \in \mathbb{R}^{m \times n}$.

Fractional Programming

$$\min \quad f(x) = \sum_{i=1}^{T} \frac{c_i x + \alpha_i}{q_i x + \beta_i}$$
$$\text{s.t.} \quad l_r \leq Ax \leq u_r$$
$$l_c \leq x \leq u_c$$
$$x_i \in \mathbb{Z} \text{ for } i = 1, \ldots, k$$
$$x_i \in \mathbb{R} \text{ for } i = k+1, \ldots, n$$

where $x \in \mathbb{Z}^k \times \mathbb{R}^{n-k}$; $c_i \in \mathbb{R}^n$, $q_i \in \mathbb{R}^n$, $\alpha_i \in \mathbb{R}$ and $\beta_i \in \mathbb{R}$ ($i = 1, \ldots, T$); $l_r \in \mathbb{R}^m$, $u_r \in \mathbb{R}^m$, $l_c \in \mathbb{R}^n$, $u_c \in \mathbb{R}^n$, and $A \in \mathbb{R}^{m \times n}$. We assume $q_i x + \beta_i > 0$ ($i = 1, \ldots, T$) for any $x \in X := \{x \mid l_r \leq Ax \leq u_r \text{ and } l_c \leq x \leq u_c\}$, and X compact.

Univariate Polynomial Programming

$$\min \quad f(x) = \sum_{i=0}^{k} c_i x^i$$
$$\text{s.t.} \quad l_c \leq x \leq u_c$$

where $x \in \mathbb{R}$, $c \in \mathbb{R}^k$, $l_c \in \mathbb{R}$, $u_c \in \mathbb{R}$.

Linear Multiplicative Programming

$$\min \quad f(x) = \prod_{i=1}^{p} f_i(x) = \prod_{i=1}^{p} \left(c_i^t x + c_{i0} \right)$$
$$\text{s.t.} \quad l_r \leq Ax \leq u_r$$
$$l_c \leq x \leq u_c$$
$$x_i \in \mathbb{Z} \text{ for } i = 1, \ldots, k$$
$$x_i \in \mathbb{R} \text{ for } i = k+1, \ldots, n$$

where $x \in \mathbb{Z}^k \times \mathbb{R}^{n-k}$, $c_i \in \mathbb{R}^n$ and $c_{i0} \in \mathbb{R}$ ($i = 1, \ldots, p$), $l_r \in \mathbb{R}^m$, $u_r \in \mathbb{R}^m$, $l_c \in \mathbb{R}^n$, $u_c \in \mathbb{R}^n$, and $A \in \mathbb{R}^{m \times n}$. We assume $c_i^t x + c_{i0} \geq 0$ ($i = 1, \ldots, p$) for $x \in \{x \mid l_r \leq Ax \leq u_r \text{ and } l_c \leq x \leq u_c\}$.

11.2. TYPES OF PROBLEMS GAMS/BARON CAN SOLVE

General Linear Multiplicative Programming

$$\min \ f(x) = \sum_{i=1}^{T} \prod_{j=1}^{p_i} (c_{ij}^0 + c_{ij}^t x)$$
$$\text{s.t.} \quad l_r \leq Ax \leq u_r$$
$$l_c \leq x \leq u_c$$
$$x_i \in \mathbb{Z} \text{ for } i = 1, \ldots, k$$
$$x_i \in \mathbb{R} \text{ for } i = k+1, \ldots, n$$

where $x \in \mathbb{Z}^k \times \mathbb{R}^{n-k}$, $c_{ij} \in \mathbb{R}^n$ and $c_{ij}^0 \in \mathbb{R}$ ($i = 1, \ldots, T; j = 1, \ldots, p_i$), $l_r \in \mathbb{R}^m$, $u_r \in \mathbb{R}^m$, $l_c \in \mathbb{R}^n$, $u_c \in \mathbb{R}^n$, and $A \in \mathbb{R}^{m \times n}$.

Separable Mixed-Integer Convex Quadratic Programming

$$\min \ f(x) = \sum_{i=1}^{n} (c_i x_i + q_i x_i^2)$$
$$\text{s.t.} \quad l_r \leq Ax \leq u_r$$
$$l_c \leq x \leq u_c$$
$$x_i \in \mathbb{Z} \text{ for } i = 1, \ldots, k$$
$$x_i \in \mathbb{R} \text{ for } i = k+1, \ldots, n$$

where $x \in \mathbb{Z}^k \times \mathbb{R}^{n-k}$, $c \in \mathbb{R}^n$, $q \in \mathbb{R}_+^n$, $l_r \in \mathbb{R}^m$, $u_r \in \mathbb{R}^m$, $l_c \in \mathbb{R}^n$, $u_c \in \mathbb{R}^n$ and $A \in \mathbb{R}^{m \times n}$.

Indefinite Quadratic Programming

$$\min \ f(x) = \sum_{i=1}^{n} \sum_{j=1}^{n} q_{ij} x_i x_j + \sum_{i=1}^{n} c_i x_i$$
$$\text{s.t.} \quad l_r \leq Ax \leq u_r$$
$$l_c \leq x \leq u_c$$
$$x_i \in \mathbb{Z} \text{ for } i = 1, \ldots, k$$
$$x_i \in \mathbb{R} \text{ for } i = k+1, \ldots, n$$

where $x \in \mathbb{Z}^k \times \mathbb{R}^{n-k}$, $q \in \mathbb{R}^{n \times n}$, $c \in \mathbb{R}^n$, $l_r \in \mathbb{R}^m$, $u_r \in \mathbb{R}^m$, $l_c \in \mathbb{R}^n$, $u_c \in \mathbb{R}^n$, and $A \in \mathbb{R}^{m \times n}$.

Mixed-Integer Linear Programming

$$\begin{aligned} \min \quad & f(x) = \sum_{i=1}^{n} c_i x_i \\ \text{s.t.} \quad & l_r \leq Ax \leq u_r \\ & l_c \leq x \leq u_c \\ & x_i \in \mathbb{Z} \text{ for } i = 1, \ldots, k \\ & x_i \in \mathbb{R} \text{ for } i = k+1, \ldots, n \end{aligned}$$

where $x \in \mathbb{Z}^k \times \mathbb{R}^{n-k}$, $c \in \mathbb{R}^n$, $l_r \in \mathbb{R}^m$, $u_r \in \mathbb{R}^m$, $l_c \in \mathbb{R}^n$, $u_c \in \mathbb{R}^n$, and $A \in \mathbb{R}^{m \times n}$.

11.3 Software and Hardware Requirements

In order to use GAMS/BARON, users will need to have access to a licensed GAMS base system as well as a licensed LP solver. A licensed NLP solver is optional and may facilitate solution of certain problems. Currently, CPLEX (ILOG 2000), MINOS (Murtagh & Saunders 1995), and SNOPT (Gill et al. 1999) may be used as the LP solver, while MINOS and SNOPT may be used as the NLP solver. Hence, a minimal GAMS/BARON system requires any one of the CPLEX, MINOS, or SNOPT solvers.

Currently, BARON is a mix of about 42,000 lines in standard FORTRAN 90 and 24,000 lines in C. Therefore, it is very portable. Thus far, it has been compiled under Windows, Linux, IBM AIX, HP-UX, Sun OS, and the Sun Solaris operating systems.

11.4 Model Requirements

11.4.1 Variable and Expression Bounds

All nonlinear variables and expressions in the mathematical program to be solved must be bounded below and above by finite numbers. It is important that finite lower and upper bounds be provided by the user on all problem variables. Note that providing finite bounds on variables is not sufficient to guarantee finite bounds on nonlinear expressions arising in the model. For example, consider $1/x$ for $x \in [0,1]$. It is important to provide bounds for problem variables that guarantee that the problem functions are finitely valued. If the user model does not include variable bounds that guarantee that all nonlinear expressions are finitely-valued, BARON's preprocessor will attempt to infer appropriate bounds from problem constraints. If this step fails, global optimality of the solutions provided is not guaranteed. Occasionally, the lack of bounds is so severe that no numerically stable lower bounding problems can be constructed, in which case BARON may terminate immediately.

11.4.2 Allowable Nonlinear Functions

In addition to multiplication and division, BARON can handle nonlinear functions that involve $\exp(x)$, $\ln(x)$, x^α for $\alpha \in \mathbb{R}$, and β^x for $\beta \in \mathbb{R}$. Currently, there is no support for other functions, including the trigonometric functions $\sin(x)$, $\cos(x)$, etc.

11.5 How to Run GAMS/BARON

BARON is capable of solving models of the following types: LP, MIP, RMIP, NLP, DNLP, CNS, RMINLP, and MINLP. If BARON is not selected as the default solver for these models, it can be invoked by issuing the following command before the `solve` statement:

$$\text{option xxx=baron;}$$

where xxx stands for LP, MIP, RMIP, NLP, DNLP, CNS, RMINLP, or MINLP. The solver choice can also be specified on the command level as follows:

$$\text{gams myfile xxx=baron}$$

where `myfile` is the file containing the GAMS model and xxx stands for LP, MIP, RMIP, NLP, DNLP, CNS, RMINLP, or MINLP.

11.6 System Output

11.6.1 System Log

Consider the following concave minimization problem from Floudas & Pardalos (1990):

$$\begin{aligned}
\min \quad & 42x_1 - 50x_1^2 + 44x_2 - 50x_2^2 + 45x_3 - 50x_3^2 \\
& +47x_4 - 50x_4^2 + 47.5x_5 - 50x_5^2 \\
\text{s.t.} \quad & 20x_1 + 12x_2 + 11x_3 + 7x_4 + 4x_5 \leq 40 \\
& 0 \leq x_i \leq 1, \quad i = 1,\ldots,5.
\end{aligned}$$

The global minimum for this problem occurs at $x = (1, 1, 0, 1, 0)$, with an objective function value of -17.

The GAMS model for solving this problem is as follows:

```
POSITIVE VARIABLES x1, x2, x3, x4, x5;

VARIABLE f;

x1.up = 1; x2.up = 1; x3.up = 1; x4.up = 1; x5.up = 1;

EQUATIONS e1, obj;

e1 ..  20*x1 + 12*x2 + 11*x3 + 7*x4 + 4*x5 =L= 40;

obj .. f =E= 42*x1 - 50*x1**2 + 44*x2 - 50*x2**2 + 45*x3 - 50*x3**2
         + 47*x4 - 50*x4**2 + 47.5*x5 - 50*x5**2;

model logex /all/;

option nlp = baron;

solve logex minimizing f using nlp;
```

The following is the system log for this model.

```
===============================================================================
                        Welcome to BARON v. 5.0
                 Global Optimization by BRANCH-AND-REDUCE
               Parts of the BARON software were created at the
                  University of Illinois at Urbana-Champaign.
```

11.6. SYSTEM OUTPUT

```
===============================================================================
                    Factorable Non-Linear Programming
===============================================================================
Starting solution is feasible with a value of      0.000000D+00
Preprocessing found feasible solution with value  -0.800000D+01
Preprocessing found feasible solution with value  -0.170000D+02
===============================================================================
            We have space for       28097     nodes in the tree
===============================================================================

    Itn. no.      Open Nodes       Total Time     Lower Bound     Upper Bound

        1              1            000:00:00    -0.250000D+03   -0.170000D+02
        1              1            000:00:00    -0.189000D+02   -0.170000D+02
        9              0            000:00:00    -0.170000D+02   -0.170000D+02


                       *** Successful Termination ***


    Total time elapsed      :   000:00:00,   in seconds:        0.08
          on parsing        :   000:00:00,   in seconds:        0.01
          on preprocessing:     000:00:00,   in seconds:        0.06
          on navigating     :   000:00:00,   in seconds:        0.00
          on relaxed        :   000:00:00,   in seconds:        0.01
          on local          :   000:00:00,   in seconds:        0.00
          on tightening     :   000:00:00,   in seconds:        0.00
          on marginals      :   000:00:00,   in seconds:        0.00
          on probing        :   000:00:00,   in seconds:        0.00


    Total no. of BaR iterations:         9
    Best solution found at node:        -1
    Max. no. of nodes in memory:         2

All done with problem
===============================================================================
```

The solver first tests feasibility of the user-supplied starting point. In this instance, no starting point is provided. Thus, GAMS initializes the starting point at zero. This point is found to be feasible with an objective function value of 0. Subsequently, BARON's preprocessor identifies two feasible solutions with objective function values of -8 and -17. BARON then reports that the supplied memory (default of 16 MB) provides enough space for storing up to 28097 branch-and-reduce "open" nodes for this problem. Then, the iteration log provides (every 100 iterations unless termination occurs or a feasible solution is found): the iteration number, number of open branch-and-bound nodes, the CPU time taken thus far, the lower bound, and the upper bound for the problem. At the end of the run, a breakdown of times in different parts of the branch-and-reduce algorithm is provided. Finally, the total number of branch-and-reduce iterations (number of search tree nodes) is reported, followed by the node where the best solution was identified (-1 stands for preprocessing as explained below).

11.6.2 Termination Messages, Model and Solver Status

Upon termination, BARON will report the node where the optimal solution was found. We refer to this node as `nodeopt`. This quantity has the following interpretation:

$$\texttt{nodeopt} = \begin{cases} -3, & \text{no feasible solution found,} \\ -2, & \text{the best solution found was the user-supplied point,} \\ -1, & \text{the best solution was found during preprocessing,} \\ i, & \text{the best solution was found in the } i\text{th node of the tree.} \end{cases}$$

In addition to reporting `nodeopt`, upon termination, BARON will issue one of the following statements:

- *** Successful Termination ***. This is the desirable termination status. The problem has been solved within tolerances in this case. If nodeopt$= -3$, the problem is infeasible.

- *** Max. Allowable Nodes in Memory Reached ***. The user will need to make more memory available to BARON or change algorithmic options to reduce the size of the search tree and memory required for storage.

- *** Max. Allowable BaR Iterations Reached ***. The user will need to increase the maximum number of allowable iterations (`maxiter`).

- *** Max. Allowable CPU Time Exceeded ***. The user will need to increase the maximum number of allowable CPU time (`maxtime`).

- *** Numerical Difficulties Encountered ***. This case should be reported to the developers.

- *** Search Interrupted by User ***. The run was interrupted by the user (Ctrl-C).

- *** Insufficient Memory for Data Structures ***. More memory is needed to set up the problem data structures.

- *** Search Terminated by BARON ***. This will happen if the required variable bounds are not provided in the input model. The user will need to read the BARON output for likely pointers to variables and expressions with missing bounds and fix the formulation, or be content with the solution provided, which may not be globally optimal.

11.7 Algorithmic and System Options

For a detailed description of the algorithmic and other options that are available to the user, readers are referred to the GAMS/BARON manual (Sahinidis & Tawarmalani 2002). All options are initialized in BARON at certain default values. If the user wishes to modify any options, this must be communicated to BARON through a GAMS options command, a GAMS model suffix, or a BARON options file. The use of these options is illustrated through examples in the remainder of this chapter.

11.8 Application to Multiplicative Programs

In this section, we address the solution of the class of linear multiplicative problems of Section 10.3 with no integer variables:

$$(\text{LMP}): \quad \min \quad \prod_{i=1}^{p} (c_i^t x + f_i)$$
$$\text{s.t.} \quad l_r \leq Ax \leq u_r$$
$$l_c \leq x \leq u_c,$$

where $x \in \mathbb{R}^n$, $c_i \in \mathbb{R}^n$ and $f_i \in \mathbb{R}$ for $i \in \{1, \ldots, p\}$, $l_r \in \mathbb{R}^m$, $u_r \in \mathbb{R}^m$, $l_c \in \mathbb{R}^n$, $u_c \in \mathbb{R}^n$, and $A \in \mathbb{R}^{m \times n}$.

LMPs and slight generalizations thereof have attracted considerable attention in the literature because of their large number of practical applications in many fields of study, including microeconomics (Henderson & Quandt 1971), financial optimization (Konno, Shirakawa & Yamazaki 1993, Maranas, Androulakis, Floudas, Berger & Mulvey 1997), VLSI chip design (Dorneich

& Sahinidis 1995, Maling, Mueller & Heller 1982), decision tree optimization (Bennett 1994), portfolio optimization (Konno, Shirakawa & Yamazaki 1993, Markowitz 1991), plant layout design (Quesada & Grossmann 1996), multi-criteria optimization problems (Keeney & Raiffa 1993), robust optimization (Mulvey, Vanderbei & Zenios 1995), and data mining/pattern recognition (Bennett & Mangasarian 1994).

Observe that LMP is linearly constrained. The only nonlinearities appear in the objective function. As posed above, the number of nonlinearities is a function of the number of variables, n, as well as the number of products, p, in the objective. Depending on the value of p, this representation implies a large number of nonlinear nonzeros in the model. Alternatively, we may rewrite LMP as:

$$(\text{LMP}_y): \quad \min \quad \prod_{i=1}^{p} y_i$$
$$\text{s.t.} \quad y_i = c_i^t x + f_i \quad i = 1, \ldots, p$$
$$l_r \leq Ax \leq u_r$$
$$l_c \leq x \leq u_c,$$

where $y \in \mathbb{R}^p$. At the expense of introducing p linear variables, the number of nonlinearities is now a function of p alone. For any fixed value of y, LMP is linear in x. This makes it possible to focus the search for a global optimum in $\mathbb{R}^p$ alone.

11.8.1 LMPs of Type 1

We first consider LMPs of "Type 1," defined as:

$$(\text{LMP1}): \quad \min \quad \prod_{i=1}^{p} c_i^t x$$
$$\text{s.t.} \quad b \leq Ax$$
$$0 \leq x,$$

where the elements of c, A, and b are generated randomly in the range [0,100] as was initially proposed by Konno & Kuno (1992) and later used for computations by Kuno et al. (1993) and Ryoo & Sahinidis (2002). These LMPs can be generated and solved in GAMS as follows:

11.8. APPLICATION TO MULTIPLICATIVE PROGRAMS

Model lmp1.gms

```
$ontext
Filename: LMP1.gms
Author: Nick Sahinidis, August 2002

Purpose:
Generate and solve random linear multiplicative models of
"Type 1." Problem instances are generated as proposed by:
   H. Konno and T. Kuno, "Linear multiplicative programming,"
   Mathematical Programming, 56(51-64), 1992.
$offtext

options optcr=0, optca=1.e-6,
    limrow=0, limcol=0,
    solprint=off,
    reslim = 10000;

sets m constraints  /1*220/
     n variables    /1*200/
     p products     /1*5/
     c cases        /1*10/
     i instances    /1*5/ ;

*for each case to be solved, we have a
*different (m,n,p) triplet
table cases(c,*)
     1    2   3
1    20   30  2
2    120  100 2
3    220  200 2
4    20   30  3
5    120  120 3
6    200  180 3
7    20   30  4
8    100  100 4
9    200  200 4
10   200  200 5 ;

parameters cc(p,n)   cost coefficients
           A(m,n)    constraint coefficients
           b(m)      left-hand-side
           rep(c,*)  summary report ;

parameters mactual, nactual, pactual,
```

```
                ResMin, Resmax, NodMin, Nodmax;

variables y(p), x(n), obj ;
x.lo(n) = 0;

equations objective, constraints(m), products(p);

objective .. obj =E= prod(p $ (ord(p) le pactual), y(p));

products(p) $ (ord(p) le pactual) ..
         y(p) =E= sum(n $ (ord(n) le nactual), cc(p,n)*x(n));

constraints(m) $ (ord(m) le mactual) ..
         b(m) =L= sum(n $ (ord(n) le nactual), A(m,n)*x(n)) ;

model lmp1 /all/;
lmp1.workspace = 32;

rep(c,'AvgResUsd') = 0;
rep(c,'AvgNodUsd')= 0;
loop (c,

  mactual = cases(c,'1');
  nactual = cases(c,'2');
  pactual = cases(c,'3');

  ResMin  = inf;
  Resmax  = 0;
  NodMin  = inf;
  Nodmax  = 0;

  loop(i,

    cc(p,n) = uniform(0,100) $ (ord(p) le pactual and ord(n) le nactual);
    A(m,n)  = uniform(0,100) $ (ord(m) le mactual and ord(n) le nactual);
    b(m)    = uniform(0,100) $ (ord(m) le mactual);

*   make sure all problems have the same zero starting point
    x.l(n)=0; y.l(p)=0;
    solve lmp1 minimizing obj using nlp;
    rep(c,'AvgResUsd')  = rep(c,'AvgResUsd') + lmp1.resusd;
    rep(c,'AvgNodUsd')  = rep(c,'AvgNodUsd') + lmp1.nodusd;
    ResMin = min(ResMin, lmp1.resusd);
    NodMin = min(NodMin, lmp1.nodusd);
    ResMax = max(ResMax, lmp1.resusd);
```

11.8. APPLICATION TO MULTIPLICATIVE PROGRAMS

```
      NodMax = max(NodMax, lmp1.nodusd);
   );
   rep(c,'MinResUsd') = ResMin;
   rep(c,'MaxResUsd') = ResMax;
   rep(c,'MinNodUsd')= NodMin;
   rep(c,'MaxNodUsd')= NodMax;
);
rep(c,'AvgResUsd') = rep(c,'AvgResUsd')/card(i);
rep(c,'AvgNodUsd')= rep(c,'AvgNodUsd')/card(i);

display rep;
```

Ten different problem sizes are considered corresponding to the ten cases of the GAMS array cases(c,*). For each problem size, five different random instances are generated and solved.

Table 11.1 presents branch-and-bound requirements for these problems with default GAMS/BARON options. For each problem size, the table shows the total number of branch-and-bound nodes (N_{tot}), the maximum number of nodes in memory (N_{mem}), and the CPU time in seconds. The minimum, maximum, and average of these quantities over the five randomly generated instances for each problem size are presented. Clearly, computational requirements increase with problem size, in particular with the number of products (p).

11.8.2 Controlling Local Search Requirements

It is instructive to take a close look at the breakdown of computational times provided by GAMS/BARON. Upon execution of the fifth instance of the largest problem, the following CPU times are reported:

```
Total time elapsed    :  000:18:21,   in seconds:    1101.46
      on parsing      :  000:00:01,   in seconds:       1.32
      on preprocessing:  000:00:09,   in seconds:       8.55
      on navigating   :  000:00:01,   in seconds:       1.38
      on relaxed      :  000:03:01,   in seconds:     180.89
      on local        :  000:14:15,   in seconds:     854.70
      on tightening   :  000:00:55,   in seconds:      54.62
```

Case	Size m n p	N_{tot} Min Max Ave.	N_{mem} Min Max Ave.	CPU s Min Max Ave.
1	20 30 2	15 19 16.6	2 4 2.8	0.22 0.32 0.26
2	120 100 2	13 25 17.0	2 5 3.2	3.62 8.52 5.61
3	220 200 2	13 27 18.2	2 4 2.8	39.15 65.84 46.90
4	20 30 3	41 82 61.6	6 12 9.6	0.31 0.45 0.37
5	120 120 3	77 162 114.4	13 21 16.8	6.81 19.18 12.63
6	200 180 3	93 179 143.8	15 21 17.2	42.92 77.21 54.01
7	20 30 4	93 155 126.6	12 22 17.2	0.36 0.75 0.54
8	100 100 4	245 1194 619.2	26 120 68.8	9.99 48.16 25.57
9	200 200 4	835 1540 1142.0	85 160 127.0	178.87 644.33 346.47
10	200 200 5	2785 4597 3404.8	288 487 385.0	675.57 1102.2 930.33

Table 11.1: Computational requirements of branch-and-bound for LMP1 with default GAMS/BARON options

```
on marginals   :   000:00:00,   in seconds:    0.00
on probing     :   000:00:00,   in seconds:    0.00
```

As 77% of the CPU time is spent in local search, this indicates that changing the BARON options to reduce local search activities may be beneficial. For this reason, we next experiment with the following GAMS/BARON options file:

File baron.opt

```
numloc 0
dolocal 0
```

The `numloc` option sets the number of local searches that will be performed during preprocessing of the root node of the tree. The `dolocal` option sets the frequency at which local searches are done in the course of the tree search. Setting these options to zero entirely eliminates the use of a local NLP solver by GAMS/BARON. The computational requirements of GAMS/BARON with this set of options are shown in Table 11.2. Compared to performance of the code under default options, the number of total nodes occasionally increases. This is expected because without local search it may

11.8. APPLICATION TO MULTIPLICATIVE PROGRAMS

Case	N_{tot} Min	Max	Ave.	N_{mem} Min	Max	Ave.	CPU s Min	Max	Ave.
1	11	19	15.4	2	4	2.8	0.08	0.1	0.09
2	13	26	17.8	2	6	3.8	1.17	1.48	1.40
3	13	27	19.2	2	4	3.2	6.07	6.84	6.55
4	45	81	62.6	7	17	11.4	0.12	0.19	0.16
5	85	163	118.2	13	24	17.6	2.86	3.83	3.35
6	111	199	166.4	17	25	21.2	8.16	11.26	9.81
7	93	157	125.6	12	21	16.2	0.23	0.32	0.28
8	251	1071	597.6	34	124	71.2	3.94	12.94	7.86
9	874	1568	1090.2	86	140	119.6	45.29	78.24	55.10
10	2591	5494	3684.8	331	490	393.6	133.47	306.09	203.56

Table 11.2: Branch-and-bound requirements for LMP1 with `baron.opt` options

take longer to identify good solutions and eliminate certain parts of the tree. On the other hand, local search is much more time consuming compared to solving the linear programming relaxations generated by BARON. As a result, the total CPU time requirements using this set of options are about 20% of the CPU requirements with the default options.

11.8.3 Reducing Memory Requirements via Branching Options

Next, we would like to discuss memory control issues with GAMS/BARON. By default, BARON stores the lower and upper bounds on all problem variables as well as a number of additional variables that define the linear programming relaxation at every node in the search tree. After the preprocessing phase, BARON informs the user how many branch-and-bound nodes it can store in the available memory. If the number of nodes in memory approaches the maximum possible, the default best-bound strategy switches to the usually inferior last-in-first-out (LIFO) option. In order to save space in such situations, the user is provided with the `numstore` and `numbranch` options, which denote the number of variables to be stored at each node and considered for branching, respectively. Only stored variables are considered for branching. Hence, `numbranch` can not exceed `numstore`. The use of these options requires the user to declare the stored variables before any other problem variables in the GAMS model. Furthermore, the branching variables should be declared before any other stored variables.

Consider lmp1.gms with lmp1.workspace=32; and the default set of options. The opening screen of BARON informs the user how many branch-and-bound nodes can be stored in 32 MB of memory. In particular, for the tenth case of the above problems, the following is reported:

```
===========================================================================
              We have space for       7279      nodes in the tree
===========================================================================
```

All variables are stored and considered for branching by default. Observe, however, that once the y variables are fixed in LMP_y, the remaining program is linear. Hence, it suffices to branch on the y variables alone. We next consider the following set of options:

File baron.op2

```
numloc 0
dolocal 0
numbranch 5
numstore 5
```

With lmp1.workspace=32, BARON reports the following for the tenth case of the problems in lmp1.gms:

```
===========================================================================
              We have space for      103716     nodes in the tree
===========================================================================
```

This particular set of options instructs BARON to store and branch on the y variables only. The memory savings are significant, although not surprising considering that there are only five y variables in this problem while there are an additional 200 continuous variables in the problem formulation. Computational requirements with this option are detailed in Table 11.3. Compared to BARON's performance with the previously considered options, CPU and node requirements are now slightly larger. However, there is sufficient space in memory for much larger search trees. This is an important feature in order to solve larger and more difficult models.

11.8. APPLICATION TO MULTIPLICATIVE PROGRAMS

Case	N_{tot} Min	Max	Ave.	N_{mem} Min	Max	Ave.	CPU s Min	Max	Ave.
1	11	19	15.4	2	4	2.8	0.08	0.09	0.09
2	13	28	18.2	2	6	3.6	1.17	1.48	1.38
3	13	27	19.2	2	4	3.0	6.13	6.63	6.44
4	47	83	62.6	6	12	10.2	0.13	0.19	0.16
5	85	181	121.8	13	23	16.8	2.79	3.96	3.32
6	106	198	159.8	18	22	19.6	7.48	10.76	9.55
7	97	159	126.6	11	23	16.6	0.22	0.31	0.27
8	281	1061	599.6	30	109	63.6	4.10	12.37	7.88
9	854	1333	1041.2	81	132	116.6	49.51	65.94	55.83
10	2973	5917	3898.0	340	465	403.4	164.63	324.89	218.88

Table 11.3: Branch-and-bound requirements for LMP1 with `baron.op2` options

11.8.4 Controlling Memory Requirements via Probing

As explained in Section 1.3.3 and Chapter 5, a variety of strategies are possible for reducing ranges of variables based on duality principles. In its simplest form, these strategies use the Lagrange multipliers of the relaxation to reduce ranges of variables whose bounds are active at the solution of the relaxed problem at any node of the tree (see Section 1.3.3). This marginals-based strategy is performed by default in BARON. A slightly more elaborate strategy, that is not carried out by default, is to probe the bounds of variables that are not at their bounds in the relaxation solution. The number of variables whose bounds are probed is determined by the BARON option `pdo`, which by default takes the value of zero. If `pdo` is nonzero, there are two probing strategies currently available in BARON. In the default strategy, variables are temporarily fixed at bounds, the resulting relaxation is solved, and the marginals thus obtained are used for range reduction. In another strategy, optimization problems are solved to minimize and maximize the probing variables over a suitable relaxation of the search space. The BARON option `pxdo` (default value is zero) can be used to specify the number of probing variables for which the minimization/maximization linear programs will be solved. In either case, at a given node BARON will rank order the problem variables for branching. The same ranking of variables is used for probing:

the **pdo** most important branching variables are chosen for probing. After probing, the relaxation is constructed and solved again, at which point the final branching decision can change.

Consider the following GAMS/BARON options file:

File baron.op3

```
numloc 0
dolocal 0
numbranch 5
numstore 5
pdo 1
pxdo 1
```

As before, we do not allow local search and have specified that only the first 5 problem variables (the y variables) be stored in every node and considered for branching. In addition, pdo = 1 specifies that probing is to performed on one variable. Further, pxdo = 1 specifies that probing is carried out by solving the optimization problems to minimize/maximize the probing variable over a relaxation of the search space. In other words, the selected branching variable is probed upon and, if sufficient range reduction occurs, the relaxation is solved again and the branching decision is reconsidered. Computational results for LMP1 with this strategy are provided in Table 11.4. While probing increases the amount of work done at a given node, it leads to considerable reduction in the ranges of variables. As a result, relaxations become tighter and the search trees much smaller. When compared with default options, probing reduces the memory requirements of the algorithm by about 70% for these problems. In some instances, this is coupled with an increase in the CPU time due to additional requirements for solving the probing subproblems.

11.8.5 Effects of Reformulation

Reformulating a problem often alters the lower bounding strategies chosen by BARON. The new lower bounding strategy could be more/less effective. For example, consider the reformulation of LMP1 where a lower bound for the y variables is introduced and the objective is changed as follows:

11.8. APPLICATION TO MULTIPLICATIVE PROGRAMS

	N_{tot}			N_{mem}			CPU s		
Case	Min	Max	Ave.	Min	Max	Ave.	Min	Max	Ave.
1	1	5	3.0	1	2	1.8	0.07	0.10	0.09
2	3	7	4.2	2	4	2.4	1.13	1.38	1.24
3	1	11	5.4	1	4	2.2	5.42	6.59	6.03
4	11	19	15.0	3	6	4.4	0.12	0.20	0.16
5	17	33	27.6	4	9	6.8	1.91	4.10	3.10
6	21	64	45.2	4	12	9.2	6.70	13.79	10.75
7	21	29	24.6	4	8	6.0	0.18	0.29	0.24
8	93	315	202.6	9	43	24.4	4.27	17.67	9.76
9	179	415	258.0	21	63	37.8	43.71	98.84	59.80
10	625	1630	1010.6	75	195	117.0	163.49	369.60	236.63

Table 11.4: Branch-and-bound requirements for LMP1 with `baron.op3` options

```
y.lo(p) = 0.000001;
objective .. obj =E= sum(p $ (ord(p) le pactual), log(y(p)));
```

In the new formulation, solutions with objective function value of 0 are chopped off. However, it is easy to check if such solutions are admissible to the model by minimizing each y variable individually. In the current set none of the problems admits such a solution. Therefore, the reformulation is correct. In Table 11.5, we report computational requirements of BARON to solve the reformulated problems using the options specified in `baron.op3`. Compared to the original LMP1 model, BARON is now four times faster on the largest instances. It should be noted, though, that the optimality tolerances have a different meaning for the original and reformulated problems.

11.8.6 LMPs of Type 2

LMPs of "Type 2" are defined as:

$$(\text{LMP2}): \quad \min \prod_{i=1}^{2} (c_i^t x + f_i)$$
$$\text{s.t.} \quad b \leq Ax$$

	N_{tot}			N_{mem}			CPU s		
Case	Min	Max	Ave.	Min	Max	Ave.	Min	Max	Ave.
1	1	7	3.8	1	2	1.8	0.07	0.09	0.08
2	3	7	3.8	2	3	2.2	1.01	1.33	1.21
3	3	7	4.6	2	2	2.0	5.52	6.33	5.84
4	10	19	14.4	2	4	3.0	0.12	0.17	0.14
5	7	31	15.8	2	6	3.8	1.79	3.14	2.26
6	21	41	32.6	4	7	5.8	6.95	8.79	8.06
7	11	29	20.2	2	5	3.2	0.14	0.25	0.20
8	27	165	77.8	5	20	9.4	2.05	7.53	4.20
9	103	133	116.4	13	21	16.4	24.08	31.33	26.82
10	144	435	268.2	13	46	30.8	31.33	85.90	57.48

Table 11.5: Branch-and-bound requirements for reformulated LMP1 with baron.op3 options

$$0 \leq x,$$

where all the elements of $c_i \in \mathbb{R}^n$ and $f_i \in \mathbb{R}$ ($i = 1, 2$) are random numbers in [0,1], the constraint matrix elements a_{ij} are generated in $[-1, 1]$ via $a_{ij} := 2\alpha - 1$, where α are random numbers in [0,1], and the left-hand-side values are generated via $b_i := \sum_j a_{ij} + 2\beta$, where β are random numbers in [0,1]. This construction was proposed by Thoai (1991) and used subsequently for computations by Ryoo & Sahinidis (2002). These LMPs can be generated and solved in GAMS as follows:

Model lmp2.gms

```
$ontext
Filename: lmp2.gms
Author: Nick Sahinidis, August 2002

Purpose:
Generate and solve random linear multiplicative models of "Type 2."
Problem instances are generated as proposed by:
   N. V. Thoai, "A global optimization approach for solving
   convex multiplicative programming problems,"
   Journal of Global Optimization, 1(341-357), 1991.
$offtext
```

11.8. APPLICATION TO MULTIPLICATIVE PROGRAMS

```
options optcr=0, optca=1.e-6,
        limrow=0, limcol=0,
        solprint=off,
        reslim = 10000;

sets m constraints /1*200/
     n variables   /1*200/
     p products    /1*2/
     c cases       /1*5/
     i instances   /1*5/ ;

*for each case to be solved, we have a different (m,n) pair
table cases(c,*)
    1    2
1   10   20
2   20   30
3   60   100
4   100  100
5   200  200 ;

parameters cc(p,n)   cost coefficients
           f(p)      constants
           A(m,n)    constraint coefficients
           b(m)      left-hand-side
           rep(c,*)  summary report ;

parameters mactual, nactual, ResMin, Resmax, NodMin, Nodmax;

equations objective, constraints(m), products(p);
variables y(p), x(n), obj ;
x.lo(n) = 0;

objective .. obj =E= prod(p, y(p));

products(p) .. y(p) =E= sum(n $ (ord(n) le nactual), cc(p,n)*x(n));

constraints(m) $ (ord(m) le mactual) ..
          b(m) =L= sum(n $ (ord(n) le nactual), A(m,n)*x(n)) ;

model lmp2 /all/;
lmp2.workspace = 32;

rep(c,'AvgResUsd') = 0;
rep(c,'AvgNodUsd')= 0;
```

```
loop (c,

  mactual = cases(c,'1');
  nactual = cases(c,'2');

  ResMin  = inf;
  Resmax  = 0;
  NodMin  = inf;
  Nodmax  = 0;

  loop(i,

    f(p)    = uniform(0,1);
    cc(p,n) = uniform(0,1) $ (ord(n) le nactual);
    A(m,n)  = (2*uniform(0,1)-1)
              $ (ord(m) le mactual and ord(n) le nactual);
    b(m)    = (sum(n, A(m,n)) + 2*uniform(0,1)) $ (ord(m) le mactual);

*   make sure all problems have the same zero starting point
    x.l(n)=0; y.l(p)=0;
    solve lmp2 minimizing obj using nlp;
    rep(c,'AvgResUsd')  = rep(c,'AvgResUsd') + lmp2.resusd;
    rep(c,'AvgNodUsd')  = rep(c,'AvgNodUsd') + lmp2.nodusd;
    ResMin = min(ResMin, lmp2.resusd);
    NodMin = min(NodMin, lmp2.nodusd);
    ResMax = max(ResMax, lmp2.resusd);
    NodMax = max(NodMax, lmp2.nodusd);

  );

  rep(c,'MinResUsd') = ResMin;
  rep(c,'MaxResUsd') = ResMax;
  rep(c,'MinNodUsd')= nodMin;
  rep(c,'MaxNodUsd')= nodMax;

);

rep(c,'AvgResUsd') = rep(c,'AvgResUsd')/card(i);
rep(c,'AvgNodUsd')= rep(c,'AvgNodUsd')/card(i);

display rep;
```

Five different cases are considered in terms of problem sizes. For each case, five different problems are generated randomly. Computational results

11.8. APPLICATION TO MULTIPLICATIVE PROGRAMS

with the default BARON options for these twenty five problems are shown in Table 11.6.

Looking at the results of Table 11.6, one can tell that these problems are relatively easy: they require no more than 23 branch-and-bound iterations to converge. However, the CPU times are large, especially for the larger of these problems. Once again, it is instructive to take a close look at the breakdown of computational times provided by GAMS/BARON. Upon execution of Problem 25, the following CPU times are reported:

```
Total time elapsed  :   000:03:48,  in seconds:    228.09
     on parsing     :   000:00:01,  in seconds:      1.23
     on preprocessing:  000:03:25,  in seconds:    205.12
     on navigating  :   000:00:00,  in seconds:      0.08
     on relaxed     :   000:00:01,  in seconds:      1.27
     on local       :   000:00:20,  in seconds:     20.21
     on tightening  :   000:00:00,  in seconds:      0.18
     on marginals   :   000:00:00,  in seconds:      0.00
     on probing     :   000:00:00,  in seconds:      0.00
```

Clearly, a large fraction of the CPU time is spent on local search that is done in preprocessing and elsewhere in the tree. For this reason, we use the following set of options:

File baron.op4

```
numloc 0
dolocal 0
```

According to these settings, local search is not permitted during the course of the algorithm. Computational results with these settings are shown in Table 11.7. Compared to the default settings, there is a 25% reduction in the CPU time of the large problem instances. Nonetheless, the breakdown of CPU times spent on Problem 25, once again, indicates that almost 99% of the time is still spent in preprocessing.

11.8.7 Controlling Time Spent on Preprocessing LPs

In addition to local search that was turned off in baron.op4, the BARON preprocessor by default solves minimization and maximization linear programs for each individual problem variable. These LPs are constructed by

Problem	m	n	p	N_{tot}	N_{mem}	CPU s
1	10	20	2	23	4	0.11
2	10	20	2	1	1	0.08
3	10	20	2	15	2	0.10
4	10	20	2	15	2	0.13
5	10	20	2	13	2	0.13
6	20	30	2	19	2	0.22
7	20	30	2	11	2	0.29
8	20	30	2	11	2	0.30
9	20	30	2	19	2	0.25
10	20	30	2	13	2	0.26
11	60	100	2	23	3	5.49
12	60	100	2	23	3	4.37
13	60	100	2	15	2	5.91
14	60	100	2	13	3	5.00
15	60	100	2	19	2	4.98
16	100	100	2	13	2	16.57
17	100	100	2	11	2	15.35
18	100	100	2	15	2	13.75
19	100	100	2	15	2	13.73
20	100	100	2	19	4	14.76
21	200	200	2	13	3	214.33
22	200	200	2	15	2	203.58
23	200	200	2	13	2	201.88
24	200	200	2	15	2	215.89
25	200	200	2	15	2	228.09

Table 11.6: Computational requirements of branch-and-bound for LMP2 with default GAMS/BARON options

11.8. APPLICATION TO MULTIPLICATIVE PROGRAMS

Problem	m	n	p	N_{tot}	N_{mem}	CPU s
1	10	20	2	23	4	0.06
2	10	20	2	1	1	0.05
3	10	20	2	15	2	0.05
4	10	20	2	15	2	0.06
5	10	20	2	13	2	0.05
6	20	30	2	19	2	0.12
7	20	30	2	11	2	0.13
8	20	30	2	11	2	0.12
9	20	30	2	19	2	0.11
10	20	30	2	13	2	0.16
11	60	100	2	23	3	2.96
12	60	100	2	21	2	2.66
13	60	100	2	15	2	3.61
14	60	100	2	13	3	3.23
15	60	100	2	19	2	3.10
16	100	100	2	13	2	9.99
17	100	100	2	11	2	9.87
18	100	100	2	15	2	9.58
19	100	100	2	15	2	10.25
20	100	100	2	19	4	9.24
21	200	200	2	13	3	142.45
22	200	200	2	15	2	166.57
23	200	200	2	13	2	147.95
24	200	200	2	15	2	160.79
25	200	200	2	15	2	162.50

Table 11.7: Computational requirements of branch-and-bound for LMP2 with `baron.op4` options

outer-approximating the search space. Their solution serves to provide starting points for local search and, most importantly, to tighten variable bounds before the search begins. For problem 25, two linear programs are solved for each of the problem's 200 x variables in addition to the two y variables. This kind of effort is not worth undertaking for problems where either the user-specified bounds are tight, or the problem is relatively easy. To turn off preprocessing LPs, the following options file is used:

File baron.op5

```
numloc 0
dolocal 0
numbranch 2
prelpdo 2
```

The BARON option `prelpdo` may be set to $-n$ to restrict the number of variables that are preprocessed to n. By default, `prelpdo` is set to 1, which allows preprocessing on all problem variables. Setting `prelpdo` to 2 or 3, respectively, restricts preprocessing to the first `numbranch` variables, or the original problem variables only (*i.e.*, the reformulation variables are not preprocessed). As there are only two nonlinear variables in this model (y_1 and y_2), we have set `prelpdo` = 2 and `numbranch` = 2. Results under these options are shown in Table 11.8 and show that restricting the number of preprocessing LPs can significantly reduce solution times. As a result of reduced preprocessing, the number of nodes in the search tree will, in general, increase. However, empirical evidence indicates that for this easy class of problems, derivation of tight bounds is not essential.

11.8.8 LMPs of Type 3

LMPs of "Type 3" are defined as:

$$(\text{LMP3}): \quad \min \ \prod_{i=1}^{p} c_i^t x$$
$$\text{s.t.} \quad b \leq Ax$$
$$1 \leq x \leq u,$$

11.8. APPLICATION TO MULTIPLICATIVE PROGRAMS

Problem	m	n	p	N_{tot}	N_{mem}	CPU s
1	10	20	2	23	4	0.03
2	10	20	2	1	1	0.02
3	10	20	2	17	2	0.03
4	10	20	2	15	2	0.04
5	10	20	2	15	2	0.03
6	20	30	2	19	2	0.05
7	20	30	2	13	2	0.05
8	20	30	2	19	2	0.05
9	20	30	2	21	2	0.06
10	20	30	2	13	2	0.05
11	60	100	2	25	3	0.37
12	60	100	2	23	2	0.34
13	60	100	2	23	2	0.36
14	60	100	2	21	4	0.37
15	60	100	2	21	2	0.34
16	100	100	2	13	2	0.66
17	100	100	2	19	2	0.70
18	100	100	2	15	2	0.65
19	100	100	2	15	2	0.66
20	100	100	2	27	4	0.80
21	200	200	2	17	3	3.44
22	200	200	2	23	2	4.03
23	200	200	2	19	2	3.59
24	200	200	2	23	2	3.94
25	200	200	2	21	2	3.90

Table 11.8: Computational requirements of branch-and-bound for LMP2 with `baron.op5` options

where the elements of $c_i \in \mathbb{R}^n$ ($i = 1, \ldots, n$) and A are chosen randomly from the set of integers $\{1, 2, \ldots, 10\}$, $b_i = \sum_j a_{ij}^2$ for $i = 1, \ldots, m$, and $u = \max_{i=1,\ldots,m} \{b_i\}$, as was initially proposed by Benson & Boger (1997) and later used for computations by Liu et al. (1999) and Ryoo & Sahinidis (2002). These LMPs can be generated and solved in GAMS as follows:

Model lmp3.gms

```
$ontext
Filename: lmp3.gms
Author: Nick Sahinidis, August 2002

Purpose:
Generate and solve random linear multiplicative models of "Type 3."
Problem instances are generated as proposed by:
   H. P. Benson and G. M. Boger, "Multiplicative programming problems:
   Analysis and efficient point search heuristic",
   Journal of Optimization Theory and Applications, 94(487-510), 1997.
$offtext

options optcr=0, optca=1.e-6,
        limrow=0, limcol=0,
        solprint=off,
        reslim = 10000;

sets m constraints  /1*220/
     n variables    /1*200/
     p products     /1*4/
     c cases        /1*9/
     i instances    /1*5/ ;

*for each case to be solved, we have a different (m,n,p) triplet
table cases(c,*)
       1    2    3
1     20   30    2
2    120  100    2
3    220  200    2
4     20   30    3
5    120  120    3
6    200  180    3
7     20   30    4
8    100  100    4
9    200  200    4  ;
```

11.8. APPLICATION TO MULTIPLICATIVE PROGRAMS

```
parameters cc(p,n)   cost coefficients
           A(m,n)    constraint coefficients
           b(m)      left-hand-side
           rep(c,*)  summary report ;

parameters mactual, nactual, pactual,
           ResMin, Resmax, NodMin, Nodmax;

variables y(p), x(n), obj ;
x.lo(n) = 1;

equations objective, constraints(m), products(p);

objective .. obj =E= prod(p $ (ord(p) le pactual), y(p));

products(p) $ (ord(p) le pactual) ..
        y(p) =E= sum(n $ (ord(n) le nactual), cc(p,n)*x(n));

constraints(m) $ (ord(m) le mactual) ..
        b(m) =L= sum(n $ (ord(n) le nactual), A(m,n)*x(n)) ;

model lmp3 /all/;
lmp3.workspace = 32;

rep(c,'AvgResUsd') = 0;
rep(c,'AvgNodUsd')= 0;
loop (c,

  mactual = cases(c,'1');
  nactual = cases(c,'2');
  pactual = cases(c,'3');

  ResMin  = inf;
  Resmax  = 0;
  NodMin  = inf;
  Nodmax  = 0;

  loop(i,

    cc(p,n) = round(uniform(1,10))
              $(ord(p) le pactual and ord(n) le nactual);
    A(m,n)  = round(uniform(1,10))
              $(ord(m) le mactual and ord(n) le nactual);
    b(m)    = sum(n $ (ord(n) le nactual), A(m,n)**2)
              $ (ord(m) le mactual);
```

```
         x.up(n) = smax(m, b(m));

*        make sure all problems have the same zero starting point
         x.l(n)=0; y.l(p)=0;
         solve lmp3 minimizing obj using nlp;
         rep(c,'AvgResUsd')   = rep(c,'AvgResUsd') + lmp3.resusd;
         rep(c,'AvgNodUsd')   = rep(c,'AvgNodUsd') + lmp3.nodusd;
         ResMin = min(ResMin, lmp3.resusd);
         NodMin = min(NodMin, lmp3.nodusd);
         ResMax = max(ResMax, lmp3.resusd);
         NodMax = max(NodMax, lmp3.nodusd);

  );

  rep(c,'MinResUsd') = ResMin;
  rep(c,'MaxResUsd') = ResMax;
  rep(c,'MinNodUsd')= NodMin;
  rep(c,'MaxNodUsd')= NodMax;

);

rep(c,'AvgResUsd') = rep(c,'AvgResUsd')/card(i);
rep(c,'AvgNodUsd')= rep(c,'AvgnodUsd')/card(i);

display rep;
```

Nine different cases are considered with varying problem dimensions. Five random problems are generated in each case, bringing the total number of problems solved to 45. Table 11.9 presents computational results for this class of problems. Compared to LMP1 and LMP2, LMP3 is clearly an easier class of problems. However, based on the insights developed for multiplicative programs from the results for LMP1 and LMP2, we anticipate that performance of the algorithm will improve for LMP3 when the following set of options are used:

File baron.op6

```
numloc 0
dolocal 0
numbranch 4
prelpdo 2
```

11.8. APPLICATION TO MULTIPLICATIVE PROGRAMS

	Size			CPU s		
Case	m	n	p	Min	Max	Ave.
1	20	30	2	0.16	0.28	0.22
2	120	100	2	2.96	5.82	4.86
3	220	200	2	15.27	25.74	23.90
4	20	30	3	0.27	0.43	0.40
5	120	120	3	5.64	6.84	6.90
6	200	180	3	16.24	26.03	25.11
7	20	30	4	0.40	0.70	0.66
8	100	100	4	1.86	13.37	6.12
9	200	200	4	10.53	16.28	13.67

Table 11.9: Computational requirements of branch-and-bound for LMP3 with default GAMS/BARON options

Indeed, as Table 11.10 demonstrates, experience gained with classes LMP1 and LMP2 can be employed gainfully to solve other multiplicative programs: the results with `baron.op6` are up to an order of magnitude faster than those with the default options.

11.8.9 Comparison with Local Search

Comparisons between local and global solvers aim to answer the following fundamental questions:

1. What is the risk that a local solver terminates with a solution that is significantly inferior to a global one?

2. What is the computational penalty incurred by using a global solver instead of a local solver?

3. If one is not willing to pay the premium required to use a global solver, is it possible to enhance the performance of local search solvers through the use of global search elements?

Limitations of Local Search

To answer the first question, we provide a number of graphs where we plot the ratio of the objective function value obtained by a local solver/heuristic

	Size			CPU s		
Case	m	n	p	Min	Max	Ave.
1	20	30	2	0.04	0.07	0.05
2	120	100	2	0.54	0.66	0.60
3	220	200	2	2.43	3.38	2.79
4	20	30	3	0.10	0.15	0.12
5	120	120	3	1.93	2.97	2.38
6	200	180	3	6.04	9.21	7.23
7	20	30	4	0.31	0.40	0.36
8	100	100	4	0.26	10.70	2.66
9	200	200	4	1.04	1.13	1.08

Table 11.10: Computational requirements of branch-and-bound for LMP3 with `baron.op6` options

to that of a globally optimal solution identified by GAMS/BARON. Figures 11.1, 11.2, and 11.3 present, respectively, these ratios for GAMS/CONOPT2, GAMS/MINOS, and the best of GAMS/CONOPT2 and GAMS/MINOS for the 50 LMP1 problems. For graphing purposes, we set this ratio to 10 for problems for which local search failed to identify a feasible solution. The same `lmp1.gms` file was used to solve all problems and the same starting point of zero was used with all solvers.

As seen in Figure 11.1, GAMS/CONOPT2 did not return a feasible solution for five out of the 50 LMP1 problems. For 25 out of the remaining 45 problems, GAMS/CONOPT2 provided solutions with objective function values that ranged from 1.00062 to 5.0755 times the value of the corresponding globally optimal objective. The average error of these solutions was approximately 75%. For 20 of the 50 problems, GAMS/CONOPT2 found a globally optimal solution.

Figure 11.2 shows that GAMS/MINOS returned feasible solutions to all 50 problems. For 27 out of these 50 problems, GAMS/MINOS provided solutions with objective function values that ranged from 1.00062 to 6.1756 times the corresponding global optima. The average error of these solutions was approximately 89%. For 23 of the 50 problems, GAMS/MINOS found a globally optimal solution.

Comparing Figures 11.1 and 11.2, we observe that the spikes do not correspond to the same problems. In particular, there are many problems for

11.8. APPLICATION TO MULTIPLICATIVE PROGRAMS

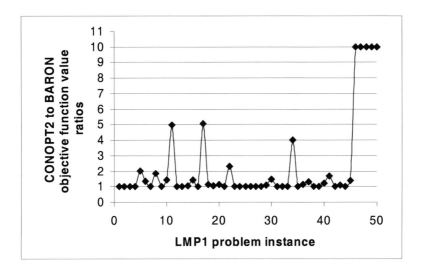

Figure 11.1: GAMS/CONOPT2 to GAMS/BARON objective function ratios for LMP1 problems

which GAMS/CONOPT2 finds a global solution and GAMS/MINOS terminates with a nonoptimal solution and *vice versa*. For this reason, a simple local search heuristic would be to run both solvers and use the best found solution. This improves the local search results as shown in Figure 11.3. The combined local search heuristic misses the global optimum in 24 out of these 50 problems. The average error of these solutions was still approximately 60%. In 26 of the 50 problems, a global optimum was identified through this approach.

Limitations of Global Search

To answer the second question above regarding the relative solution time requirements of local and global search, we plot the ratio of CPU time taken by BARON to the time taken by a local search solver/heuristic. In Figures 11.4, 11.5, and 11.6, respectively, we present these ratios for GAMS/CONOPT2, GAMS/MINOS, and the combined GAMS/CONOPT2 and GAMS/MINOS for the 10 cases of LMP1. The average CPU times of Table 11.4 were used for this purpose along with the corresponding times of the local solvers. The

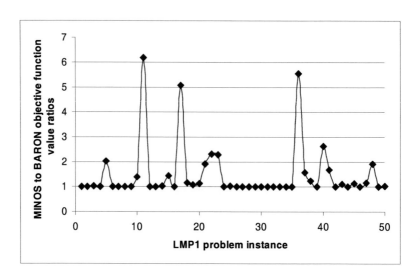

Figure 11.2: GAMS/MINOS to GAMS/BARON objective function ratios for LMP1 problems

CPU times of GAMS/CONOPT2 and GAMS/MINOS were added in order to produce the last figure. As shown in these figures, GAMS/CONOPT2 and GAMS/MINOS are, on average, 47 and 20 times faster than GAMS/BARON for these problems. The heuristic that combines GAMS/CONOPT2 and GAMS/MINOS is 14 times faster than GAMS/BARON. The periodicity in these figures indicates that the two local search solvers have a similar behavior on this set of problems.

Additional Comparative Results

For LMP2 problems, GAMS/CONOPT2 as well as GAMS/MINOS identify globally optimal solutions for all instances. This is not surprising as these problems involve very mild nonlinearities (a single product of two variables in the objective). For the same reason, the CPU times of GAMS/BARON shown in Table 11.8 are comparable to those of the two local search solvers. As shown in Figure 11.7, BARON's LP-based branch-and-bound becomes faster than local search for the largest of these problems.

Finally, for LMP3, local search results are mixed. On one hand, us-

11.8. APPLICATION TO MULTIPLICATIVE PROGRAMS

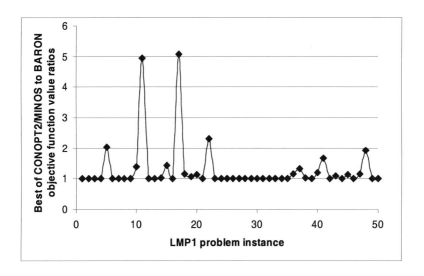

Figure 11.3: Objective function ratios of the best of GAMS/CONOPT2 and GAMS/MINOS to GAMS/BARON for LMP1 problems

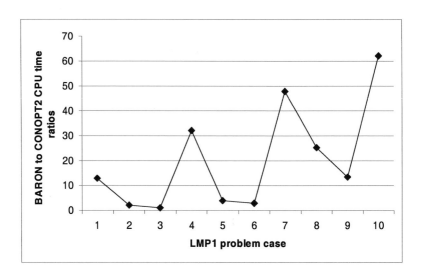

Figure 11.4: GAMS/BARON to GAMS/CONOPT2 solution time ratios for LMP1 problems

ing GAMS/MINOS provides good solutions within 1.5% of optimality in all cases with a global solution identified in 32 of the 45 problems that were solved. GAMS/BARON (Table 11.10) takes an average of 11 times more than GAMS/MINOS to solve these problems. On the other hand, GAMS/CONOPT2 did not provide feasible solutions for the last 25 of these 45 problems. For the first 20, it provided solutions within 3% of optimality; in 17 cases, a globally optimal solution was identified. GAMS/BARON took five times more CPU time than GAMS/CONOPT2 to solve these 20 problems.

Combining Local and Global Search Elements

The above results illustrate that, for multiplicative problems, local search missed the global minimum by a significant margin (approximately 60%) in a significant fraction of the problems solved (approximately 50%). On the other hand, global search required one to two orders of magnitude more computational time than local search. In an attempt to provide a rigorous method that meets the needs of practitioners, GAMS/BARON provides an

11.8. APPLICATION TO MULTIPLICATIVE PROGRAMS

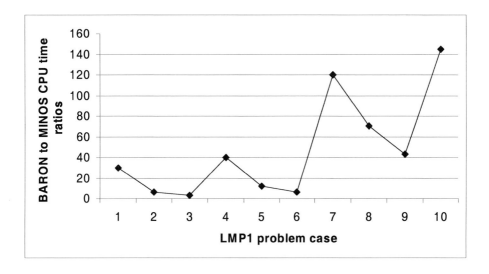

Figure 11.5: GAMS/BARON to GAMS/MINOS solution time ratios for LMP1 problems

integrated local search routine that also yields a rigorous lower bound upon termination. In particular, consider the following BARON options file:

File `baron.op7`

```
maxiter 1
dolocal 0
```

The `maxiter` option sets the number of iterations to 1. Hence, the run will terminate immediately after the root node is solved. The `dolocal` option signifies that no local search is permitted in the course of the tree, *i.e.*, after the root node relaxation is solved. However, BARON performs a number of local searches in the problem preprocessing step. The number of these local searches is controlled using the `numloc` option. By default, ten local searches are performed. The BARON preprocessor starts with a local search from the user supplied starting point. It also solves a number of preprocessing LPs that minimize/maximize the problem variables over an outer-approximation of the search space. Dual solutions from these LPs are used to further reduce

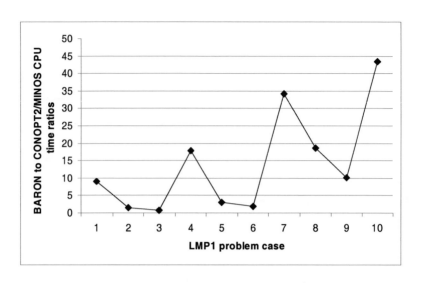

Figure 11.6: GAMS/BARON to GAMS/CONOPT2/GAMS/MINOS solution time ratios for LMP1 problems

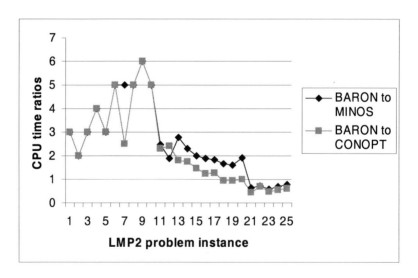

Figure 11.7: Time ratios of global to local solvers for LMP2 problems

11.8. APPLICATION TO MULTIPLICATIVE PROGRAMS

the ranges of variables. In concert, interval arithmetic operations on the nonlinear part of the model are utilized to contract variable ranges even more. Subsequently, numloc − 1 additional local searches are performed, this time from randomly generated starting points over the contracted region. Having reduced the space through optimality- and feasibility-based range reduction techniques increases the likelihood that local search will yield solutions of high quality. At the end of this process, the root node relaxation is constructed and solved over the restricted region to determine a lower bound. The relaxation solution is tested for feasibility and may further improve the incumbent.

The preprocessor of GAMS/BARON utilizes GAMS/MINOS for local search and CPLEX for solving the preprocessing LPs. Upon return to GAMS, execution goes to the default NLP solver for an additional local search. In our implementation, GAMS/CONOPT2 is used as the default GAMS NLP solver. Hence, this preprocessing strategy combines the best of NLP and LP technology currently available with multistart stochastic global optimization and deterministic range reduction and lower bounding techniques. At the end of this process, rigorous lower and upper bounds on the global optimum are reported.

Results with the GAMS/BARON preprocessor on the 50 LMP1 problems are shown in Figure 11.8 and Table 11.11. Figure 11.8 presents the objective function value of the solutions returned by the presolver as a multiple of the globally optimal objective function value. This figure can be compared to Figures 11.1, 11.2, and 11.3 that present the same ratios for GAMS/CONOPT2, GAMS/MINOS, and the best of GAMS/CONOPT2 and GAMS/MINOS for the same set of LMP1 problems. Clearly, the preprocessor of GAMS/BARON provides solutions of much better quality than any of the stand alone local solvers. The preprocessor identifies globally optimal solutions for 33 of the 50 test problems. The solutions provided for the remainder 17 problems, show errors of no more than 92%, averaging 17% as compared to 60% for the standard local search solvers. Considering the entire data set, the solutions provided by the preprocessor average within 6%. As Table 11.11 indicates, the preprocessor takes about 40 times less the CPU time required by the exact algorithm (Table 11.4). As Figures 11.4, 11.5, and 11.6 show, the preprocessor's CPU time requirements are comparable to those of the stand alone local solvers. In particular, for the most difficult of these problems (cases 3, 6, 9, and 10), the preprocessor takes no more than twice the time taken by GAMS/CONOPT2 and no more than three times

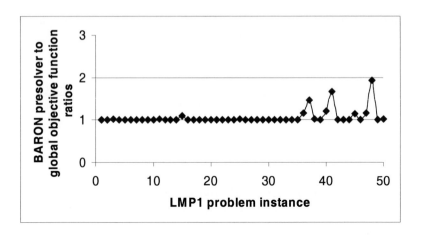

Figure 11.8: GAMS/BARON presolver solution to global objective function ratios for LMP1 problems

the time taken by GAMS/MINOS.

11.9 Application to Pooling Problems

In this section, we use the GAMS/BARON solver to address the solution of the pooling problem that was discussed at length in Chapter 9. We model pooling problems using the pq-formulation:

$$\text{(PQ)} \quad \min \sum_{j=1}^{J} \left(\sum_{l=1}^{L} y_{lj} \sum_{i=1}^{I} c_i q_{il} - d_j \sum_{l=1}^{L} y_{lj} + \sum_{i=1}^{I} c_i z_{ij} - \sum_{i=1}^{I} d_j z_{ij} \right)$$

$$\text{s.t.} \ A_i^L \leq \sum_{l=1}^{L} \sum_{j=1}^{J} q_{il} y_{lj} + \sum_{j=1}^{J} z_{ij} \leq A_i^U \quad i = 1, \ldots, I$$

$$\sum_{j=1}^{J} q_{il} y_{lj} \leq q_{il} S_l \qquad i = 1, \ldots, I; l = 1, \ldots, L$$

$$D_j^L \leq \sum_{l=1}^{L} y_{lj} + \sum_{i=1}^{I} z_{ij} \leq D_j^U \qquad j = 1, \ldots, J$$

11.9. APPLICATION TO POOLING PROBLEMS

	CPU s		
Case	Min	Max	Ave.
1	0.07	0.09	0.08
2	1.52	2.33	1.97
3	10.37	14.29	12.20
4	0.07	0.08	0.08
5	1.68	2.20	2.01
6	6.60	8.51	7.70
7	0.07	0.08	0.08
8	0.79	1.21	1.01
9	7.96	11.55	9.17
10	5.94	8.92	7.41

Table 11.11: CPU requirements for LMP1 with `baron.op7` options

$$\sum_{l=1}^{L}\left(\sum_{i=1}^{I} C_{ik}q_{il} - P_{jk}^{L}\right) y_{lj} + \sum_{i=1}^{I}(C_{ik} - P_{jk}^{L})z_{ij} \geq 0$$
$$k = 1, \ldots, K; \; l = 1, \ldots, L$$

$$\sum_{l=1}^{L}\left(\sum_{i=1}^{I} C_{ik}q_{il} - P_{jk}^{U}\right) y_{lj} + \sum_{i=1}^{I}(C_{ik} - P_{jk}^{U})z_{ij} \leq 0$$
$$k = 1, \ldots, K; \; l = 1, \ldots, L$$

$$\sum_{i=1}^{I} q_{il} = 1 \qquad l = 1, \ldots, L$$

$$\sum_{i=1}^{I} q_{il}y_{lj} = y_{lj} \qquad l = 1, \ldots, L; \; j = 1 \ldots, J$$

$$q_{il} \geq 0, \forall(i,l); \quad y_{lj} \geq 0, \forall(l,j); \quad z_{ij} \geq 0, \forall(i,j).$$

The variables, indices, and parameters used in the model are detailed in Table 11.12.

The pooling tables of Table 9.4 can be solved in GAMS as follows. Consider the Haverly 1 problem (see Figure 9.4). This problem instance is provided in GAMS format below:

Indices	i	raw materials, $i = 1, \ldots, I$
	j	products, $j = 1, \ldots, J$
	k	qualities, $k = 1, \ldots, K$
	l	pools, $l = 1, \ldots, L$
Variables	q_{il}	the fraction of raw material i in pool l
	y_{lj}	total flow from pool l to product j
	z_{ij}	direct flow of raw material i to product j
Parameters	c_i	unit cost of the ith raw material
	d_j	price of jth product
	A_i^L	lower bound on the availability of ith raw material
	A_i^U	upper bound on the availability of ith raw material
	C_{ik}	k^{th} quality of raw material i
	D_j^L	lower bound on the demand of jth product
	D_j^U	upper bound on the demand of jth product
	P_{jk}^L	lower bound on k^{th} quality of jth product
	P_{jk}^U	upper bound on k^{th} quality of jth product
	S_l	l^{th} pool capacity

Table 11.12: Indices, variables, and parameters in the pq-formulation

11.9. APPLICATION TO POOLING PROBLEMS

File haverly1.gms

```
$ontext
    Gams Model for the Pooling Problem
    Author: Mohit Tawarmalani
    model: haverly1
$offtext

$eolcom #

# Set Declarations
    set comp /1*3/;
    set pro  /1*2/;
    set qual /1*1/;
    set pool /1*1/;

# components related parameters
table compparams(comp,*)
          1     2     3
    1     0    300    6
    2     0    300   16
    3     0    300   10  ;

parameters cl(comp), cu(comp), cprice(comp);
cl(comp)     = compparams(comp,'1');
cu(comp)     = compparams(comp,'2');
cprice(comp) = compparams(comp,'3');

table cqual(comp,qual)
          1
    1     3
    2     1
    3     2 ;

# pool related parameters
parameters psize(pool);
psize(pool) = 300;

# product related parameters
table prodparams(pro,*)
          1     2     3
    1     0    100    9
    2     0    200   15  ;

parameters prl(pro), pru(pro), pprice(pro);
```

```
prl(pro) = prodparams(pro,'1');
pru(pro) = prodparams(pro,'2');
pprice(pro) = prodparams(pro,'3');

parameter pqlbd(pro, qual);
pqlbd(pro, qual) = 0.0;

table pqubd(pro, qual)
        1
   1    2.5
   2    1.5 ;

# network related parameters
table ubq(comp, pool)
        1
   1    1
   2    1
   3    0 ;

table uby(pool, pro)
        1     2
   1    100   200 ;

table ubz(comp, pro)
        1     2
   3    100   200 ;

$include pool.gms
```

The GAMS models for the entire collection of pooling problems solved in Chapter 9 are provided in Appendix A. Using these problem data files, the following model generates the pq-formulation for the respective pooling problem:

File pool.gms

```
$ontext

Filename: pool.gms
Author: Mohit Tawarmalani, August 2002

Purpose:
To encode the pq-formulation of the pooling problem described in:
M. Tawarmalani and N. V. Sahinidis, "Convexification and
```

11.9. APPLICATION TO POOLING PROBLEMS

```
Global Optimization of the Pooling Problem," May 2002,
Mathematical Programming, submitted.

$offtext

options optcr=0, optca=1.e-6,
    limrow=0, limcol=0,
    reslim = 10000;

positive variables q(comp, pool), y(pool, pro), z(comp, pro);
q.up(comp, pool) = ubq(comp, pool);
y.up(pool, pro) = uby(pool, pro);
z.up(comp, pro) = ubz(comp, pro);

variable cost;

equations obj                   objective function,
          clower(comp)          lower bound component availability
          cupper(comp)          upper bound component availability
          pszrlt(comp,pool)     ss-rlt on pool size constraints
          plower(pro)           minimum product production
          pupper(pro)           maximum product demand
          pqlower(pro,qual)     minimum product quality requirement
          pqupper(pro,qual)     maximum product quality
          fraction(pool)        fractions sum to one
          extensions(pool, pro) convexification constraints;

obj.. cost =e= sum(pro, sum(pool$(uby(pool,pro) > 0),
               sum(comp$(ubq(comp, pool) > 0),
                 cprice(comp)*y(pool,pro)*q(comp,pool)))
               - pprice(pro)*sum(pool$(uby(pool,pro) > 0),
               y(pool, pro))
               + sum(comp$(ubz(comp,pro)>0),
               (cprice(comp)-pprice(pro))*z(comp, pro))
              );

clower(comp).. sum(pool$(ubq(comp,pool)>0),
                   sum(pro$(uby(pool,pro)>0),
                     q(comp,pool)*y(pool, pro)))
               + sum(pro$(ubz(comp,pro)>0), z(comp, pro))
               =g= cl(comp);

cupper(comp).. sum(pool$(ubq(comp,pool)>0),
                   sum(pro$(uby(pool,pro)>0),
                     q(comp,pool)*y(pool, pro)))
```

```
                    + sum(pro$(ubz(comp,pro)>0), z(comp, pro))
                    =l= cu(comp);

pszrlt(comp,pool)$(ubq(comp,pool)>0)..
            sum(pro$(uby(pool,pro)>0), q(comp,pool)*y(pool,pro))
            =l= q(comp,pool)*psize(pool);

plower(pro).. sum(pool$(uby(pool,pro)>0), y(pool,pro))
            + sum(comp$(ubz(comp, pro)>0), z(comp, pro)) =g= prl(pro);

pupper(pro).. sum(pool$(uby(pool,pro)>0), y(pool,pro))
            + sum(comp$(ubz(comp, pro)>0), z(comp, pro)) =l= pru(pro);

pqlower(pro, qual).. sum(pool$(uby(pool,pro)>0),
                       sum(comp$(ubq(comp,pool)>0),
                         cqual(comp, qual)*q(comp,pool)*y(pool,pro)))
                   + sum(comp$(ubz(comp, pro)>0),
                         cqual(comp, qual)*z(comp, pro))  =g=
                     sum(pool$(uby(pool,pro)>0),
                         pqlbd(pro, qual)*y(pool,pro))
                   + sum(comp$(ubz(comp, pro)>0),
                         pqlbd(pro, qual)*z(comp, pro));

pqupper(pro, qual).. sum(pool$(uby(pool,pro)>0),
                       sum(comp$(ubq(comp,pool)>0),
                         cqual(comp, qual)*q(comp,pool)*y(pool,pro)))
                   + sum(comp$(ubz(comp, pro)>0),
                         cqual(comp, qual)*z(comp, pro))  =l=
                     sum(pool$(uby(pool,pro)>0),
                         pqubd(pro, qual)*y(pool,pro))
                   + sum(comp$(ubz(comp, pro)>0),
                         pqubd(pro, qual)*z(comp, pro));

fraction(pool).. sum(comp$(ubq(comp,pool)>0), q(comp, pool)) =e= 1;

extensions(pool, pro)$(uby(pool,pro)>0)..
            sum(comp$(ubq(comp,pool)>0), q(comp, pool)*y(pool, pro))
            =e= y(pool, pro);

model poolprob /all/;

$if %gams.nlp% == baron $include baronoptions.gms

solve poolprob minimizing cost using nlp;
```

11.9. APPLICATION TO POOLING PROBLEMS

When BARON is specified as the NLP solver, `baronoptions.gms` is included by the above model. This GAMS code automatically generates the appropriate option file and is explained later in Section 11.9.6. We first use the default BARON options to solve Haverly 1. The global optimum is found by BARON in 3 iterations as shown below:

```
===============================================================================
                        Welcome to BARON v. 5.0
                 Global Optimization by BRANCH-AND-REDUCE
           BARON is a product of The Optimization Firm, LLC.
           Parts of the BARON software were created at the
                 University of Illinois at Urbana-Champaign.
===============================================================================
                      Factorable Non-Linear Programming
===============================================================================
Preprocessing found feasible solution with value   -0.100000D+03
Preprocessing found feasible solution with value   -0.133333D+03
Preprocessing found feasible solution with value   -0.400000D+03
===============================================================================
           We have space for      27715     nodes in the tree
===============================================================================

   Itn. no.     Open Nodes     Total Time     Lower Bound     Upper Bound

      1             1           000:00:00    -0.500000D+03   -0.400000D+03
      1             1           000:00:00    -0.500000D+03   -0.400000D+03
      3             0           000:00:00    -0.400000D+03   -0.400000D+03


                     *** Successful Termination ***


   Total time elapsed     :   000:00:00,     in seconds:      0.05
         on parsing       :   000:00:00,     in seconds:      0.00
         on preprocessing:   000:00:00,     in seconds:      0.04
         on navigating    :   000:00:00,     in seconds:      0.00
         on relaxed       :   000:00:00,     in seconds:      0.00
         on local         :   000:00:00,     in seconds:      0.01
         on tightening    :   000:00:00,     in seconds:      0.00
         on marginals     :   000:00:00,     in seconds:      0.00
         on probing       :   000:00:00,     in seconds:      0.00


   Total no. of BaR iterations:         3
   Best solution found at node:         1
   Max. no. of nodes in memory:         2

All done with problem
===============================================================================
```

Note that BARON's preprocessor finds progressively better solutions and eventually a global optimum. It subsequently takes three iterations to prove globality of this solution.

11.9.1 Controlling Time Spent in Preprocessing

Consider the Adhya 1 problem presented in Section A.1. When solved using the default options, BARON terminates in 337 iterations as shown below:

```
===============================================================================
                       Welcome to BARON v. 5.0
                  Global Optimization by BRANCH-AND-REDUCE
              BARON is a product of The Optimization Firm, LLC.
                  Parts of the BARON software were created at the
                     University of Illinois at Urbana-Champaign.
===============================================================================
                      Factorable Non-Linear Programming
===============================================================================
 Preprocessing found feasible solution with value    0.000000D+00
 Preprocessing found feasible solution with value   -0.852381D+01
 Preprocessing found feasible solution with value   -0.549803D+03
===============================================================================
            We have space for      11192     nodes in the tree
===============================================================================

   Itn. no.    Open Nodes    Total Time    Lower Bound      Upper Bound

         1          1        000:00:00     -0.852730D+03    -0.549803D+03
         1          1        000:00:00     -0.745375D+03    -0.549803D+03
       100         15        000:00:01     -0.550765D+03    -0.549803D+03
       200         19        000:00:01     -0.549955D+03    -0.549803D+03
       300         17        000:00:01     -0.549806D+03    -0.549803D+03
*      330          7        000:00:01     -0.549803D+03    -0.549803D+03
*      332          3        000:00:01     -0.549803D+03    -0.549803D+03
       337          0        000:00:01     -0.549803D+03    -0.549803D+03


                    *** Successful Termination ***


   Total time elapsed    :    000:00:01,    in seconds:     1.46
           on parsing    :    000:00:00,    in seconds:     0.02
           on preprocessing:  000:00:00,    in seconds:     0.22
           on navigating :    000:00:00,    in seconds:     0.05
           on relaxed    :    000:00:01,    in seconds:     0.59
           on local      :    000:00:00,    in seconds:     0.33
           on tightening :    000:00:00,    in seconds:     0.25
           on marginals  :    000:00:00,    in seconds:     0.00
           on probing    :    000:00:00,    in seconds:     0.00


   Total no. of BaR iterations:       337
   Best solution found at node:       332
   Max. no. of nodes in memory:        21

 All done with problem
===============================================================================
```

11.9. APPLICATION TO POOLING PROBLEMS

By default, BARON performs ten local searches in the preprocessing phase of the algorithm. It is often beneficial to change the number of local searches performed to gain a time advantage or to get better upper bounding solutions. In this example, we restrict the number of local searches to one by using the following options file:

```
numloc    1
```

If the user supplies an initial starting point, then this point is used for local search. If no starting point is provided, BARON generates an arbitrary starting point for local search. As seen below, the above option file considerably reduces the time spent in pre-processing.

```
===============================================================================
                         Welcome to BARON v. 5.0
                   Global Optimization by BRANCH-AND-REDUCE
             BARON is a product of The Optimization Firm, LLC.
               Parts of the BARON software were created at the
                   University of Illinois at Urbana-Champaign.
===============================================================================
                         Factorable Non-Linear Programming
===============================================================================
 Preprocessing found feasible solution with value    0.000000D+00
 Preprocessing found feasible solution with value   -0.852381D+01
===============================================================================
              We have space for      11192      nodes in the tree
===============================================================================

    Itn. no.      Open Nodes      Total Time      Lower Bound      Upper Bound

        1              1           000:00:00      -0.852730D+03    -0.852381D+01
*       1              1           000:00:00      -0.745375D+03    -0.566667D+02
*       1              1           000:00:00      -0.745375D+03    -0.549803D+03
        1              1           000:00:00      -0.745375D+03    -0.549803D+03
      100             15           000:00:00      -0.550765D+03    -0.549803D+03
      200             12           000:00:01      -0.549943D+03    -0.549803D+03
      300             15           000:00:01      -0.549806D+03    -0.549803D+03
*     331              4           000:00:01      -0.549803D+03    -0.549803D+03
      340              0           000:00:01      -0.549803D+03    -0.549803D+03


                         *** Successful Termination ***


    Total time elapsed      :    000:00:01,    in seconds:        1.25
             on parsing     :    000:00:00,    in seconds:        0.02
             on preprocessing:   000:00:00,    in seconds:        0.02
             on navigating  :    000:00:00,    in seconds:        0.05
             on relaxed     :    000:00:01,    in seconds:        0.70
             on local       :    000:00:00,    in seconds:        0.27
             on tightening  :    000:00:00,    in seconds:        0.19
```

```
             on marginals     :   000:00:00,   in seconds:     0.00
             on probing       :   000:00:00,   in seconds:     0.00


        Total no. of BaR iterations:      340
        Best solution found at node:      331
        Max. no. of nodes in memory:       19

   All done with problem
   ============================================================================
```

However, notice that the global optimum is not found during the preprocessing phase with this option setting. Instead, it is identified at the root node. Also, note that the number of iterations increases slightly.

For the sake of illustration, we solve Foulds 4 (see Section A.3) with numloc = 0. Here are the results:

```
   ============================================================================
                          Welcome to BARON v. 5.0
                    Global Optimization by BRANCH-AND-REDUCE
              BARON is a product of The Optimization Firm, LLC.
                  Parts of the BARON software were created at the
                     University of Illinois at Urbana-Champaign.
   ============================================================================
                        Factorable Non-Linear Programming
   ============================================================================
                 We have space for        459    nodes in the tree
   ============================================================================

      Itn. no.     Open Nodes      Total Time      Lower Bound     Upper Bound

          1             1           000:00:03      -0.800000D+01    0.500000D+00
   *      1             0           000:00:04      -0.800000D+01   -0.800000D+01
          1             0           000:00:04      -0.800000D+01   -0.800000D+01


                          *** Successful Termination ***


        Total time elapsed   :   000:00:04,   in seconds:     4.24
             on parsing      :   000:00:00,   in seconds:     0.15
             on preprocessing:   000:00:03,   in seconds:     2.66
             on navigating   :   000:00:00,   in seconds:     0.01
             on relaxed      :   000:00:00,   in seconds:     0.10
             on local        :   000:00:01,   in seconds:     1.32
             on tightening   :   000:00:00,   in seconds:     0.00
             on marginals    :   000:00:00,   in seconds:     0.00
             on probing      :   000:00:00,   in seconds:     0.00


        Total no. of BaR iterations:        1
        Best solution found at node:        1
        Max. no. of nodes in memory:        1
```

11.9. APPLICATION TO POOLING PROBLEMS

```
All done with problem
================================================================================
```

In this case, the time spent in preprocessing increases as compared to the corresponding run with the default value of `numloc = 10`. This happens because BARON determines that branching will be needed to solve this problem. It therefore sets the ground by performing extensive range reduction on problem variables through the solution of appropriately constructed preprocessing LPs. These preprocessing LPs can be turned off using the option `prelpdo = 0`. When the local searches and preprocessing LPs are turned off, BARON solves the model quickly as shown below:

```
================================================================================
                        Welcome to BARON v. 5.0
                 Global Optimization by BRANCH-AND-REDUCE
               BARON is a product of The Optimization Firm, LLC.
                Parts of the BARON software were created at the
                    University of Illinois at Urbana-Champaign.
================================================================================
                        Factorable Non-Linear Programming
================================================================================
              We have space for        459      nodes in the tree
================================================================================

   Itn. no.    Open Nodes     Total Time     Lower Bound       Upper Bound

*     1            1           000:00:00     -0.100000D+52     -0.323611D+00
*     1            0           000:00:01     -0.800000D+01     -0.800000D+01
      1            0           000:00:01     -0.800000D+01     -0.800000D+01


                        *** Successful Termination ***

       Total time elapsed    :  000:00:01,   in seconds:        0.94
              on parsing     :  000:00:00,   in seconds:        0.15
              on preprocessing: 000:00:00,   in seconds:        0.00
              on navigating  :  000:00:00,   in seconds:        0.00
              on relaxed     :  000:00:00,   in seconds:        0.06
              on local       :  000:00:01,   in seconds:        0.73
              on tightening  :  000:00:00,   in seconds:        0.00
              on marginals   :  000:00:00,   in seconds:        0.00
              on probing     :  000:00:00,   in seconds:        0.00

       Total no. of BaR iterations:      1
       Best solution found at node:      1
       Max. no. of nodes in memory:      1

All done with problem
================================================================================
```

11.9.2 Reducing Memory Requirements

In the previous runs for Foulds 4, BARON informs the user that "We have space for 459 nodes in memory." The default memory allocation for BARON's data structures allows for 16 MB, which for this problem provides for only 459 nodes in memory. By default, BARON stores bounds on all original problem variables and variables it introduces for constructing the relaxation. In the case of pooling problems, the variables z_{ij}, which denote the flow from components to pools, appear linearly in the formulation. Therefore, it is not necessary to branch on these variables and they do not need to be stored for branching purposes in the search tree. The number of branching variables can therefore be reduced to the cardinality of the nonzero q_{il} and y_{lj} variables in the pooling formulation. In this example, there are 160 such variables. Setting numbranch = numstore = 160 increases the capacity to 1560 nodes. This value can be automatically computed and used in GAMS/BARON by including the following GAMS code:

```
scalar numbranch, numstore;
numbranch = sum(pool, sum(comp$(ubq(comp,pool)>0), 1) +
             sum(pro$(uby(pool,pro)>0), 1));

numstore = numbranch;

file opt6 /baron.op6/;
if (%gams.optfile% eq 6,
put opt6;
put 'numbranch ', numbranch:4:0 /;
put 'numstore ', numstore:4:0 /;
putclose opt6;
);
```

A more detailed options file for the pooling problem will be developed later (see Section 11.9.6).

11.9.3 Controlling the Size of the Search Tree

BARON implements many range reduction techniques to reduce the size of the search tree. The reader is referred to Section 1.3.3, Chapter 5, and Section 11.8.4 for a more detailed discussion. Here, we illustrate the use of probing on the pooling problem Adhya 1. When this problem is solved

11.9. APPLICATION TO POOLING PROBLEMS

using numloc = 1, numbranch = numstore = 13 (see Sections 11.9.1 and 11.9.2), BARON proves global optimality of the solution in 1.36 seconds after performing 361 iterations:

```
===============================================================================
                          Welcome to BARON v. 5.0
                   Global Optimization by BRANCH-AND-REDUCE
                 BARON is a product of The Optimization Firm, LLC.
                  Parts of the BARON software were created at the
                    University of Illinois at Urbana-Champaign.
===============================================================================
                        Factorable Non-Linear Programming
===============================================================================
 Preprocessing found feasible solution with value    0.000000D+00
 Preprocessing found feasible solution with value   -0.852381D+01
===============================================================================
              We have space for      20717      nodes in the tree
===============================================================================

     Itn. no.      Open Nodes      Total Time      Lower Bound      Upper Bound

        1              1           000:00:00      -0.852730D+03    -0.852381D+01
   *    1              1           000:00:00      -0.745375D+03    -0.566667D+02
   *    1              1           000:00:00      -0.745375D+03    -0.549803D+03
        1              1           000:00:00      -0.745375D+03    -0.549803D+03
      100             16           000:00:00      -0.552235D+03    -0.549803D+03
      200              8           000:00:01      -0.549996D+03    -0.549803D+03
      300             15           000:00:01      -0.549811D+03    -0.549803D+03
   *  328             13           000:00:01      -0.549806D+03    -0.549803D+03
   *  352              6           000:00:01      -0.549803D+03    -0.549803D+03
      361              0           000:00:01      -0.549803D+03    -0.549803D+03


                         *** Successful Termination ***


     Total time elapsed    :   000:00:01,    in seconds:    1.36
              on parsing   :   000:00:00,    in seconds:    0.02
              on preprocessing: 000:00:00,   in seconds:    0.02
              on navigating :  000:00:00,    in seconds:    0.06
              on relaxed   :   000:00:01,    in seconds:    0.67
              on local     :   000:00:00,    in seconds:    0.42
              on tightening :  000:00:00,    in seconds:    0.17
              on marginals :   000:00:00,    in seconds:    0.00
              on probing   :   000:00:00,    in seconds:    0.00


     Total no. of BaR iterations:     361
     Best solution found at node:     352
     Max. no. of nodes in memory:      22

 All done with problem
===============================================================================
```

However, when probing is turned on by setting pdo = pxdo = 5, BARON is able to prove globality in just five iterations:

```
================================================================
                       Welcome to BARON v. 5.0
              Global Optimization by BRANCH-AND-REDUCE
            BARON is a product of The Optimization Firm, LLC.
            Parts of the BARON software were created at the
                 University of Illinois at Urbana-Champaign.
================================================================
                   Factorable Non-Linear Programming
================================================================
 Preprocessing found feasible solution with value    0.000000D+00
 Preprocessing found feasible solution with value   -0.852381D+01
================================================================
          We have space for      20715    nodes in the tree
================================================================

     Itn. no.    Open Nodes    Total Time    Lower Bound    Upper Bound

         1           1         000:00:00    -0.852730D+03   -0.852381D+01
*        1           1         000:00:00    -0.745375D+03   -0.566667D+02
*        1           1         000:00:00    -0.745375D+03   -0.549803D+03
         1           1         000:00:00    -0.554966D+03   -0.549803D+03
*        5           1         000:00:00    -0.549810D+03   -0.549803D+03
*        5           1         000:00:00    -0.549810D+03   -0.549803D+03
*        5           1         000:00:00    -0.549810D+03   -0.549803D+03
         5           0         000:00:00    -0.549803D+03   -0.549803D+03


                     *** Successful Termination ***


      Total time elapsed    :    000:00:00,    in seconds:      0.49
           on parsing       :    000:00:00,    in seconds:      0.01
           on preprocessing:    000:00:00,    in seconds:      0.02
           on navigating    :    000:00:00,    in seconds:      0.00
           on relaxed       :    000:00:00,    in seconds:      0.04
           on local         :    000:00:00,    in seconds:      0.09
           on tightening    :    000:00:00,    in seconds:      0.01
           on marginals     :    000:00:00,    in seconds:      0.00
           on probing       :    000:00:00,    in seconds:      0.32


      Total no. of BaR iterations:        5
      Best solution found at node:        5
      Max. no. of nodes in memory:        2

 All done with problem
================================================================
```

11.9.4 Controlling Local Search Time During Navigation

In the case of the benchmark pooling problems, BARON does not spend an extensive amount of time in local searches. If local search time needs to be controlled, it is possible to change `dolocal` and `maxheur` to increase/decrease the time spent in performing heuristic local searches. More specifically, `maxheur` changes the maximum number of stochastic searches at a node and `dolocal` determines the nodes at which such a heuristic is performed.

11.9.5 Reduced Branching Space

As described in Section 9.4.1, branching can be restricted to the q_{il} variables without affecting the convergence property of branch-and-bound. However, restricting the branching space does not always lead to improved performance. For example, consider restricting branching to the q_{il} variables by using the following options file:

```
numloc     1
numbranch  5
numstore   13
pdo     5
pxdo    5
```

In this case, BARON takes 1435 iterations instead of 5 (as in Section 11.9.3) to prove optimality of the globally optimal solution for Adhya 1:

```
===========================================================================
                     Welcome to BARON v. 5.0
              Global Optimization by BRANCH-AND-REDUCE
         BARON is a product of The Optimization Firm, LLC.
         Parts of the BARON software were created at the
              University of Illinois at Urbana-Champaign.
===========================================================================
                  Factorable Non-Linear Programming
===========================================================================
Preprocessing found feasible solution with value    0.000000D+00
Preprocessing found feasible solution with value   -0.852381D+01
===========================================================================
           We have space for      20715     nodes in the tree
===========================================================================

  Itn. no.    Open Nodes    Total Time    Lower Bound    Upper Bound

       1            1        000:00:00   -0.852730D+03  -0.852381D+01
```

```
  *     1        1       000:00:00   -0.745375D+03   -0.566667D+02
  *     1        1       000:00:00   -0.745375D+03   -0.549803D+03
        1        1       000:00:00   -0.557025D+03   -0.549803D+03
      100       26       000:00:04   -0.549935D+03   -0.549803D+03
      200       33       000:00:06   -0.549866D+03   -0.549803D+03
      300       38       000:00:09   -0.549839D+03   -0.549803D+03
  *   315       32       000:00:09   -0.549829D+03   -0.549803D+03
  *   330       42       000:00:10   -0.549826D+03   -0.549803D+03
  *   330       42       000:00:10   -0.549826D+03   -0.549803D+03
  *   330       42       000:00:10   -0.549826D+03   -0.549803D+03
  *   332       43       000:00:10   -0.549826D+03   -0.549803D+03
  *   332       43       000:00:10   -0.549826D+03   -0.549803D+03
      400       67       000:00:12   -0.549816D+03   -0.549803D+03
      500       89       000:00:15   -0.549811D+03   -0.549803D+03
      600      111       000:00:18   -0.549809D+03   -0.549803D+03
      700      129       000:00:21   -0.549808D+03   -0.549803D+03
      800      141       000:00:24   -0.549807D+03   -0.549803D+03
      900      139       000:00:26   -0.549806D+03   -0.549803D+03
     1000      132       000:00:29   -0.549806D+03   -0.549803D+03
     1100      125       000:00:32   -0.549805D+03   -0.549803D+03
     1200      107       000:00:34   -0.549805D+03   -0.549803D+03
     1300       78       000:00:37   -0.549804D+03   -0.549803D+03
     1400       28       000:00:38   -0.549804D+03   -0.549803D+03
     1435        0       000:00:38   -0.549803D+03   -0.549803D+03

            *** Successful Termination ***

  Total time elapsed      :   000:00:38,   in seconds:   38.45
       on parsing         :   000:00:00,   in seconds:    0.02
       on preprocessing   :   000:00:00,   in seconds:    0.02
       on navigating      :   000:00:00,   in seconds:    0.18
       on relaxed         :   000:00:03,   in seconds:    2.62
       on local           :   000:00:02,   in seconds:    1.84
       on tightening      :   000:00:01,   in seconds:    0.80
       on marginals       :   000:00:00,   in seconds:    0.00
       on probing         :   000:00:33,   in seconds:   32.97

  Total no. of BaR iterations:    1435
  Best solution found at node:     332
  Max. no. of nodes in memory:     143

All done with problem
==========================================================================
```

11.9.6 Pooling Problem Computations

In this section, we provide computational results using different BARON options. These options are generated on the fly by the following GAMS file which is included by pool.gms when BARON is invoked as the GAMS NLP

11.9. APPLICATION TO POOLING PROBLEMS 373

solver:

File baronoptions.gms

```
scalar numbranch, numstore, dolocal,
       numloc, pdo, pxdo, maxheur;

numloc = 1;

file opt1 /baron.opt/;
if (%gams.optfile% eq 1,
put opt1;
put 'numloc ', numloc:4:0 /;
putclose opt1;
);

numbranch = sum(pool, sum(comp$(ubq(comp,pool)>0), 1) +
            sum(pro$(uby(pool,pro)>0), 1));
numstore = numbranch;

file opt2 /baron.op2/;
if (%gams.optfile% eq 2,
put opt2;
put 'numloc ', numloc:4:0 /;
put 'numbranch ', numbranch:4:0 /;
put 'numstore ', numstore:4:0 /;
putclose opt2;
);

pdo = 5;
pxdo = 5;

file opt3 /baron.op3/;
if (%gams.optfile% eq 3,
put opt3;
put 'numloc ', numloc:4:0 /;
put 'numbranch ', numbranch:4:0 /;
put 'numstore ', numstore:4:0 /;
put 'pdo ', pdo:4:0 /;
put 'pxdo', pxdo:4:0 /;
putclose opt3;
);

dolocal = -999;
maxheur = 3;
```

```
file opt4 /baron.op4/;
if (%gams.optfile% eq 4,
put opt4;
put 'numloc ', numloc:4:0 /;
put 'dolocal ', dolocal:4:0 /;
put 'maxheur ', maxheur:4:0 /;
put 'numbranch ', numbranch:4:0 /;
put 'numstore ', numstore:4:0 /;
put 'pdo ', pdo:4:0 /;
put 'pxdo', pxdo:4:0 /;
putclose opt4;
);

numbranch = sum(pool, sum(comp$(ubq(comp,pool)>0), 1));

file opt5 /baron.op5/;
if (%gams.optfile% eq 5,
put opt5;
put 'numloc ', numloc:4:0 /;
put 'dolocal ', dolocal:4:0 /;
put 'numbranch ', numbranch:4:0 /;
put 'numstore ', numstore:4:0 /;
put 'pdo ', pdo:4:0 /;
put 'pxdo', pxdo:4:0 /;
putclose opt5;
);
```

A brief summary of these options is as follows:

> 0: defaults;
> 1: numloc = 1;
> 2: numloc = 1, numbranch = numstore = number of nonzero q and y variables;
> 3: numloc = 1, numbranch = numstore = number of nonzero q and y variables, pdo = pxdo = 5;
> 4: numloc = 1, numbranch = numstore = number of nonzero q and y variables, pdo = pxdo = 5, dolocal = −999, maxheur = 3;
> 5: numloc = 1, numbranch = numstore = number of nonzero q and y variables, pdo = pxdo = 5, dolocal = −999.

11.9. APPLICATION TO POOLING PROBLEMS

As seen, these options set the number of stochastic random searches in preprocessing to 1 (`numloc` option), the number of branching and stored variables to the number of nonzero q and y variables (options `numbranch` and `numstore`), the number of probing LPs in every node to five (`pdo` and `pxdo` options), the frequency at which local search is done to once in 999 nodes (`dolocal` option), and the number of repeated applications of stochastic search on a node to three (`maxheur` option).

In Tables 11.13 and 11.14, we provide the number of iterations BARON requires for proving global optimality under each of these option settings. As seen in these tables, Option Setting 5 requires many iterations to prove optimality. This is because, under this option, BARON is not allowed to branch on the y_{lj} variables. Options 3 and 4 are the most suitable for pooling problems. We will make use of Option Setting 3 in our later comparisons with local solvers.

In Table 11.15, we present the objective function values of the solutions found and CPU time requirements for GAMS/CONOPT2, GAMS/MINOS, and GAMS/BARON. As seen, GAMS/CONOPT2 misses the global optimum in 13 of these 14 pooling problems. This solver is able to find the global solution only for Foulds 5. For problem RT 2, GAMS/CONOPT2 is not able to find any solution with a negative objective function value. GAMS/MINOS does better than GAMS/CONOPT2 for these problems. In particular, it misses the global solution in ten of the problems and identifies a global optimum in the four remaining cases. As seen in this table, the suboptimal solutions obtained by local solvers are very far from the global solutions.

In Figure 11.9, we provide a pictorial representation of the global optimality gap when the pooling models are solved using GAMS/CONOPT2. In this graph, the optimal objective for each problem has been scaled to -1. The problems are labelled using the first letter of the problem names used in previous tables followed by the problem instance. The horizontal line in each bar indicates the scaled objective function value obtained by using GAMS/CONOPT2. The darker shaded area in each bar is proportional to the error gap of the local solver. In Figure 11.10, we present a similar comparison between GAMS/BARON and GAMS/MINOS. Finally, Figure 11.11 compares the best local search results obtained by GAMS/CONOPT2 and GAMS/MINOS to the global solutions obtained by GAMS/BARON. Even when both of the local search solvers are combined, the local solver misses

	Option Setting					
Problem	0	1	2	3	4	5
Adhya 1	337	340	361	5	5	1435
Adhya 2	325	624	302	11	11	5403
Adhya 3	423	423	483	5	5	41
Adhya 4	35	35	45	1	1	5
Bental 4	3	3	3	1	1	1
Bental 5	-1	1	1	1	1	1
Foulds 2	-1	-1	-1	-1	-1	-1
Foulds 3	-1	-1	-1	-1	-1	-1
Foulds 4	-1	-1	-1	-1	-1	-1
Foulds 5	-1	-1	-1	-1	-1	-1
Haverly 1	3	3	3	1	1	1
Haverly 2	3	3	3	1	1	1
Haverly 3	3	3	3	1	1	1
RT 2	177	177	183	3	3	9

Table 11.13: BARON iterations (nodes) using different option settings

the global solution in ten out of the 14 pooling problems.

11.10 Problems from `globallib` and `minlplib`

The following miscellaneous benchmarks solved in Section 10.6 can be found in `globallib` and `minlplib` under different names:

e10: appears in `globallib` as ex4_1_8;
e14: appears in `minlplib` as ex1223b;
e15: appears in `minlplib` as ex1221;
e20: appears in `globallib` as ex8_1_8;
e28: appears in `globallib` as himmel11;
e29: appears in `minlplib` as ex1224;
e39: appears in `globallib` as ex8_1_6.

11.10. PROBLEMS FROM GLOBALLIB AND MINLPLIB

	Option Setting					
Problem	0	1	2	3	4	5
Adhya 1	1.47	1.26	1.35	0.5	0.5	36.13
Adhya 2	1.27	2.04	0.94	0.74	0.74	53.17
Adhya 3	2.67	2.39	2.86	0.66	0.67	1.56
Adhya 4	1.08	0.46	0.55	0.46	0.45	0.5
Bental 4	0.06	0.03	0.03	0.03	0.03	0.03
Bental 5	0.12	0.12	0.13	0.13	0.13	0.13
Foulds 2	0.01	0.02	0.02	0.02	0.02	0.03
Foulds 3	1.9	1.89	1.89	1.86	1.86	1.88
Foulds 4	1.67	1.68	1.68	1.68	1.68	1.68
Foulds 5	1.04	1.04	1.03	1.04	1.07	1.05
Haverly 1	0.05	0.03	0.02	0.02	0.03	0.03
Haverly 2	0.05	0.03	0.02	0.03	0.03	0.03
Haverly 3	0.07	0.02	0.03	0.03	0.03	0.03
RT 2	0.76	0.52	0.49	0.4	0.4	0.39

Table 11.14: BARON CPU time (seconds) using different option settings

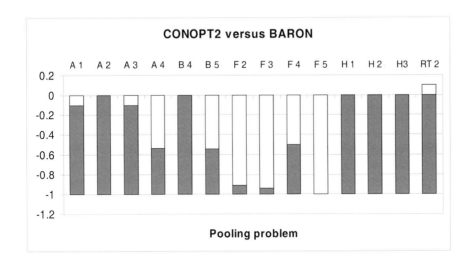

Figure 11.9: GAMS/CONOPT2 versus GAMS/BARON solutions

	GAMS/CONOPT		GAMS/MINOS		GAMS/BARON	
	CPU s	Obj	CPU s	Obj	CPU s	Obj
Adhya 1	0.01	-56.67	0.01	0	0.50	-549.80
Adhya 2	0.01	0	0.01	0	0.74	-549.80
Adhya 3	0.02	-57.74	0.01	0	0.66	-561.05
Adhya 4	0.01	-470.83	0.04	-470.83	0.46	-877.65
Bental 4	0.01	0	0	-100	0.03	-450
Bental 5	0.01	-1900	0.01	-1900	0.13	-3500
Foulds 2	0.01	-1000	0.01	-700	0.02	-1100
Foulds 3	0.06	-7.5	1.69	-8	1.86	-8
Foulds 4	0.05	-4	2.01	-8	1.68	-8
Foulds 5	0.05	-8	1.24	-8	1.04	-8
Haverly 1	0.01	0	0.01	-100	0.02	-400
Haverly 2	0.01	0	0.01	-600	0.03	-600
Haverly 3	0.01	0	0.01	-125	0.03	-750
RT 2	0.01	458.46	0.02	-1728.17	0.40	-4391.83

Table 11.15: CPU time requirements and objective function values for GAMS/CONOPT2, GAMS/MINOS, and GAMS/BARON

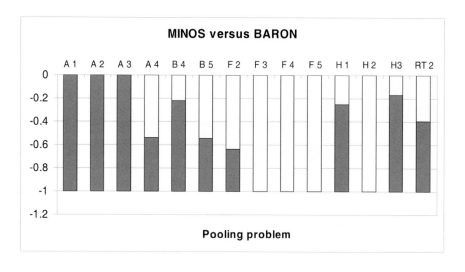

Figure 11.10: GAMS/MINOS versus GAMS/BARON solutions

11.10. PROBLEMS FROM GLOBALLIB AND MINLPLIB

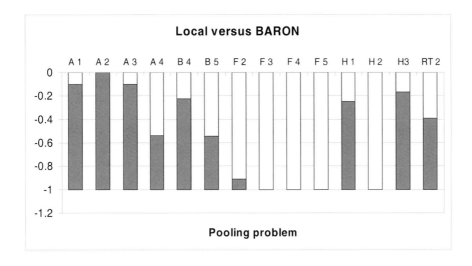

Figure 11.11: Local search versus GAMS/BARON solutions

In addition, the Gupta-Ravindran problems of Section 10.7.2 can be found in minlplib under the names nvs01 through nvs24.

In the course of writing this book, we contributed the following problems to gamslib and minlplib:

- the refrigerant design problem of Chapter 8 was contributed under the name primary.gms;

- the just-in-time system design problem of Section 10.7.1 was contributed as jit.gms.

In addition, we contributed to gamslib or minlplib the following problems from Chapter 10:

- the separable concave quadratic programs of Tables 10.1 and 10.2;

- the indefinite quadratic programs of Tables 10.4 and 10.5;

- the generalized linear multiplicative programs of Table 10.10;

- the miscellaneous engineering design and optimization benchmarks of Tables 10.12 and 10.13.

Each of these problems can be found in either `globallib` or `minlplib` under the name `st_xxx`, where `xxx` is the name used in Chapter 10. These problems were contributed to `globallib` with the exception of the following problems that were contributed to `minlplib`:

- problems miqp1, miqp2, miqp3, miqp4, miqp5 from Table 10.4;
- problems test1, test2, test3, test4, test5, test6, test8, testGR1, testGR3, and testPh14 from Table 10.5;
- problem e13 from Table 10.12;
- problems e27, e31, e32, e35, e36, e38, and e40 from Tables 10.12 and 10.13.

As Chapter 10 provides the values of the optimal solutions for all the above problems, readers can download the models from the corresponding libraries and use GAMS/BARON to reproduce the results. The default options of the software will suffice to solve all the above problems in modest computing times, with the following exceptions. In problems e32 and e35, branching preference should be given to the integer variables by setting their corresponding `BRANCHING_PRIORITIES` to, say, 100. For the Gupta-Ravindran problems 5, 22, and 24, branching is more efficient when restricted to the original problem variables and the largest edge is bisected in each node as discussed in Section 10.7.2.

11.11 Local Landscape Analyzer

In this section, we describe a facility that allows a user to conveniently perform multiple local searches on a nonlinear optimization problem. The local landscape analyzer is often helpful in diagnosing nonconvexity in nonlinear programs. If different local searches lead to significantly different local optima, there is a strong indication that the model is nonconvex.

We illustrate the performance of the local analyzer on the pooling problems Adhya 3 and 4. The following option file invokes BARON's local analyzer:

11.11. LOCAL LANDSCAPE ANALYZER

```
numloc    100
locres    1
mdo       0
tdo       0
prelpdo   0
maxiter   0
```

When the above options are used, BARON does not perform any preprocessing LPs, marginals-based range reduction, or reduction based on interval arithmetic operations. Instead, BARON carries out 100 local searches from random starting points. The `locres` option forces BARON to produce a succinct report of the different local optima identified (distinguished only by their objective function value). For example, when Adhya 3 is run with these options, BARON identifies eleven distinct solutions:

```
Local search found the following    11  distinct solution value(s):
   0.000000D+00         3  times
  -0.512635D+02        12  times
  -0.566667D+02         2  times
  -0.577436D+02         4  times
  -0.650000D+02        37  times
  -0.203514D+03         2  times
  -0.234561D+03         1  times
  -0.422456D+03         2  times
  -0.549803D+03         7  times
  -0.559615D+03        21  times
  -0.561045D+03         9  times
```

Next, we allow BARON to perform range reduction using the following option file:

```
numloc    100
locres    1
maxiter   0
```

In this case, BARON produces the following local solution report:

```
Local search found the following    5  distinct solution value(s):
   0.000000D+00         2  times
  -0.422456D+03         2  times
```

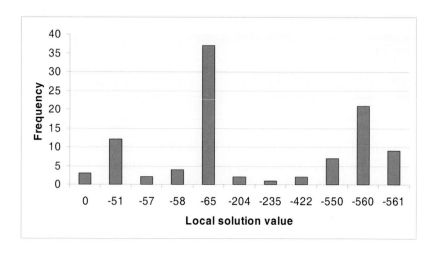

Figure 11.12: Landscape analysis with stochastic search

```
-0.549803D+03      5 times
-0.559615D+03     76 times
-0.561045D+03     15 times
```

In Figure 11.12 and Figure 11.13, we plot the objective function values versus the number of times they are found during 100 local searches with and without range reduction, respectively. Observe that the frequencies with which solutions close to the global minimum are identified increases considerably when range reduction is invoked. This is because BARON eliminates parts of the feasible region based on optimality arguments after the first local search. Thereafter, local search is restricted to parts of the search space more likely to contain good quality solutions.

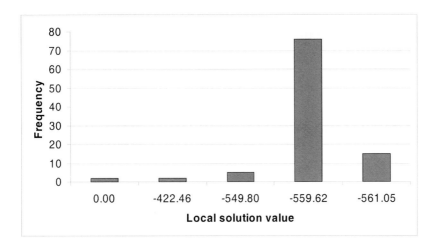

Figure 11.13: Landscape analysis with stochastic search and range reduction

11.12 Finding the K Best or All Feasible Solutions

11.12.1 Motivation and Alternative Approaches

It is not uncommon that the second best solution to an optimization problem comes with only a slight degradation of the objective while it leads to significant improvement of a secondary objective. For this reason, modelers frequently seek to identify not only the best solution to an optimization problem but the best few solutions. Secondary objectives are then used in order to compare the best few solutions before selecting one to be applied.

We first describe a technique that is commonly used in integer programming to find alternate feasible solutions. Consider an optimization problem with n integer variables, x_i, $i = 1, \ldots, n$. For simplicity, we assume $0 \leq x_i \leq x_i^u$, $i = 1, \ldots, n$. First, the following binary expansions are introduced in the model:

$$x_i = \sum_{j=1}^{\lfloor \log_2(x_i^u) \rfloor} 2^{j-1} y_{ij}, \quad i = 1, \ldots, n.$$

Once the problem is solved in terms of binary variables, it is straightforward to exclude a binary combination y^* by introducing the well-known *linear* integer cut:

$$\sum_{(i,j)\in\mathcal{B}^*} y_{ij}^* - \sum_{(i,j)\in\mathcal{N}^*} y_{ij}^* \leq \|\mathcal{B}^*\| - 1$$

where $\mathcal{B}^* = \{(i,j)|y_{ij}^* = 1\}$, $\mathcal{N}^* = \{(i,j)|y_{ij}^* = 0\}$, and $\|\mathcal{B}^*\|$ is the cardinality of $\mathcal{B}^*$. Alternatively, one can work directly in the space of the integer variables by introducing the following more intuitive, yet *nonlinear*, cut:

$$\sum_{i=1}^{n}(x_i^* - x_i)^2 \geq 1.$$

An absolute value or other norm may also be used instead. These nonlinear cuts may be linearized at the expense of introducing additional problem variables and constraints. Several solutions of the optimization formulation can then be obtained by solving a series of models in which integer cuts are successively introduced to exclude the previous models' optimal solutions from further consideration. This allows the identification of as many feasible solutions as desired. However, such an approach requires the search of a number of branch-and-bound trees.

Instead of the successive introduction of any of the integer cuts above when multiple solutions are sought, BARON modifies the standard rule for fathoming a node in the course of a branch-and-bound algorithm. As detailed in Section 7.6.1, instead of deleting all inferior nodes when a feasible solution is found, we delete only the current node when it becomes infeasible or a point. Nodes where feasible solutions are identified are branched further until they become points in the search space or infeasible. An obvious advantage of this scheme versus using the integer cuts is that all feasible solutions can be identified through a single application of branch-and-bound, thus avoiding a great deal of duplicate work in subsequent search trees. Below, we demonstrate additional advantages of this scheme.

It is straightforward to modify this algorithm to provide only the K best solutions of the model for any prespecified integer value of K. In every branch-and-bound iteration, one would need to maintain the (up to) K best solutions found thus far, and delete nodes provably worse than the Kth best known solution.

11.12.2 Finding All Solutions to Combinatorial Optimization Problems

The `numsol` option of BARON can be used to specify the desired number of best solutions to be returned. To demonstrate this feature of BARON, consider the following problem from `gamslib`:

$$\min \sum_{i=1}^{4} 10^{4-i} x_i$$
$$\text{s.t.} \quad 2 \leq x_1 \leq 4$$
$$x_2 = 3$$
$$2 \leq x_3 \leq 4$$
$$2 \leq x_4 \leq 3.$$

This problem has 18 feasible integer points. The following code is from `gamslib` with the exception of a single line which has been added to append to the `report` array the number of nodes taken by the MIP solver for solving the problem:

Model `icut.gms`

```
$title Integer Cut Example (ICUT,SEQ=160)
$eolcom !
$Ontext
  Sometimes it may be required to exclude certain integer solutions.
  Additional constraints, called cuts, can be added to exclude such
  solutions. To exclude the k'th integer solution we can write:

     cut(k)..   sum(i, abs(x(i)-xsol(i,k))) =g= 1;

  The absolute function has to be simulated using 0/1 variables
  and some additional constraints. When the solution to be excluded
  is at lower or upper bound, we do not need additional 0/1 variables.

  In this example we simply show how to enumerate all possible
  combinations of four integer variables.


GAMS Development Coorporation, Formulation and Language Example.

$Offtext
```

```
sets  i         index on integer variable  / 1 * 4 /
      ie(i)     variables fixed
      in(i)     not fixed
      il(i)     solutions at lower bound
      iu(i)     solutions at upper bound
      ib(i)     solution between bounds

      kk        cut identification set  / 1 * 100 /
      k(kk)     dynamic subset of k
      bb(kk,i)  cut memory
      bl(kk,i)  cut memory
      bu(kk,i)  cut memory


variables  x(i)      test variable
           z         some objective variable
           b(kk,i)   flip-flop for in between solutions
           u(kk,i)   changes up
           l(kk,i)   changes down

integer variable x;
binary variable b; positive variable u,l;

equations cut(kk)       main cut equations
          cutu(kk,i)    upper bound limit for inbetween integers
          cutl(kk,i)    lower bound limit for inbetween integers
          cutul(kk,i)   definition of positive and negative deviations
          obj           obj definition;

parameters  cutrhs(kk)    cut RHS value
            cutlx(kk,i)   cut lower bound
            cutux(kk,i)   cut upper bound
            cuts(kk,i)    cut solution value

            report(kk,*)  cut report variabl
            whatnext      loop control variable;

* pick an objective function which will order the solutions

obj..    z =e= sum(i, power(10,card(i)-ord(i))*x(i));

cut(k)..      - sum(bu(k,i), x(i)) + sum(bl(k,i),x(i))
              + sum(bb(k,i), l(bb) + u(bb)) =g= cutrhs(k);

cutu(bb(k,i)).. u(bb) =l= cutux(bb)*b(bb);
```

11.12. FINDING THE K BEST OR ALL FEASIBLE SOLUTIONS

```
cutl(bb(k,i)).. l(bb) =l= cutlx(bb)*(1-b(bb));

cutul(bb(k,i)).. x(i)  =e= cuts(bb) + u(bb) - l(bb);

model enum / all /;

* get an initial solution and set bounds

x.lo(i)    = 2;
x.up(i)    = 4;
x.fx('2') = 3;       ! fix one variable
x.up('4') = 3;       ! only two values

x.l(i)     = x.lo(i);

k(kk) = no;                     ! make cut set empty
ie(i) = yes$(x.lo(i)=x.up(i));  ! find fixed variables
in(i) = yes - ie(i);            ! find free variables

whatnext      = 1;  ! initial loop control
enum.resusd = 0;    ! inital CPU used

enum.reslim = 60;  ! dont spend more than 60 seconds on on problem

* We enumerate all solutions so we are happy with the first solution
* the solver finds.

enum.optcr = 0; enum.optca = 1e06;

loop(kk$whatnext,
   il(in) = yes$(x.l(in)=x.lo(in));        ! find variables at lower
   iu(in) = yes$(x.l(in)=x.up(in));        ! find variables at upper
   ib(in) = yes - ie(in) - iu(in) - il(in); ! find variables between
   k(kk)          = yes;    ! add
   bl(kk,il)      = yes;    !    cut
   bu(kk,iu)      = yes;    !    information
   bb(kk,ib)      = yes;    !    as needed
   cutux(kk,ib) = x.up(ib) - x.l(ib);
   cutlx(kk,ib) = x.l(ib)  - x.lo(ib);
   cuts(kk,ib)  = x.l(ib);
   cutrhs(kk)   = 1 + sum(il, x.l(il)) - sum(iu, x.l(iu));
   report(kk,i) = x.l(i);                   ! save previous solution
   report(kk,'binaries') = card(bb);        ! remember binaries
   report(kk,'CPU time') = enum.resusd;     ! remember time
```

388   CHAPTER 11. GAMS/BARON: A TUTORIAL

```
    report(kk,'Nodes used') = enum.nodusd;    ! remember nodes used
    solve enum min z us mip;
    enum.limcol = 0;     ! turn off
    enum.limrow = 0;     !    all
    enum.solprint = 2;   !        output
    whatnext = enum.modelstat=1 or enum.modelstat=8 );

display enum.solvestat, enum.modelstat, report;
```

Note that the implementation of integer cuts is nontrivial and requires repetitive `solve` statements, *i.e.*, repetitive calls to a MIP solver. On the other hand, the same model can be solved using BARON through the following, much simpler, GAMS code:

Model `iall.gms`

```
$title All Integer Solutions Example
$eolcom !
$Ontext
  Purpose: demonstrate use of BARON option 'numsol' to obtain
  the best numsol solutions of an optimization problem in a
  single branch-and-bound search tree.

  Author: Nick Sahinidis, August 2002

  The model is the one solved in ICUT.GMS

$Offtext
set i index on integer variable  / 1 * 4 /

variables x(i) variables
          z    objective variable

integer variable x;

x.lo(i) = 2;
x.up(i) = 4;
x.fx('2') = 3;    ! fix one variable
x.up('4') = 3;    ! only two values

equation obj obj definition;

* pick an objective function which will order the solutions
```

```
obj .. z =e= sum(i, power(10,card(i)-ord(i))*x(i));

model enum / all /;

solve enum minimizing z using mip;
```

We use an option file to ask BARON to provide the best 50 solutions:

```
numsol 50
```

In addition to entirely eliminating the burden on the user to encode the cutting plane generation scheme, BARON's capability to return multiple solutions has significant CPU time savings as well. This is demonstrated in Figures 11.14, 11.15, and 11.16. As more feasible solutions are identified, additional integer cuts are added in icut.gms. This is shown in Figure 11.14. As a result of the increased model complexity, the number of nodes and CPU seconds required by CPLEX increases as shown in Figures 11.15 and 11.16, respectively. The CPU times of Figure 11.16 are small but indicate an increasing trend as more integer cuts are added. In total, CPLEX requires 0.27 seconds to solve these problems. On the other hand, BARON returns all 18 solutions after searching a single branch-and-bound tree with 63 nodes in 0.08 seconds.

The above example involved a feasible space with only 18 solutions and very small computational requirements. It is instructive to consider a more difficult problem. For this, we consider the same problem with a somewhat enlarged feasible space:

$$\min \sum_{i=1}^{4} 10^{4-i} x_i$$
$$\text{s.t.} \quad 2 \leq x_i \leq 4, \quad i = 1, \ldots, 4.$$

The problem now has 81 feasible solutions. As a result, the traditional method for identifying all feasible solutions requires 81 calls to a MIP solver. Doing so with icut.gms, one obtains the results shown in Figures 11.17, 11.18, and 11.19. The impact of integer cuts is dramatic in this case. As more cuts are added (Figure 11.17), CPLEX requires the solution of much larger trees (Figure 11.18) requiring increasingly more CPU times (Figure 11.19). Apparently, this is due to the structure of the integer cuts. Each of

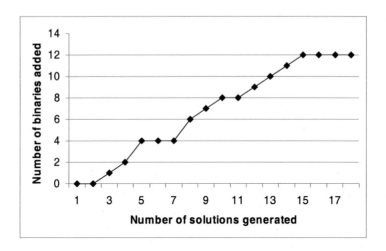

Figure 11.14: Number of binaries required for integer cuts in icut.gms

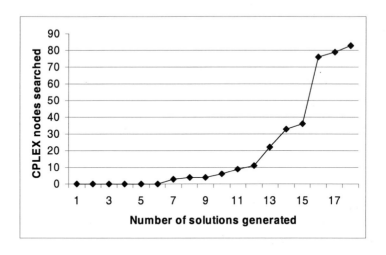

Figure 11.15: CPLEX nodes to solve icut.gms MIPs

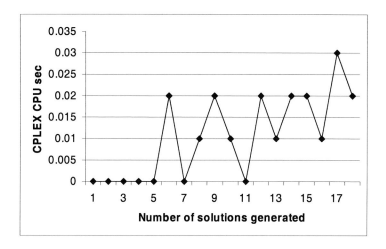

Figure 11.16: CPLEX CPU seconds to solve icut.gms MIPs

these cuts excludes only one integer solution and, for this formulation, the state-of-the-art MIP code requires exponentially increasing CPU time for a problem for which explicit enumeration requires only sorting of 81 items. The last MIP requires CPLEX to expand over 140,000 nodes in the tree and take over 100 CPU seconds. On the other hand, BARON with numsol set to 100, requires the enumeration of a tree with only 511 nodes in 0.56 seconds of CPU time.

It should be emphasized that the above results do not represent a comparison between CPLEX and BARON. Instead, they represent a comparison between two different approaches to finding all (or multiple) solutions to an integer program. The fact that BARON required several orders of magnitude less CPU time than CPLEX to identify all solutions to the above small MIP is a consequence of the fact that the formulation solved by BARON includes only bounds on the integer variables and none of the integer cuts that were part of the formulation solved by CPLEX.

11.12.3 Refrigerant Design Problem

BARON's technology for identifying all or the best K solutions of an optimization problem grew out of a need to find all solutions to the refrigerant

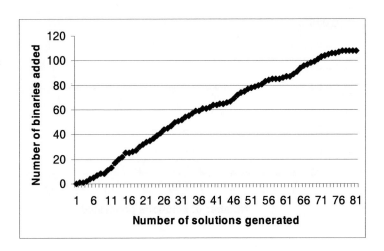

Figure 11.17: Number of binaries required for integer cuts in icut.gms with enlarged search space

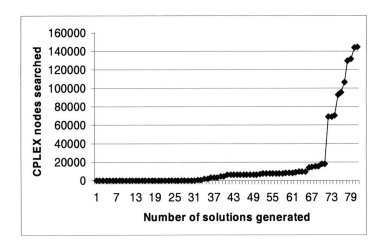

Figure 11.18: CPLEX nodes to solve icut.gms MIPs with enlarged search space

11.12. FINDING THE K BEST OR ALL FEASIBLE SOLUTIONS

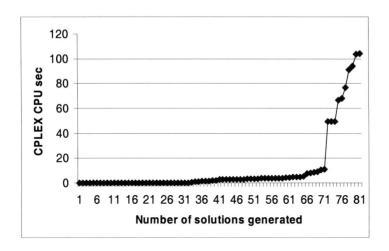

Figure 11.19: CPLEX CPU seconds to solve `icut.gms` MIPs with enlarged search space

design problem of Chapter 8. There, the need for a clever way to enumerate the search tree for all feasible solutions was further amplified by the presence of nonlinearities. The model of Chapter 8 is available in `gamslib` under the name `primary.gms` and includes a significant number of nonlinearities, some of which are shown below:

```
e5  .. Tbr-Tb/Tc =E= 0;

e6  .. Tavgr-294.25/Tc =E= 0;

e7  .. Tcndr-316.45/Tc =E= 0;

e8  .. Tevpr-272.05/Tc =E= 0;

e9  .. alpha+log(Pc/1.013)-6.09648/Tbr-1.28862*log(Tbr)+0.169347*Tbr**6
       =E= -5.97214;

e10 .. beta+15.6875/Tbr+13.4721*log(Tbr)-0.43577*Tbr**6 =E= 15.2518;

e11 .. omega*beta-alpha =E= 0;

e12 ..
```

```
 4.1868 * Cpla
 - ( Cpl0 + 3.7413/(1-Tavgr) + 35.563135 * omega
     + 52.3782 * omega * (1-Tavgr)**(1/3) / Tavgr
     + 3.620747 * omega / (1-Tavgr)
   ) =E= 12.0553 ;

 e14 .. dHve-dHvb*((1-Tevpr)/(1-Tbr))**0.38 =E= 0;

 e15 .. h-Tbr*log(Pc/1.013)/(1-Tbr) =E= 0;

 e16 .. G-0.4605*h =E= 0.4835;

 e17 .. k-(h-(1+Tbr)*G)/G/(3+Tbr)/(1-Tbr)**2 =E= 0;

 e18 .. Pvpcr-exp(-G*(1-Tcndr**2+k*(3+Tcndr)*(1-Tcndr)**3)/Tcndr) =E= 0;

 e19 .. Pvpc-Pvpcr*Pc =E= 0;

 e20 .. Pvper-exp(-G*(1-Tevpr**2+k*(3+Tevpr)*(1-Tevpr)**3)/Tevpr) =E= 0;

 e21 .. Pvpe-Pvper*Pc =E= 0;

 objdef ..   obj =E= -dHve/Cpla;
```

Note that there are no GAMS **parameters** in the above model, *i.e.*, Tbr, Tb, Tc, Tavgr, etc. are all variables that appear nonlinearly in this formulation, which contains a total of 58 integer variables and 22 continuous variables. The degree of nonlinearity in this model is so severe that GAMS/SBB and GAMS/DICOPT do not even find a single feasible solution. The algorithm outlined in this section yields all feasible solutions shown in Tables 8.4 and 8.5.

11.12.4 Finding All Solutions to Systems of Nonlinear Equations

A large body of literature addresses the problem of finding all solutions to systems of nonlinear equations. This problem is significant in many scientific, engineering, and management fields. For recent works on the development of mathematical software for this problem, we refer the reader to Verschelde (1999) and Wise, Sommese & Watson (2000), as well as references therein. Here, we use the **numsol** option of BARON to find all solutions to two specific

11.12. FINDING THE K BEST OR ALL FEASIBLE SOLUTIONS

problems from the literature.

The Robot Problem

We consider a common problem in kinematic analysis of robot manipulators, the so-called *indirect-position* or *inverse kinematics* problem, in which the desired position and orientation of the robot hand is given and the relative robot joint displacements are to be found. In particular, we consider the following set of equations for the PUMA robot (Tsai & Morgan 1985):

$$\gamma_1 x_1 x_3 + \gamma_2 x_2 x_3 + \gamma_3 x_1 + \gamma_4 x_2 + \gamma_5 x_4 + \gamma_6 x_7 + \gamma_7 = 0$$
$$\gamma_8 x_1 x_3 + \gamma_9 x_2 x_3 + \gamma_{10} x_1 + \gamma_{11} x_2 + \gamma_{12} x_4 + \gamma_{13} = 0$$
$$\gamma_{14} x_6 x_8 + \gamma_{15} x_1 + \gamma_{16} x_2 = 0$$
$$\gamma_{17} x_1 + \gamma_{18} x_2 + \gamma_{19} = 0$$
$$x_1^2 + x_2^2 - 1 = 0$$
$$x_3^2 + x_4^2 - 1 = 0$$
$$x_5^2 + x_6^2 - 1 = 0$$
$$x_7^2 + x_8^2 - 1 = 0$$
$$-1 \le x_i \le 1, \quad i = 1, \ldots, 8$$

where

$\gamma_1 = 0.004731$	$\gamma_6 = 1$	$\gamma_{11} = -0.07745$	$\gamma_{16} = 0.004731$
$\gamma_2 = -0.3578$	$\gamma_7 = -0.3571$	$\gamma_{12} = -0.6734$	$\gamma_{17} = -0.7623$
$\gamma_3 = -0.1238$	$\gamma_8 = 0.2238$	$\gamma_{13} = -0.6022$	$\gamma_{18} = 0.2238$
$\gamma_4 = -001637$	$\gamma_9 = 0.7638$	$\gamma_{14} = 1$	$\gamma_{19} = 0.3461$
$\gamma_5 = -0.9338$	$\gamma_{10} = 0.2638$	$\gamma_{15} = 0.3578$	

The first four equations of this problem are bilinear while the last four are generalized cylinders. The same problem is available in a somewhat different form as model ex14_1_6 in `globallib`. As opposed to the previous examples of this section, this example is continuous. However, we note that BARON's scheme for finding all feasible solutions will work well in continuous spaces as long as the sought-after solutions are isolated (separated by a certain distance). The BARON option `isoltol` (default value of 10^{-4}) allows the user to specify the isolation tolerance to be used to discriminate among different solutions. In order for two feasible solution vectors to be considered different, at least one of their coordinates must differ by `isoltol`.

The robot problem can be modeled in GAMS as follows:

File robot.gms

```
$ontext

Filename: robot.gms

Author: Nick Sahinidis, August 2002

Purpose:  Find all solutions of the PUMA robot problem
  L.-W. Tsai and A. P. Morgan, "Solving the kinematics of the
  most general six- and five-degree-of-freedom manipulators by
  continuation methods," ASME J. Mech. Transm. Automa. Des.,
  107, 189-200, 1985.

$offtext

options limrow=0,
        limcol=0,
        solprint=off,
        nlp = baron,
        optca = 1e-6,
        optcr = 0;

VARIABLES   x1, x2, x3, x4, x5, x6, x7, x8, obj;

x1.lo = -1; x2.lo = -1; x3.lo = -1; x4.lo = -1;
x5.lo = -1; x6.lo = -1; x7.lo = -1; x8.lo = -1;

x1.up = 1; x2.up = 1; x3.up = 1; x4.up = 1;
x5.up = 1; x6.up = 1; x7.up = 1; x8.up = 1;

EQUATIONS e1, e2, e3, e4, e5, e6, e7, e8, objective;

e1..  0.004731*x1*x3 - 0.1238*x1 - 0.3578*x2*x3
      - 0.001637*x2 - 0.9338*x4 + x7 =E= 0.3571;

e2..  0.2238*x1*x3 + 0.2638*x1 + 0.7623*x2*x3
      - 0.07745*x2 - 0.6734*x4 - x7 =E= 0.6022;

e3..  x6*x8 + 0.3578*x1 + 0.004731*x2  =E= 0;

e4..  - 0.7623*x1 + 0.2238*x2 =E= -0.3461;

e5..  power(x1,2) + power(x2,2)   =E= 1;
```

11.12. FINDING THE K BEST OR ALL FEASIBLE SOLUTIONS

```
e6..  power(x3,2) + power(x4,2)   =E= 1;

e7..  power(x5,2) + power(x6,2)   =E= 1;

e8..  power(x7,2) + power(x8,2)   =E= 1;

objective..  obj =E= 0;

model robot /all/;

robot.optfile = 1;

file opt /baron.opt/;

set i /1*20/;
parameter report(i,*);

loop (i,
   put opt;
   put 'numsol ', ord(i):4:0 /;
   putclose opt;

   solve robot using nlp minimizing obj;
   report(i,'Nodes Used') = robot.nodusd;
   report(i,'CPU Time')  = robot.resusd;
);

display report;
```

Using this model, we require BARON to solve this model for 1, 2, ..., 20 feasible solutions. In Figures 11.20 and 11.21, we plot the number of nodes and CPU seconds taken, respectively, per number of solutions requested for each of the first 16 runs. These numbers initially increase but flatten once numsol exceeds five. When numsol was set to 17 or above, BARON returned only 16 feasible solutions, which shows that this problem has exactly 16 feasible solutions within the default isolation tolerance.

The Boon Problem

The Boon problem from the field of neurophysiology is as follows (Boon 1992, Verschelde 1999):
$$x_1^2 + x_3^2 - 1 = 0$$

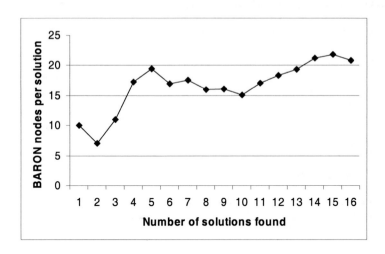

Figure 11.20: Nodes per solution found for the robot problem

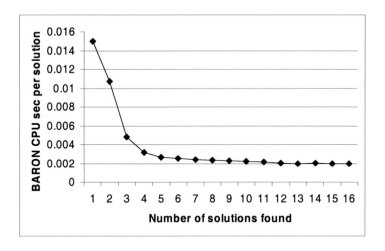

Figure 11.21: CPU times per solution found for the robot problem

11.12. FINDING THE K BEST OR ALL FEASIBLE SOLUTIONS

$$x_2^2 + x_4^2 - 1 = 0$$
$$x_5 x_3^3 + x_6 x_4^3 - 1.2 = 0$$
$$x_5 x_1^3 + x_6 x_2^3 - 1.2 = 0$$
$$x_5 x_3^2 x_1 + x_6 x_4^2 x_2 - 0.7 = 0$$
$$x_5 x_3 x_1^2 + x_6 x_4 x_2^2 - 0.7 = 0$$

for which we will seeks solutions over the box $[-10, 10]^6$ using the following GAMS model:

File boon.gms

```
$ontext

Filename: boon.gms

Author: Nick Sahinidis, August 2002

Purpose:  Find all solutions of the Boon problem from neurophysiology

References:
  S. Boon, "Solving systems of equations," Sci. Math. Num-Analysis,
  Newsgroup Article, 1992.

  J. Verschelde, "Algorithm 795: PHCpack: A general-purpose solver
  for polynomial systems by homotopy continuation," ACM Transactions
  on Mathematical  Softqare, 25(2), 251-276, 1999.

$offtext

options limrow=0,
        limcol=0,
        solprint=off,
        nlp = baron,
        optca = 1e-6,
        optcr = 0;

VARIABLES   x1, x2, x3, x4, x5, x6, obj;

x1.lo = -10; x2.lo = -10; x3.lo = -10;
x4.lo = -10; x5.lo = -10; x6.lo = -10;

x1.up = 10; x2.up = 10; x3.up = 10;
x4.up = 10; x5.up = 10; x6.up = 10;
```

```
EQUATIONS e1, e2, e3, e4, e5, e6, objective;

e1.. power(x1,2) + power(x3,2) =E= 1;

e2.. power(x2,2) + power(x4,2) =E= 1;

e3.. x3 * x5 * power(x3,2) + x4 * x6 * power(x4,2) =E= 1.2;

e4.. x1 * x5 * power(x1,2) + x2 * x6 * power(x2,2) =E= 1.2;

e5.. x1 * x5 * power(x3,2) + x2 * x6 * power(x4,2) =E= 0.7;

e6.. x3 * x5 * power(x1,2) + x4 * x6 * power(x2,2) =E= 0.7;

objective.. obj =E= 0;

model boon /all/;

boon.optfile = 1;

file opt /baron.opt/;

set i /1*20/;
parameter report(i,*);

loop (i,

  put opt;
  put 'numsol ', ord(i):4:0 /;
  putclose opt;

  solve boon using nlp minimizing obj;
  report(i,'Nodes Used') = boon.nodusd;
  report(i,'CPU Time') = boon.resusd;
);

display report;
```

Note that this problem involves fourth order polynomial equality constraints. In the GAMS model, we recast the problem in terms of quadratics and products. For numsol equal to 20, BARON enumerates 1965 nodes in 3.11 seconds, and returns exactly eight solutions, which implies that the problem has only eight solutions within the default isolation tolerance.

11.12. FINDING THE K BEST OR ALL FEASIBLE SOLUTIONS

In closing this section, it should be noted that finding all solutions to systems of nonlinear equations has direct applications in many fields, including chemical equilibrium, economic equilibrium, complementarity problems, stability analysis of dynamical systems, and many other areas in sciences and engineering. Thus, we expect this and the global optimization features of BARON to find many interesting uses in applied science and engineering.

Appendix A

GAMS Models for Pooling Problems

In this appendix, we provide the GAMS formulations for the pooling problems solved in Chapter 9 and Section 11.9. The main GAMS code that takes these files as input can be found in page 360.

A.1 Problems Adhya 1, 2, 3, and 4

These models originate from Adhya et al. (1999).

File adhya1.gms

```
$ontext
    Gams Model for the Pooling Problem
    Author: Mohit Tawarmalani
    model: adhya1
$offtext

$eolcom #

# Set Declarations
    set comp /1*5/;
    set pro  /1*4/;
    set qual /1*4/;
    set pool /1*2/;

# components related parameters
```

404 APPENDIX A. GAMS MODELS FOR POOLING PROBLEMS

```
table compparams(comp,*)
        1   2   3
    1   0   75  7
    2   0   75  3
    3   0   75  2
    4   0   75  10
    5   0   75  5 ;

parameters cl(comp), cu(comp), cprice(comp);
cl(comp) = compparams(comp,'1');
cu(comp) = compparams(comp,'2');
cprice(comp) = compparams(comp,'3');

table cqual(comp,qual)
        1   2    3   4
    1   1   6    4   0.5
    2   4   1    3   2
    3   4   5.5  3   0.9
    4   3   3    3   1
    5   1   2.7  4   1.6 ;

# pool related parameters
parameters psize(pool);
psize(pool) = 75;

# product related parameters
table prodparams(pro,*)
        1   2   3
    1   0   10  16
    2   0   25  25
    3   0   30  15
    4   0   10  10 ;

parameters prl(pro), pru(pro), pprice(pro);
prl(pro) = prodparams(pro,'1');
pru(pro) = prodparams(pro,'2');
pprice(pro) = prodparams(pro,'3');

parameter pqlbd(pro, qual);
pqlbd(pro, qual) = 0.0;

table pqubd(pro, qual)
        1   2    3     4
    1   3   3    3.25  0.75
    2   4   2.5  3.5   1.5
```

A.1. PROBLEMS ADHYA 1, 2, 3, AND 4

```
     3   1.5  5.5  3.9  0.8
     4    3    4    4   1.8 ;

# network related parameters
table ubq(comp, pool)
          1    2
    1     1    0
    2     1    0
    3     0    1
    4     0    1
    5     0    1 ;

table uby(pool, pro)
          1    2    3    4
    1    10   25   30   10
    2    30   10   10   25 ;

parameter ubz(comp, pro);
ubz(comp, pro) = 0.0;

$include pool.gms
```

File adhya2.gms

```
$ontext
    Gams Model for the Pooling Problem
    Author: Mohit Tawarmalani
    model: adhya2
$offtext

$eolcom #

# Set Declarations
    set comp /1*5/;
    set pro  /1*4/;
    set qual /1*6/;
    set pool /1*2/;

# components related parameters
table compparams(comp,*)
          1    2    3
    1     0   75    7
    2     0   75    3
```

```
         3      0     75     2
         4      0     75    10
         5      0     75     5 ;

parameters cl(comp), cu(comp), cprice(comp);
cl(comp)     = compparams(comp,'1');
cu(comp)     = compparams(comp,'2');
cprice(comp) = compparams(comp,'3');

table cqual(comp,qual)
           1    2    3    4    5    6
    1      1    6    4  0.5    5    9
    2      4    1    3    2    4    4
    3      4  5.5    3  0.9    7   10
    4      3    3    3    1    3    4
    5      1  2.7    4  1.6    3    7 ;

# pool related parameters
parameters psize(pool);
psize(pool) = 75;

# product related parameters
table prodparams(pro,*)
           1    2    3
    1      0   10   16
    2      0   25   25
    3      0   30   15
    4      0   10   10 ;

parameters prl(pro), pru(pro), pprice(pro);
prl(pro)    = prodparams(pro,'1');
pru(pro)    = prodparams(pro,'2');
pprice(pro) = prodparams(pro,'3');

parameter pqlbd(pro, qual);
pqlbd(pro, qual) = 0.0;

table pqubd(pro, qual)
           1    2     3    4    5    6
    1      3    3  3.25 0.75    6    5
    2      4  2.5   3.5  1.5    7    6
    3    1.5  5.5   3.9  0.8    7    6
    4      3    4     4  1.8    8    6 ;

# network related parameters
```

A.1. PROBLEMS ADHYA 1, 2, 3, AND 4

```
table ubq(comp, pool)
        1    2
   1    1    0
   2    1    0
   3    0    1
   4    0    1
   5    0    1 ;

table uby(pool, pro)
        1    2    3    4
   1   10   25   30   10
   2   30   10   10   25 ;

parameter ubz(comp, pro);
ubz(comp, pro) = 0.0;

$include pool.gms
```

File adhya3.gms

```
$ontext
    Gams Model for the Pooling Problem
    Author: Mohit Tawarmalani
    model: adhya3
$offtext

$eolcom #

# Set Declarations
    set comp /1*8/;
    set pro  /1*4/;
    set qual /1*6/;
    set pool /1*3/;

# components related parameters
table compparams(comp,*)
        1    2    3
   1    0   75    7
   2    0   75    3
   3    0   75    2
   4    0   75   10
   5    0   75    5
   6    0   75    5
```

```
         7       0     75     9
         8       0     75    11 ;

parameters cl(comp), cu(comp), cprice(comp);
cl(comp) = compparams(comp,'1');
cu(comp) = compparams(comp,'2');
cprice(comp) = compparams(comp,'3');

table cqual(comp,qual)
             1     2     3     4     5     6
     1       1     6     4   0.5     5     9
     2       4     1     3     2     4     4
     3       4   5.5     3   0.9     7    10
     4       3     3     3     1     3     4
     5       1   2.7     4   1.6     3     7
     6     1.8   2.7     4   3.5   6.1     3
     7       5     1   1.7   2.9   3.5   2.9
     8       3     3     3     1     5     2 ;

# pool related parameters
parameters psize(pool);
psize(pool) = 75;

# product related parameters
table prodparams(pro,*)
         1     2     3
     1   0    10    16
     2   0    25    25
     3   0    30    15
     4   0    10    10 ;

parameters prl(pro), pru(pro), pprice(pro);
prl(pro) = prodparams(pro,'1');
pru(pro) = prodparams(pro,'2');
pprice(pro) = prodparams(pro,'3');

parameter pqlbd(pro, qual);
pqlbd(pro, qual) = 0;

table pqubd(pro, qual)
             1     2     3      4     5     6
     1       3     3  3.25   0.75     6     5
     2       4   2.5   3.5    1.5     7     6
     3     1.5   5.5   3.9    0.8     7     6
     4       3     4     4    1.8     8     6 ;
```

A.1. PROBLEMS ADHYA 1, 2, 3, AND 4

```
# network related parameters
table ubq(comp, pool)
        1   2   3
    1   1   0   0
    2   1   0   0
    3   0   1   0
    4   0   1   0
    5   0   1   0
    6   0   0   1
    7   0   0   1
    8   0   0   1 ;

table uby(pool, pro)
        1   2   3   4
    1   10  25  30  10
    2   10  10  25  30
    3   30  10  10  25 ;

parameter ubz(comp, pro);
ubz(comp, pro) = 0;

$include pool.gms
```

File adhya4.gms

```
$ontext
    Gams Model for the Pooling Problem
    Author: Mohit Tawarmalani
    model: adhya4
$offtext

$eolcom #

# Set Declarations
    set comp /1*8/;
    set pro  /1*5/;
    set qual /1*4/;
    set pool /1*2/;

# components related parameters
table compparams(comp,*)
        1   2   3
```

```
    1    0   85   15
    2    0   85    7
    3    0   85    4
    4    0   85    5
    5    0   85    6
    6    0   85    3
    7    0   85    5
    8    0   85    5 ;

parameters cl(comp), cu(comp), cprice(comp);
cl(comp) = compparams(comp,'1');
cu(comp) = compparams(comp,'2');
cprice(comp) = compparams(comp,'3');

table cqual(comp,qual)
          1    2    3    4
    1   0.5  1.9  1.3    1
    2   1.4  1.8  1.7  1.6
    3   1.2  1.9  1.4  1.4
    4   1.5  1.2  1.7  1.3
    5   1.6  1.8  1.6    2
    6   1.2  1.1  1.4    2
    7   1.5  1.5  1.5  1.5
    8   1.4  1.6  1.2  1.6 ;

# pool related parameters
parameters psize(pool);
psize(pool) = 85;

# product related parameters
table prodparams(pro,*)
        1    2    3
    1   0   15   10
    2   0   25   25
    3   0   10   30
    4   0   20    6
    5   0   15   10 ;

parameters prl(pro), pru(pro), pprice(pro);
prl(pro) = prodparams(pro,'1');
pru(pro) = prodparams(pro,'2');
pprice(pro) = prodparams(pro,'3');

parameter pqlbd(pro, qual);
pqlbd(pro, qual) = 0.0;
```

```
table pqubd(pro, qual)
        1    2    3    4
   1   1.2  1.7  1.4  1.7
   2   1.4  1.3  1.8  1.4
   3   1.3  1.3  1.9  1.9
   4   1.2  1.1  1.7  1.6
   5   1.6  1.9   2   2.5 ;

# network related parameters
table ubq(comp, pool)
        1    2
   1    1    0
   2    1    0
   3    1    0
   4    1    0
   5    0    1
   6    0    1
   7    0    1
   8    0    1 ;

table uby(pool, pro)
        1    2    3    4    5
   1   15   25   10   20   15
   2   10   20   15   15   25 ;

parameter ubz(comp, pro);
ubz(comp, pro) = 0.0;

$include pool.gms
```

A.2 Problems Bental 4 and 5

These problems originate from Ben-Tal et al. (1994).

File bental4.gms

```
$ontext
    Gams Model for the Pooling Problem
    Author: Mohit Tawarmalani
    model: bental4
$offtext
```

```
$eolcom #

# Set Declarations
    set comp /1*4/;
    set pro /1*2/;
    set qual /1*1/;
    set pool /1*1/;

# components related parameters
table compparams(comp,*)
        1    2    3
    1   0   300   6
    2   0    50  15
    3   0   300  16
    4   0   300  10 ;

parameters cl(comp), cu(comp), cprice(comp);
cl(comp) = compparams(comp,'1');
cu(comp) = compparams(comp,'2');
cprice(comp) = compparams(comp,'3');

table cqual(comp,qual)
        1
    1   3
    2   1
    3   1
    4   2 ;

# pool related parameters
parameters psize(pool);
psize(pool) = 300;

# product related parameters
table prodparams(pro,*)
        1    2    3
    1   0   100   9
    2   0   200  15 ;

parameters prl(pro), pru(pro), pprice(pro);
prl(pro) = prodparams(pro,'1');
pru(pro) = prodparams(pro,'2');
pprice(pro) = prodparams(pro,'3');

parameter pqlbd(pro, qual);
```

A.2. PROBLEMS BENTAL 4 AND 5

```
pqlbd(pro, qual) = 0.0;

table pqubd(pro, qual)
          1
    1    2.5
    2    1.5 ;

# network related parameters
table ubq(comp, pool)
          1
    1    1
    2    1
    3    1
    4    0 ;

table uby(pool, pro)
          1     2
    1    100   200 ;

table ubz(comp, pro)
          1     2
    1    0     0
    2    0     0
    3    0     0
    4    100   200 ;

$include pool.gms
```

File bental5.gms

```
$ontext
    Gams Model for the Pooling Problem
    Author: Mohit Tawarmalani
    model: bental5
$offtext

$eolcom #

# Set Declarations
    set comp /1*13/;
    set pro  /1*5/;
    set qual /1*2/;
    set pool /1*3/;
```

```
# components related parameters
table compparams(comp,*)
            1       2       3
    1       0      600      6
    2       0      600     16
    3       0      600     15
    4       0      600     12
    5       0      600      6
    6       0      600     16
    7       0      600     15
    8       0      600     12
    9       0      600      6
   10       0      600     16
   11       0      600     15
   12       0      600     12
   13       0      600     10 ;

parameters cl(comp), cu(comp), cprice(comp);
cl(comp) = compparams(comp,'1');
cu(comp) = compparams(comp,'2');
cprice(comp) = compparams(comp,'3');

table cqual(comp,qual)
            1       2
    1       3       1
    2       1       3
    3      1.2      5
    4      1.5     2.5
    5       3       1
    6       1       3
    7      1.2      5
    8      1.5     2.5
    9       3       1
   10       1       3
   11      1.2      5
   12      1.5     2.5
   13       2      2.5 ;

# pool related parameters
parameters psize(pool);
psize(pool) = 600;

# product related parameters
table prodparams(pro,*)
```

A.2. PROBLEMS BENTAL 4 AND 5

```
              1     2     3
      1       0    100    18
      2       0    200    15
      3       0    100    19
      4       0    100    16
      5       0    100    14 ;

parameters prl(pro), pru(pro), pprice(pro);
prl(pro) = prodparams(pro,'1');
pru(pro) = prodparams(pro,'2');
pprice(pro) = prodparams(pro,'3');

parameter pqlbd(pro, qual);
pqlbd(pro, qual) = 0.0;

table pqubd(pro, qual)
              1     2
      1      2.5    2
      2      1.5   2.5
      3       2    2.6
      4       2     2
      5       2     2 ;

# network related parameters
table ubq(comp, pool)
              1     2     3
      1       1     0     0
      2       1     0     0
      3       1     0     0
      4       1     0     0
      5       0     1     0
      6       0     1     0
      7       0     1     0
      8       0     1     0
      9       0     0     1
     10       0     0     1
     11       0     0     1
     12       0     0     1
     13       0     0     0 ;

table uby(pool, pro)
              1     2     3     4     5
      1      100   200   100   100   100
      2      100   100   100   200   100
      3      200   100   100   100   100 ;
```

```
table ubz(comp, pro)
            1       2       3       4       5
   13     100     200     100     100     100 ;

$include pool.gms
```

A.3 Problems Foulds 2, 3, 4, and 5

These problems originate from Foulds et al. (1992).

File foulds2.gms

```
$ontext
    Gams Model for the Pooling Problem
    Author: Mohit Tawarmalani
    model: foulds2
$offtext

$eolcom #

# Set Declarations
    set comp /1*6/;
    set pro /1*4/;
    set qual /1*1/;
    set pool /1*2/;

# components related parameters
table compparams(comp,*)
         1    2    3
   1     0   600    6
   2     0   600   16
   3     0   600   10
   4     0   600    3
   5     0   600   13
   6     0   600    7 ;

parameters cl(comp), cu(comp), cprice(comp);
cl(comp)     = compparams(comp,'1');
cu(comp)     = compparams(comp,'2');
cprice(comp) = compparams(comp,'3');
```

A.3. PROBLEMS FOULDS 2, 3, 4, AND 5

```
table cqual(comp,qual)
        1
    1   3
    2   1
    3   2
    4   3.5
    5   1.5
    6   2.5 ;

# pool related parameters
parameters psize(pool);
psize(pool) = 600;

# product related parameters
table prodparams(pro,*)
        1    2    3
    1   0   100   9
    2   0   200  15
    3   0   100   6
    4   0   200  12 ;

parameters prl(pro), pru(pro), pprice(pro);
prl(pro) = prodparams(pro,'1');
pru(pro) = prodparams(pro,'2');
pprice(pro) = prodparams(pro,'3');

parameter pqlbd(pro, qual);
pqlbd(pro, qual) = 0.0;

table pqubd(pro, qual)
        1
    1   2.5
    2   1.5
    3   3
    4   2 ;

# network related parameters
table ubq(comp, pool)
        1    2
    1   1    0
    2   1    0
    3   0    0
    4   0    1
    5   0    1
    6   0    0 ;
```

```
table uby(pool, pro)
         1    2    3    4
    1   100  200  100  200
    2   100  200  100  200 ;

table ubz(comp, pro)
         1    2    3    4
    3   100  200  100  200
    6   100  200  100  200 ;

$include pool.gms
```

File foulds3.gms

```
$ontext
    Gams Model for the Pooling Problem
    Author: Mohit Tawarmalani
    model: foulds3
$offtext

$eolcom #

# Set Declarations
    set comp /1*32/;
    set pro  /1*16/;
    set qual /1*1/;
    set pool /1*8/;

# components related parameters
table compparams(comp,*)
         1    2    3
    1    0   16   20
    2    0   16   19
    3    0   16   18
    4    0   16   17
    5    0   16   19
    6    0   16   18
    7    0   16   17
    8    0   16   16
    9    0   16   18
   10    0   16   17
   11    0   16   16
```

A.3. PROBLEMS FOULDS 2, 3, 4, AND 5

```
    12    0    16    15
    13    0    16    17
    14    0    16    16
    15    0    16    15
    16    0    16    14
    17    0    16    16
    18    0    16    15
    19    0    16    14
    20    0    16    13
    21    0    16    15
    22    0    16    14
    23    0    16    13
    24    0    16    12
    25    0    16    14
    26    0    16    13
    27    0    16    12
    28    0    16    11
    29    0    16    13
    30    0    16    12
    31    0    16    11
    32    0    16    10 ;

parameters cl(comp), cu(comp), cprice(comp);
cl(comp)     = compparams(comp,'1');
cu(comp)     = compparams(comp,'2');
cprice(comp) = compparams(comp,'3');

table cqual(comp,qual)
            1
     1      1
     2      1.1
     3      1.2
     4      1.3
     5      1.1
     6      1.2
     7      1.3
     8      1.4
     9      1.2
    10      1.3
    11      1.4
    12      1.5
    13      1.3
    14      1.4
    15      1.5
    16      1.6
```

```
        17    1.4
        18    1.5
        19    1.6
        20    1.7
        21    1.5
        22    1.6
        23    1.7
        24    1.8
        25    1.6
        26    1.7
        27    1.8
        28    1.9
        29    1.7
        30    1.8
        31    1.9
        32     2  ;

# pool related parameters
parameters psize(pool);
psize(pool) = 16;

# product related parameters
table prodparams(pro,*)
            1    2     3
      1     0    1     20
      2     0    1   19.5
      3     0    1     19
      4     0    1   18.5
      5     0    1     18
      6     0    1   17.5
      7     0    1     17
      8     0    1   16.5
      9     0    1     16
     10     0    1   15.5
     11     0    1     15
     12     0    1   14.5
     13     0    1     14
     14     0    1   13.5
     15     0    1     13
     16     0    1   12.5 ;

parameters prl(pro), pru(pro), pprice(pro);
prl(pro) = prodparams(pro,'1');
pru(pro) = prodparams(pro,'2');
pprice(pro) = prodparams(pro,'3');
```

A.3. PROBLEMS FOULDS 2, 3, 4, AND 5

```
parameter pqlbd(pro, qual);
pqlbd(pro, qual) = 0.0;

table pqubd(pro, qual)
          1
   1    1.05
   2    1.1
   3    1.15
   4    1.2
   5    1.25
   6    1.3
   7    1.35
   8    1.4
   9    1.45
  10    1.5
  11    1.55
  12    1.6
  13    1.65
  14    1.7
  15    1.75
  16    1.8  ;

# network related parameters
table ubq(comp, pool)
         1    2    3    4    5    6    7    8
   1     1    0    0    0    0    0    0    0
   2     1    0    0    0    0    0    0    0
   3     1    0    0    0    0    0    0    0
   4     1    0    0    0    0    0    0    0
   5     0    1    0    0    0    0    0    0
   6     0    1    0    0    0    0    0    0
   7     0    1    0    0    0    0    0    0
   8     0    1    0    0    0    0    0    0
   9     0    0    1    0    0    0    0    0
  10     0    0    1    0    0    0    0    0
  11     0    0    1    0    0    0    0    0
  12     0    0    1    0    0    0    0    0
  13     0    0    0    1    0    0    0    0
  14     0    0    0    1    0    0    0    0
  15     0    0    0    1    0    0    0    0
  16     0    0    0    1    0    0    0    0
  17     0    0    0    0    1    0    0    0
  18     0    0    0    0    1    0    0    0
  19     0    0    0    0    1    0    0    0
```

```
    20    0    0    0    0    1    0    0    0
    21    0    0    0    0    0    1    0    0
    22    0    0    0    0    0    1    0    0
    23    0    0    0    0    0    1    0    0
    24    0    0    0    0    0    1    0    0
    25    0    0    0    0    0    0    1    0
    26    0    0    0    0    0    0    1    0
    27    0    0    0    0    0    0    1    0
    28    0    0    0    0    0    0    1    0
    29    0    0    0    0    0    0    0    1
    30    0    0    0    0    0    0    0    1
    31    0    0    0    0    0    0    0    1
    32    0    0    0    0    0    0    0    1 ;

parameter uby(pool, pro);
uby(pool, pro) = 16;

parameter ubz(comp, pro);
ubz(comp, pro) = 0;

$include pool.gms
```

File foulds4.gms

```
$ontext
    Gams Model for the Pooling Problem
    Author: Mohit Tawarmalani
    model: foulds4
$offtext

$eolcom #

# Set Declarations
    set comp /1*11/;
    set pro  /1*16/;
    set qual /1*1/;
    set pool /1*8/;

# components related parameters
table compparams(comp,*)
            1    2    3
    1       0   16   20
    2       0   16   19
```

A.3. PROBLEMS FOULDS 2, 3, 4, AND 5

```
     3    0   16   18
     4    0   16   17
     5    0   16   16
     6    0   16   15
     7    0   16   14
     8    0   16   13
     9    0   16   12
    10    0   16   11
    11    0   16   10 ;

parameters cl(comp), cu(comp), cprice(comp);
cl(comp) = compparams(comp,'1');
cu(comp) = compparams(comp,'2');
cprice(comp) = compparams(comp,'3');

table cqual(comp,qual)
          1
    1     1
    2    1.1
    3    1.2
    4    1.3
    5    1.4
    6    1.5
    7    1.6
    8    1.7
    9    1.8
   10    1.9
   11     2 ;

# pool related parameters
parameters psize(pool);
psize(pool) = 16;

# product related parameters
table prodparams(pro,*)
          1    2    3
    1     0    1   20
    2     0    1 19.5
    3     0    1   19
    4     0    1 18.5
    5     0    1   18
    6     0    1 17.5
    7     0    1   17
    8     0    1 16.5
    9     0    1   16
```

```
      10    0    1 15.5
      11    0    1   15
      12    0    1 14.5
      13    0    1   14
      14    0    1 13.5
      15    0    1   13
      16    0    1 12.5 ;

parameters prl(pro), pru(pro), pprice(pro);
prl(pro) = prodparams(pro,'1');
pru(pro) = prodparams(pro,'2');
pprice(pro) = prodparams(pro,'3');

parameter pqlbd(pro, qual);
pqlbd(pro, qual) = 0.0;

table pqubd(pro, qual)
           1
    1   1.05
    2   1.1
    3   1.15
    4   1.2
    5   1.25
    6   1.3
    7   1.35
    8   1.4
    9   1.45
   10   1.5
   11   1.55
   12   1.6
   13   1.65
   14   1.7
   15   1.75
   16   1.8 ;

# network related parameters
table ubq(comp, pool)
           1    2    3    4    5    6    7    8
    1      1    0    0    0    0    0    0    0
    2      0    1    1    0    0    0    0    0
    3      0    0    1    1    1    0    0    0
    4      1    0    0    1    0    1    1    0
    5      0    1    1    0    1    0    0    1
    6      0    0    1    1    1    1    0    0
    7      1    0    0    1    0    1    1    0
```

A.3. PROBLEMS FOULDS 2, 3, 4, AND 5

```
    8   0   1   0   0   1   0   1   1
    9   0   0   0   0   0   1   1   1
   10   1   0   0   0   0   0   0   1
   11   0   1   0   0   0   0   0   0 ;

parameter uby(pool, pro);
uby(pool, pro) = 16;

parameter ubz(comp, pro);
ubz(comp, pro) = 0.0;

$include pool.gms
```

File foulds5.gms

```
$ontext
    Gams Model for the Pooling Problem
    Author: Mohit Tawarmalani
    model: foulds5
$offtext

$eolcom #

# Set Declarations
    set comp /1*11/;
    set pro  /1*16/;
    set qual /1*1/;
    set pool /1*8/;

# components related parameters
table compparams(comp,*)
             1      2      3
    1        0     16     20
    2        0     16     19
    3        0     16     18
    4        0     16     17
    5        0     16     16
    6        0     16     15
    7        0     16     14
    8        0     16     13
    9        0     16     12
   10        0     16     11
   11        0     16     10 ;
```

```
parameters cl(comp), cu(comp), cprice(comp);
cl(comp) = compparams(comp,'1');
cu(comp) = compparams(comp,'2');
cprice(comp) = compparams(comp,'3');

table cqual(comp,qual)
        1
   1    1
   2    1.1
   3    1.2
   4    1.3
   5    1.4
   6    1.5
   7    1.6
   8    1.7
   9    1.8
  10    1.9
  11    2  ;

# pool related parameters
parameters psize(pool);
psize(pool) = 16;

# product related parameters
table prodparams(pro,*)
        1    2    3
   1    0    1    20
   2    0    1    19.5
   3    0    1    19
   4    0    1    18.5
   5    0    1    18
   6    0    1    17.5
   7    0    1    17
   8    0    1    16.5
   9    0    1    16
  10    0    1    15.5
  11    0    1    15
  12    0    1    14.5
  13    0    1    14
  14    0    1    13.5
  15    0    1    13
  16    0    1    12.5 ;

parameters prl(pro), pru(pro), pprice(pro);
```

A.3. PROBLEMS FOULDS 2, 3, 4, AND 5

```
prl(pro)    = prodparams(pro,'1');
pru(pro)    = prodparams(pro,'2');
pprice(pro) = prodparams(pro,'3');

parameter pqlbd(pro, qual);
pqlbd(pro, qual) = 0.0;

table pqubd(pro, qual)
         1
    1  1.05
    2  1.1
    3  1.15
    4  1.2
    5  1.25
    6  1.3
    7  1.35
    8  1.4
    9  1.45
   10  1.5
   11  1.55
   12  1.6
   13  1.65
   14  1.7
   15  1.75
   16  1.8 ;

# network related parameters
table ubq(comp, pool)
         1   2   3   4   5   6   7   8
    1    1   0   0   0   0   0   1   0
    2    1   0   1   0   0   0   1   0
    3    1   0   1   0   0   0   1   0
    4    1   0   1   0   1   0   1   0
    5    0   0   1   0   1   0   0   1
    6    0   0   0   0   1   0   0   1
    7    0   0   0   1   1   0   0   1
    8    0   1   0   1   0   1   0   1
    9    0   1   0   1   0   1   0   0
   10    0   1   0   1   0   1   0   0
   11    0   1   0   0   0   1   0   0 ;

parameter uby(pool, pro);
uby(pool, pro) = 16;

parameter ubz(comp, pro);
```

```
ubz(comp, pro) = 0;

$include pool.gms
```

A.4 Problems Haverly 1, 2, and 3

These problems originate from Haverly (1978) and Haverly (1979).

File `haverly1.gms`

This model can be found in page 359.

File `haverly2.gms`

```
$ontext
    Gams Model for the Pooling Problem
    Author: Mohit Tawarmalani
    model: haverly2
$offtext

$eolcom #

# Set Declarations
    set comp /1*3/;
    set pro  /1*2/;
    set qual /1*1/;
    set pool /1*1/;

# components related parameters
table compparams(comp,*)
          1    2    3
    1     0   800   6
    2     0   800  16
    3     0   800  10 ;

parameters cl(comp), cu(comp), cprice(comp);
cl(comp) = compparams(comp,'1');
cu(comp) = compparams(comp,'2');
cprice(comp) = compparams(comp,'3');

table cqual(comp,qual)
```

A.4. PROBLEMS HAVERLY 1, 2, AND 3

```
                1
     1          3
     2          1
     3          2 ;

# pool related parameters
parameters psize(pool);
psize(pool) = 800;

# product related parameters
table prodparams(pro,*)
            1      2      3
     1      0     600     9
     2      0     200     15 ;

parameters prl(pro), pru(pro), pprice(pro);
prl(pro) = prodparams(pro,'1');
pru(pro) = prodparams(pro,'2');
pprice(pro) = prodparams(pro,'3');

parameter pqlbd(pro, qual);
pqlbd(pro, qual) = 0.0;

table pqubd(pro, qual)
            1
     1     2.5
     2     1.5 ;

# network related parameters
table ubq(comp, pool)
            1
     1      1
     2      1
     3      0 ;

table uby(pool, pro)
            1      2
     1     600    200 ;

table ubz(comp, pro)
            1      2
     3     600    200 ;

$include pool.gms
```

430   APPENDIX A. GAMS MODELS FOR POOLING PROBLEMS

File `haverly3.gms`

```
$ontext
    Gams Model for the Pooling Problem
    Author: Mohit Tawarmalani
    model: haverly3
$offtext

$eolcom #

# Set Declarations
    set comp /1*3/;
    set pro /1*2/;
    set qual /1*1/;
    set pool /1*1/;

# components related parameters
table compparams(comp,*)
            1     2     3
    1       0    300    6
    2       0    300   13
    3       0    300   10  ;

parameters cl(comp), cu(comp), cprice(comp);
cl(comp)     = compparams(comp,'1');
cu(comp)     = compparams(comp,'2');
cprice(comp) = compparams(comp,'3');

table cqual(comp,qual)
            1
    1       3
    2       1
    3       2 ;

# pool related parameters
parameters psize(pool);
psize(pool) = 300;

# product related parameters
table prodparams(pro,*)
            1     2     3
    1       0    100    9
    2       0    200   15 ;

parameters prl(pro), pru(pro), pprice(pro);
```

```
prl(pro) = prodparams(pro,'1');
pru(pro) = prodparams(pro,'2');
pprice(pro) = prodparams(pro,'3');

parameter pqlbd(pro, qual);
pqlbd(pro, qual) = 0.0;

table pqubd(pro, qual)
          1
    1    2.5
    2    1.5 ;

# network related parameters
table ubq(comp, pool)
          1
    1    1
    2    1
    3    0 ;

table uby(pool, pro)
          1      2
    1    100    200 ;

table ubz(comp, pro)
          1      2
    3    100    200 ;

$include pool.gms
```

A.5 Problem RT 2

This problem is due to Rehfeldt and Tisljar and has been taken from Audet et al. (2000).

File rt2.gms

```
$ontext
    Gams Model for the Pooling Problem
    Author: Mohit Tawarmalani
    model: rt2
$offtext
```

```
$eolcom #

# Set Declarations
    set comp /1*3/;
    set pro  /1*3/;
    set qual /1*4/;
    set pool /1*2/;

# components related parameters
table compparams(comp,*)
            1        2        3
    1       0     60.9756   49.2
    2       0    161.29     62
    3       0      5       300 ;

parameters cl(comp), cu(comp), cprice(comp);
cl(comp)     = compparams(comp,'1');
cu(comp)     = compparams(comp,'2');
cprice(comp) = compparams(comp,'3');

table cqual(comp,qual)
            1       2       3       4
    1     0.82      3     99.2    90.5
    2     0.62      0     87.9    83.5
    3     0.75      0    114      98.7 ;

# pool related parameters
table poolparams(pool,*)
            1
    1     12.5
    2     17.5
;

parameters psize(pool);
psize(pool) = poolparams(pool,'1');

# product related parameters
table prodparams(pro,*)
            1       2       3
    1       5     300     190
    2       5     300     230
    3       5     300     150 ;

parameters prl(pro), pru(pro), pprice(pro);
prl(pro) = prodparams(pro,'1');
```

A.5. PROBLEM RT 2

```
pru(pro) = prodparams(pro,'2');
pprice(pro) = prodparams(pro,'3');

table pqlbd(pro, qual)
           1     2     3     4
    1    0.74    0    95    85
    2    0.74    0    96    88
    3    0.74    0    91   83.5 ;

table pqubd(pro, qual)
           1     2     3     4
    1    0.79    3   114   98.7
    2    0.79   0.9  114   98.7
    3    0.79    3   114   98.7 ;

# network related parameters
parameter ubq(comp, pool);
ubq(comp, pool) = 1;

table uby(pool, pro)
          1     2     3
    1   12.5  12.5  12.5
    2   17.5  17.5  17.5 ;

table ubz(comp, pro)
           1      2      3
    1      0     7.5     0
    2   161.29    0   161.29
    3    7.5      0      0  ;

$include pool.gms
```

Bibliography

Adhya, N., Tawarmalani, M. & Sahinidis, N. V. (1999). A Lagrangian approach to the pooling problem, *Industrial & Engineering Chemistry* **38**: 1956–1972.

Adjiman, C. S., Dallwig, S., Floudas, C. A. & Neumaier, A. (1998). A global optimization method, αBB, for general twice-differentiable constrained NLPs–I. Theoretical advances, *Computers & Chemical Engineering* **22**: 1137–1158.

Aggarwal, S. C. (1977). An alternative method of integer solutions to linear fractional functionals by a branch and bound technique, *Z. Angew. Math. Mech.* **57**: 52–53.

Ahmed, S. (2000). *Strategic Planning under Uncertainity: Stochastic Programming Approaches*, PhD thesis, Department of Mechanical & Industrial Engineering, University of Illinois, Urbana, IL.

Ahmed, S., Tawarmalani, M. & Sahinidis, N. V. (2000). A finite branch and bound algorithm for two-stage stochastic integer programs. *Mathematical Programming*. Submitted.

Aho, A. V., Sethi, R. & Ullman, J. D. (1986). *Compilers: Principles, Techniques and Tools*, Addison-Wesley.

Ahuja, R. K., Magnanti, T. L. & Orlin, J. B. (1993). *Network Flows. Theory, Algorithms and Applications*, Prentice Hall, Englewood Cliffs, NJ.

Al-Khayyal, F. A. & Falk, J. E. (1983). Jointly constrained biconvex programming, *Mathematics of Operations Research* **8**: 273–286.

Al-Khayyal, F. A. & Sherali, H. D. (2000). On finitely terminating branch-and-bound algorithms for some global optimization problems, *SIAM Journal on Optimization* **10**: 1049–1057.

Alizadeh, F. (1995). Interior point methods in semidefinite programming with applications to combinatorial optimization, *SIAM Journal of Optimization* **5**: 13–51.

Andersen, D. E. & Andersen, K. D. (1995). Presolving in linear programming, *Mathematical Programming* **71A**: 221–245.

Androulakis, I. P., Visweswaran, V. & Floudas, C. A. (1996). Distributed decomposition-based approaches in global optimization, In C. A. Floudas and P. M. Pardalos (eds.), *State of the Art in Global Optimization,* Kluwer, Boston, MA pp. 285–301.

Arora, S. R., Swarup, K. & Puri, M. C. (1977). The set covering problem with linear fractional functional, *Indian Journal of Pure and Applied Mathematics* **8**: 578–588.

Ashrafi, N. & Berman, O. (1992). Optimization models for selection of programs, considering cost & reliability, *IEEE Trans. Reliab.* **41**: 281–287.

Audet, C., Brimberg, J., Hansen, P. & Mladenovic, N. (2000). Pooling problem: Alternate formulations and solution methods, *G-2000-23*, Les Cashiers Du GERAD, Montréal.

Balas, E. (1985). Disjunctive programming and hierarchy of relaxations for discrete optimization problems, *SIAM Journal Alg. Disc. Meth.* **6**: 466–486.

Balas, E. (1988). Lecture Notes on Integer Programming. GSIA, Carnegie-Melon University.

Balas, E. (1998). Disjunctive programming: Properties of the convex hull of feasible points, *Discrete Applied Mathematics* **89**(1-3): 3–44. Original manuscript was published as a technical report in 1974.

Balas, E., Ceria, S. & Cornuèjols, G. (1993). A lift and project cutting plane algorithm for mixed 0−1 programs, *Mathematical Programming* **58**: 295–324.

Balas, E. & Mazzola, J. B. (1984). Nonlinear 0−1 programming: I. Linearization techniques, *Mathematical Programming* **30**: 1–21.

Bazaraa, M. S., Sherali, H. D. & Shetty, C. M. (1993). *Nonlinear Programming, Theory and Algorithms*, 2nd edn, Wiley Interscience, Series in Discrete Mathematics and Optimization.

Beale, E. M. L. (1979). Branch and bound methods for mathematical programming systems, *Annals of discrete mathematics* **5**: 201–219.

Beale, E. M. L. & Forrest, J. J. H. (1976). Global optimization using special ordered sets, *Mathematical Programming* **10**: 52–69.

Beale, E. M. L. & Tomlin, J. A. (1970). Special facilities in a general mathematical programming system for nonconvex problems using ordered sets of variables, In J. Lawrence (ed.), *Proceedings of the Fifth International Conference on Operational Research,* Tavistock Publications, London pp. 447–454.

Beck, J. V. & Arnold, K. J. (1977). *Parameter Estimation in Engineering and Science*, John Wiley & Sons, New York.

Ben-Tal, A., Eiger, G. & Gershovitz, V. (1994). Global minimization by reducing the duality gap, *Mathematical Programming* **63**: 193–212.

Bennett, K. P. (1994). Global tree optimization: A non-greedy decision tree algorithm, *Computing Sciences and Statistics* **26**: 156–160.

Bennett, K. P. & Mangasarian, O. L. (1994). Bilinear separation of two sets in $n-$space, *Computational Optimization and Applications* **2**: 207–227.

Benson, H. P. & Boger, G. M. (1997). Multiplicative programming problems: Analysis and efficient point search heuristic, *Journal of Optimization Theory and Applications* **94**: 487–510.

Biegler, L. T., Grossmann, I. E. & Westerberg, A. W. (1997). *Systematic Methods of Chemical Process Design*, Prentice Hall, Upper Saddle River, New Jersey.

Biegler, L. T. & Tjoa, I. B. (1991). Catalyst mixing for packed bed reactor. In I. E. Grossmann (ed.), *CACHE Design Case Study Volume 6: Chemical*

Engineering Optimization Models with GAMS, CACHE Corporation, Austin, TX.

Bienstock, D. (1996). Computational study of a family of mixed-integer quadratic programming problems, *Mathematical Programming* **74A**: 121–140.

Bienstock, D. & Shapiro, J. F. (1988). Optimizing resource acquisition decisions by stochastic programming, *Management Science* **34**: 215–229.

Bitran, G. R., Haas, E. A. & Matsuo, H. (1986). Production planning of style goods with high setup costs and forecast revisions, *Operations Research* **34**: 226–236.

Blair, C. E. & Jeroslow, R. G. (1984). Constructive characterization of the value-function of a mixed-integer program I, *Discrete Applied Mathematics* **9**: 217–233.

Blum, M., Floyd, R. W., Pratt, V., Rivest, R. L. & Tarjan, R. E. (1973). Time bounds for selection, *Journal of Computer and System Sciences* **7**: 448–461.

Boon, S. (1992). Solving systems of equations. *Sci. Math. Num-Analysis*. Newsgroup Article 3529.

Borchers, B. & Mitchell, J. E. (1994). An improved branch and bound for mixed integer nonlinear programs, *Comput. Oper. Res.* **21**: 359–367.

Borchers, B. & Mitchell, J. E. (1997). A computational comparison of branch and bound and outer approximation algorithms for 0−1 mixed integer nonlinear programs, *Comput. Oper. Res.* **24**: 699–701.

Boyd, A. (1995). On the convergence of Fenchel cutting planes in mixed-integer programming, *SIAM Journal of Optimization* **5**: 421–435.

Bracken, J. & McCormick, G. P. (1968). *Selected Applications of Nonlinear Programming*, John Wiley & Sons, New York.

Brignole, E. A., Bottini, S. B. & Gani, R. (1986). A strategy for the design and selection of solvents for separation processes, *Fluid Phase Equilibria* **29**: 125–132.

Brook, A., Kendrick, D. & Meeraus, A. (1988). *GAMS–A User's Guide*, Scientific Press, Redwood City, CA.

Burkard, R. E., Hamacher, H. & Rote, G. (1992). Sandwich approximation of univariate convex functions with an application to separable convex programming, *Naval Research Logistics* **38**: 911–924.

Bussieck, M. R. (2002). MINLP World. http://www.gamsworld.org/minlp/index.htm.

Carøe, C. C. (1998). *Decomposition in Stochastic Integer Programming*, PhD thesis, University of Copenhagen, Denmark.

Carøe, C. C., Ruszczyński, A. & Schultz, R. (1997). Unit commitment under uncertainty via two-stage stochastic programming, In *Proceedings of NOAS 97*, Carøe et al. (eds.), Department of Computer Science, University of Copenhagen pp. 21–30.

Carøe, C. C. & Schultz, R. (1999). Dual decomposition in stochastic integer programming, *Operations Research Letters* **24**: 37–45.

Carøe, C. C. & Tind, J. (1998). L-shaped decomposition of two-stage stochastic programs with integer recourse, *Mathematical Programming* **83**: 451–464.

Ceria, S. & Soares, J. (1999). Convex programming for disjunctive convex optimization, *Mathematical Programming* **86A**: 595–614.

Charnes, A. & Cooper, W. W. (1962). Programming with linear fractional functionals, *Naval Research Logistics Quarterly* **9**: 181–186.

Cheung, B. K.-S., Langevin, A. & Delmaire, H. (1997). Coupling genetic algorithm with a grid search method to solve mixed integer nonlinear programming problems, *Comput. Math. Appl.* **34**: 13–23.

Churi, N. & Achenie, L. E. K. (1996). Novel mathematical programming model for computer aided molecular design, *Industrial & Engineering Chemistry Research* **35**: 3788–3794.

Churi, N. & Achenie, L. E. K. (1997a). On the use of a mixed integer nonlinear programming model for refrigerant design, *International Transactions of Operational Research* **4**: 45–54.

Churi, N. & Achenie, L. E. K. (1997b). The optimal design of refrigerant mixtures for a two-evaporator refrigeration system, *Computers & Chemical Engineering* **21**: S349–S354.

Colville, A. R. (1968). A comparative study of nonlinear programming codes, *Technical Report 320-2940*, IBM, New York.

Constantinou, L., Bagherpour, K., Gani, R., Klein, J. A. & Wu, D. T. (1996). Computer aided product design: Problem formulations, methodology and applications, *Computers & Chemical Engineering* **20**: 685–702.

Constantinou, L. & Gani, R. (1994). New group contribution method for estimating properties of pure compounds, *AIChE J.* **40**: 1697–1710.

Crama, Y. (1993). Concave extensions for non-linear 0−1 maximization problems, *Mathematical Programming* **61**: 53–60.

Dakin, R. J. (1965). A tree search algorithm for mixed integer programming problems, *Computer Journal* **8**: 250–255.

Dantzig, G. B. (1963). *Linear Programming and Extensions*, Princeton University Press, Princeton, NJ.

Dempster, M. A. H. (1982). A stochastic approach to hierarchical planning and scheduling, In *Deterministic and Stochastic Scheduling*, Dempster et al. (eds.), D. Riedel Publishing Co., Dordrecht pp. 271–296.

Devillers, J. & Putavy, C. (1996). Designing biodegradable molecules from the combined use of a backpropagation neural network and a genetic algorithm, In J. Devillers (ed.), *Genetic Algorithms in Molecular Design*, Academic Press, London pp. 303–314.

Dixon, L. C. W. & Szegoe, G. P. (eds) (1975). *Towards global optimisation. Proceedings of a Workshop at the University of Cagliari, Italy, October 1974.*, North-Holland Publishing Company, New York.

Dorneich, M. C. & Sahinidis, N. V. (1995). Global optimization algorithms for chip layout and compaction, *Engineering Optimization* **25**: 131–154.

Duran, M. A. & Grossmann, I. E. (1986). An outer-approximation algorithm for a class of mixed-integer nonlinear programs, *Mathematical Programming* **36**: 307–339.

Duvedi, A. P. & Achenie, L. E. K. (1996). Designing environmentally safe refrigerants using mathematical programming, *Chemical Engineering Science* **51**: 3727–3739.

Epperly, T. G. W. & Swaney, R. E. (1996). Branch and bound for global NLP: New bounding LP, In I. E. Grossmann (ed.), *Global Optimization in Engineering Design*, Kluwer Academic Publishers, Boston, MA pp. 1–36.

Falk, J. E. & Polocsay, S. W. (1994). Image space analysis of generalized fractional programs, *Journal of Global Optimization* **4**: 63–88.

Falk, J. E. & Soland, R. M. (1969). An algorithm for separable nonconvex programming problems, *Management Science* **15**: 550–569.

Floudas, C. A. (1999). *Deterministic Global Optimization: Theory, Algorithms and Applications*, Kluwer Academic Publishers, Dordrecht.

Floudas, C. A. & Ciric, A. R. (1989). Strategies for overcoming uncertainties in heat exchanger network synthesis, *Computers & Chemical Engineering* **13**: 1133–1152.

Floudas, C. A. & Pardalos, P. M. (1990). *A Collection of Test Problems for Constrained Global Optimization Algorithms*, Lecture Notes in Computer Science, Springer–Verlag, Berlin.

Foulds, L. R., Haugland, D. & Jornsten, K. (1992). A bilinear approach to the pooling problem, *Optimization* **24**: 165–180.

Fourer, R., Gay, D. M. & Kernighan, B. W. (1993). *AMPL: A Modeling Language for Mathematical Programming*, Duxbury Press, Pacific Grove, CA.

Fruhwirth, B., Burkard, R. E. & Rote, G. (1989). Approximation of convex curves with applications to the bicriteria minimum cost flow problem, *European Journal of Operational Research* **42**: 326–338.

Fujie, T. & Kojima, M. (1997). Semidefinite programming relaxation for nonconvex quadratic programs, *Journal of Global Optimization* **10**: 367–380.

GAMS Development Corporation (2002). GAMS World. http://www.gamsworld.org/.

Gani, R. & Fredenslund, A. (1993). Computer aided molecular and mixture design with specified constraints, *Fluid Phase Equilibria* **82**: 39–46.

Gani, R., Nielsen, B. & Fredenslund, A. (1991). A group contribution approach to computer-aided molecular design, *AIChE J.* **37**: 1318–1332.

Ghildyal, V. & Sahinidis, N. V. (2001). Solving global optimization problems with BARON, In A. Migdalas, P. Pardalos, and P. Varbrand (eds.), *From Local to Global Optimization. A Workshop on the Occasion of the 70th Birthday of Professor Hoang Tuy*, Kluwer Academic Publishers, Boston, MA pp. 205–230.

Ghosh, A. & McLafferty, S. (1987). *Location Strategies for Retail and Service Firms*, Lexington Books, Massachusetts.

Ghosh, A., McLafferty, S. & Craig, S. (1995). Multifacility retail networks, In *Facility Location: A Survey of Applications and Methods*, Z. Drezner (ed.), Springer-Verlag, New York pp. 301–330.

Gill, P. E., Murray, W. & Saunders, M. A. (1999). User's Guide for SNOPT 5.3: A FORTRAN Package for Large-Scale Nonlinear Programming, *Technical report*, University of California, San Diego and Stanford University, CA.

Gilmore, P. C. & Gomory, R. E. (1963). A linear programming approach to the cutting stock problem—Part II, *Operations Research* **11**: 52–53.

Glover, F. (1975). Improved linear integer programming formulations of nonlinear integer problems, *Management Science* **2**: 455–460.

Glover, F. & Woolsey, E. (1973). Further reduction of 0−1 polynomial programming problems to zero-one linear programming problems, *Operations Research* **21**: 156–161.

Glover, F. & Woolsey, E. (1974). Converting a 0−1 polynomial programming problem to a 0−1 linear program, *Operations Research* **22**: 180–182.

Granot, D. & Granot, F. (1976). On Solving fractional $(0-1)$ programs by implicit enumeration, *INFOR* **14**: 241–249.

Granot, D. & Granot, F. (1977). On integer and mixed integer fractional programming problems, *Annals of Discrete Mathematics* **1**: 221–231.

Greenberg, H. J. (1995). Analysis of the pooling problem, *ORSA Journal on Computing* **7**: 205–217.

Griewank, A. (2000). *Evaluating derivatives. Principles and Techniques of Algorithmic Differentiation*, Vol. 19 of *Frontiers in Applied Mathematics*, SIAM, Philadelphia, PA.

Gruber, P. M. (1993). Aspects of approximation of convex bodies, In P. M. Gruber and J. M. Wills (eds.), *Handbook of Convex Geometry*, North-Holland .

Gruber, P. M. & Kenderov, P. (1982). Approximation of convex bodies by polytopes, *Rendiconti Circ. Mat. Palermo, Serie II* **31**: 195–225.

Grunspan, M. & Thomas, M. E. (1973). Hyperbolic integer programming, *Naval Research Logistics Quarterly* **20**: 341–356.

Gunasekaran, A., Goyal, S. K., Martikainen, T. & Yli-Olli, P. (1993). Equipment selection problems in just-in-time manufacturing systems, *Journal Operational Research Society* **4**: 345–353.

Gunn, D. J. & Thomas, W. J. (1965). Mass transport and chemical reaction in multifunctional catalyst systems, *Chemical Engineering Science* **20**: 89–100.

Gupta, O. K. & Ravindran, A. (1985). Branch and bound experiments in convex nonlinear integer programming, *Management Science* **31**: 1533–1546.

Gutierrez, R. A. & Sahinidis, N. V. (1996). A branch-and-bound approach for machine selection in just-in-time manufacturing systems, *International Journal of Production Research* **34**: 797–818.

Hamed, A. S. E. & McCormick, G. P. (1993). Calculation of bounds on variables satisfying nonlinear inequality constraints, *Journal of Global Optimization* **3**: 25–47.

Hammer, P. L. & Rudeanu, S. (1968). *Boolean methods in operations research and related areas*, Springer, New York.

Hansen, E. R. (1979). Global optimization using interval analysis: The one-dimensional case, *Journal of Optimization Theory and Applications* **29**: 331–344.

Hansen, E. R. (1992). *Global optimization using interval analysis*, Pure and Applied Mathematics, Marcel Dekker, New York.

Hansen, P., Jaumard, B. & Lu, S.-H. (1991). An analytic approach to global optimization, *Mathematical Programming* **52**: 227–254.

Hansen, P., Jaumard, B. & Mathon, V. (1993). Constrained nonlinear 0−1 programming, *ORSA Journal of Computing* **5**: 87–119.

Hansen, P., Jaumard, B. & Xiong, J. (3). Decomposition and interval arithmetic applied to global minimization of polynomial and rational functions, *Journal of Global Optimization* **4**: 421–437.

Hansen, P., Poggi de Aragao, M. V. & Ribeiro, C. C. (1991). Hyperbolic 0−1 programming and query optimization in information retrieval, *Mathematical Programming* **52**: 255–263.

Haque, M. A. & Ahmed, S. (1998). p-choice facility location in discrete space. In preparation.

Hashizume, S., Fukushima, M., Katoh, N. & Ibaraki, T. (1987). Approximation algorithms for combinatorial fractional programming problems, *Mathematical Programming* **37**: 255–267.

Haverly, C. A. (1978). Studies of the behaviour of recursion for the pooling problem, *ACM SIGMAP Bulletin* **25**: 19–28.

Haverly, C. A. (1979). Behaviour of recursion model—More studies, *ACM SIGMAP Bulletin* **26**: 22–28.

Henderson, J. M. & Quandt, R. E. (1971). *Microeconomic Theory*, 2nd edn, McGraw-Hill.

Hillestad, R. J. & Jacobsen, S. E. (1980). Reverse convex programming, *Applied Mathematics and Optimization* **6**: 63–78.

Hiriart-Urruty, J.-B. & Lemaréchal, C. (1993a). *Convex Analysis and Minimization Algorithms I*, Springer-Verlag, Berlin.

Hiriart-Urruty, J.-B. & Lemaréchal, C. (1993b). *Convex Analysis and Minimization Algorithms II*, Springer-Verlag, Berlin.

Hock, W. & Schittkowski, K. (1981). *Test examples for nonlinear programming codes*, Vol. 187 of *Lecture Notes in Economics and Mathematical Systems*, Springer-Verlag, New York.

Hoffman, K. (1981). A method for globally minimizing concave functions over convex sets, *Mathematical Programming* **20**: 22–32.

Hooker, J. (2000). *Logic-Based Methods for Optimization: Combining Optimization and Constraint Satisfaction*, John Wiley & Sons, New York, NY.

Hopcroft, J. E. & Ullman, J. D. (1979). *Introduction to Automata and Formal Languages*, Addison-Wesley, Reading, MA.

Horst, R., Thoai, N. V. & Tuy, H. (1989). On an outer-approximation in global optimization, *Optimization* **20**: 255–264.

Horst, R. & Tuy, H. (1996). *Global Optimization: Deterministic Approaches*, Third edn, Springer Verlag, Berlin.

Horvath, A. L. (1992). *Molecular Design: Chemical Structure Generation from the Properties of Pure Organic Compounds*, Elsevier, New York, NY.

IBM (1995). *Optimization Subroutine Library Guide and Reference Release 2.1*, fifth edn, International Business Machines Corporation, Kingston, NY.

ILOG (1997). *CPLEX 6.0 User's Manual*, ILOG CPLEX Division, Incline Village, NV.

ILOG (2000). *CPLEX 7.0 User's Manual*, ILOG CPLEX Division, Incline Village, NV.

Isbelle, J. R. & Marlow, W. H. (1956). Attrition games, *Naval Research Logistics Quarterly* **3**: 71–93.

Jeroslow, R. G. (1989). *Logic Based Decision Support: Mixed Integer Model Formulation*, North-Holland, New York.

Joback, K. G. & Reid, R. C. (1987). Estimation of pure-component properties from group-contributions, *Chem. Eng. Comm.* **57**: 233–243.

Joback, K. G. & Stephanopoulos, G. (1990). Designing molecules possessing desired physical property values, *Proceedings of the 1989 Foundations of Computer-Aided Process Design Conference, Snowmass, CO,* Elsevier, Amsterdam pp. 195–230.

Joback, K. G. & Stephanopoulos, G. (1995). Searching spaces of discrete solutions: The design of molecules possessing desired physical properties, *Advances in Chemical Engineering* **21**: 257–311.

Jorjani, S., Scott, C. H. & Woodruff, D. L. (1995). Selection of an optimal subset of sizes, *Technical report*, University of California, Davis, CA.

Kalantari, B. & Rosen, J. B. (1987). An algorithm for global minimization of linearly constrained convex quadratic functions, *Mathematics of Operations Research* **12**: 544–561.

Kearfott, R. B. (1996). *Rigorous Global Search: Continuous Problems*, Vol. 13 of *Nonconvex Optimization and Its Applications*, Kluwer Academic Publishers, Dordrecht.

Keeney, R. L. & Raiffa, H. (1993). *Decisions with Multiple Objective*, Cambridge University Press, Cambridge, Massachusetts.

Kim, Y. S. (1998). Problems in reliability. Personal Communication.

King, A. J., Takriti, S. & Ahmed, S. (1997). Issues in risk modeling for multistage systems, *Technical Report RC-20993*, IBM Research Division.

Klein Haneveld, W. K., Stougie, L. & van der Vlerk, M. H. (1995). On the convex hull of the simple integer recourse objective function, *Annals of Operational Research* **56**: 209–224.

Klein Haneveld, W. K., Stougie, L. & van der Vlerk, M. H. (1996). An algorithm for the construction of convex hulls in simple integer recourse programming, *Annals of Operational Research* **64**: 67–81.

Kocis, G. R. & Grossmann, I. E. (1988). Global optimization of nonconvex MINLP problems in process synthesis, *Industrial and Engineering Chemistry Research* **27**: 1407–1421.

Kojima, M. & Tunçel, L. (1999). Cones of matrices and successive convex relaxations of nonconvex sets, *Technical report*, Department of Mathematical and Computing Sciences, Tokyo Institute of Technology, Japan.

Konno, H. & Kuno, T. (1989). Linear multiplicative programming, *Technical Report IHSS 89-13*, Institute of Human and Social Sciences, Tokyo Institute of Technology, Tokyo, Japan.

Konno, H. & Kuno, T. (1990). Generalized linear multiplicative and fractional programming, *Annals of Operations Research* **25**: 147–162.

Konno, H. & Kuno, T. (1992). Linear multiplicative programming, *Mathematical Programming* **56**: 51–64.

Konno, H. & Kuno, T. (1995). Multiplicative programming problems, In R. Horst and P. M. Pardalos (eds.), *Handbook of Global Optimization*, Kluwer Academic Publishers, Boston, MA pp. 369–405.

Konno, H., Kuno, T. & Yajima, Y. (1994). Global optimization of a generalized convex multiplicative function, *Journal of Global Optimization* **4**: 47–62.

Konno, H., Shirakawa, H. & Yamazaki, H. (1993). A mean-absolute deviation-skewness portfolio optimization model, *Annals of Operations Research* **45**: 205–220.

Konno, H., Yajima, Y. & Matsui, T. (1991). Parametric simplex algorithms for solving a special class of nonconvex minimization problems, *Journal of Global Optimization* **1**: 65–81.

Krarup, J. & Bilde, O. (1977). Plant location, set covering and economic lot Size: An $O(mn)$ algorithm for structured problems, In L. Collatz et al. (eds.), *International Series of Numerical Mathematics*, Vol. 36, Birkhäuser Verlag, Basel pp. 155–180.

Kuno, T., Yajima, Y. & Konno, H. (1993). An outer approximation method for minimizing the product of several convex functions on a convex set, *Journal of Global Optimization* **3**: 325–335.

Lamar, B. W. (1993). An improved branch and bound algorithm for minimum concave cost network flow problems, *Journal of Global Optimization* **3**: 261–287.

Land, A. H. & Doig, A. G. (1960). An automatic method for solving discrete programming problems, *Econometrica* **28**: 497–520.

Laporte, G. & Louveaux, F. V. (1993). The integer L-shaped method for stochastic integer programs with complete recourse, *Operations Research Letters* **13**: 133–142.

Lawler, E. L. (1978). Sequencing jobs to minimize total weighted completion time subject to precedence constraints, *Annals of Discrete Mathematics* **2**: 75–90.

Lazimy, R. (1982). Mixed-integer quadratic programming, *Mathematical Programming* **22**: 332–349.

Lazimy, R. (1985). Improved algorithm for mixed-integer quadratic programs and a computational study, *Mathematical Programming* **32**: 100–113.

Lee, E. K. & Mitchell, J. E. (1997). Computational experience in nonlinear mixed integer programming, In U. Zimmermann (ed.), *Operations Research Proceedings, Selected Papers of the Symposium, SOR'96, Braunschweig, Germany,* Springer, Berlin pp. 95–100.

Li, H.-L. (1994). A global approach for general 0−1 fractional programming, *European Journal of Operational Research* **73**: 590–596.

Li, H.-L. & Chou, C.-T. (1994). A global approach for nonlinear mixed discrete programming in design optimization, *Engineering Optimization* **22**: 109–122.

Liebman, J., Lasdon, L. S., Schrage, L. & Waren, A. D. (1986). *Modeling and Optimization with GINO,* The Scientific Press, Palo Alto, CA.

Liu, M. L., Sahinidis, N. V. & Shectman, J. P. (1996). Planning of chemical process networks via global concave minimization, In I. E. Grossmann (ed.), *Global Optimization in Engineering Design,* Kluwer Academic Publishers, Boston, MA pp. 195–230.

Liu, X. J., Umegaki, T. & Yamamoto, Y. (1999). Heuristic Methods for Linear Multiplicative Programming, *Journal of Global Optimization* **4**: 433–447.

Lobo, M. S., Vandenberghe, L., Boyd, S. & Lebret, H. (1998). Applications of second-order cone programming, *Linear Algebra Applications* **284**: 193–228.

Logsdon, J. S. & Biegler, L. T. (1989). Accurate solution of differential-algebraic optimization problems, *Industrial & Engineering Chemistry Research* **28**: 1628–1639.

Lokketangen, A. & Woodruff, D. L. (1996). Progressive hedging and tabu search applied to mixed integer (0, 1) multi-stage stochastic programming, *Journal of Heuristics* **2**: 111–128.

Lovàsz, L. & Schrijver, A. (1991). Cones of matrices and set-functions and 0−1 optimization, *SIAM Journal on Optimization* **1**: 166–190.

Macchietto, S., Odele, O. & Omatsone, O. (1990). Design of optimal solvents for liquid-liquid extraction and gas absorption processes, *Transactions of the Institute of Chemical Engineers* **68**: 429–433.

Maling, K., Mueller, S. H. & Heller, W. R. (1982). On finding most optimal rectangular package plans, *Proceedings of the 19th Design Automation Conference*, pp. 663–670.

Mangasarian, O. L. & McLinden, L. (1985). Simple bounds for solutions of monotone complementarity problems and convex programs, *Mathematical Programming* **32**: 32–40.

Manousiouthakis, V. & Sourlas, D. (1992). A global optimization approach to rationally constrained rational programming, *Chemical Engineering Communications* **115**: 127–147.

Maranas, C. D., Androulakis, I. P., Floudas, C. A., Berger, A. J. & Mulvey, J. M. (1997). Solving long-term financial planning problems via global optimization, *Journal of Economic Dynamics & Control* **21**: 1405–1425.

Marcoulaki, E. C. & Kokossis, A. C. (1998). Molecular design synthesis using stochastic optimization as a tool for scoping and screening, *Computers & Chemical Engineering* **22**: S11–S18.

Marin, R. (1998). Points of location on a fixture. Personal Communication.

Markowitz, H. M. (1991). *Portfolio Selection*, 2nd edn, Basil Blackwell Inc.

Martelli, G. (1962). *Jemmy Twitcher—A Life on the Fourth Earl of Sandwich, 1782-1792*, Jonathan Cape, London.

Martin, R. K. (1987). Generating alternative mixed-integer programming models using variable redefinition, *Operations Research* **35**: 820–831.

Matsui, T. (1996). $\mathcal{NP}$-hardness of linear multiplicative programming and related problems, *Journal of Global Optimization* **9**: 113–119.

McBride, R. D. & Yormark, J. S. (1980). An implicit enumeration algorithm for quadratic integer programming, *Management Science* **26**: 282–296.

McCormick, G. P. (1972). Converting general nonlinear programming problems to separable nonlinear programming problems, *Technical Report T-267*, The George Washington University, Washington D.C.

McCormick, G. P. (1976). Computability of global solutions to factorable nonconvex programs: Part I—Convex underestimating problems, *Mathematical Programming* **10**: 147–175.

McCormick, G. P. (1983). *Nonlinear Programming: Theory, Algorithms and Applications*, John Wiley & Sons.

Meeraus, A. (2002). GLOBAL World. http://www.gamsworld.org/global/index.htm.

Megiddo, N. (1979). Combinatorial optimization with rational objective functions, *Mathematics of Operations Research* **4**: 414–424.

Minoux, M. (1986). *Mathematical Programming. Theory and Algorithms*, John Wiley & Sons, New York.

Mittelmann, H. D. (2002). Performance World. http://www.gamsworld.org/performance/board.htm.

Mulvey, J. M., Vanderbei, R. J. & Zenios, S. A. (1995). Robust optimization of large-scale systems, *Operations Research* **43**: 264–281.

Murtagh, B. A. & Saunders, M. A. (1995). MINOS 5.5 User's Guide, *Technical Report SOL 83-20R*, Systems Optimization Laboratory, Department of Operations Research, Stanford University, CA.

Murty, K. G. & Kabadi, S. N. (1987). Some $\mathcal{NP}$-complete problems in quadratic and nonlinear programming, *Mathematical Programming* **39**: 117–129.

Nabar, S. V. & Schrage, L. (1992). Formulating and solving business problems as nonlinear integer programs, *Technical report*, Graduate School of Business, University of Chicago.

Nakanishi, M. & Cooper, L. G. (1974). Parameter estimate for multiplicative interactive choice models: Least squares approach, *Journal of Marketing Research* **11**: 303–311.

Nanda, G. (2001). *Design of Efficient Secondary Refrigerants*, Master's thesis, Department of Chemical Engineering, University of Illinois, Urbana, IL.

Nanda, G. & Sahinidis, N. V. (2002). Design of efficient secondary refrigerants. In preparation.

Naser, S. F. & Fournier, R. L. (1991). A system for the design of an optimum liquid-liquid extractant molecule, *Computers & Chemical Engineering* **15**: 397–414.

Nemhauser, G. L. & Wolsey, L. A. (1988). *Integer and Combinatorial Optimization*, Wiley Interscience, Series in Discrete Mathematics and Optimization.

Odele, O. & Macchietto, S. (1993). Computer aided molecular design: A novel method for optimal solvent selection, *Fluid Phase Equilibria* **82**: 47–54.

Ostrovskii, G. M., Volin, Y. M. & Borisov, W. W. (1971). Über die berechnung von ableitungen, *Wiss. Z. Tech. Hochshule für Chemie* **13**: 382–384.

Ourique, J. E. & Telles, A. S. (1998). Computer-aided molecular design with simulated annealing and molecular graphs, *Computers & Chemical Engineering* **22**: S615–S618.

Pardalos, P. M. (1994-2002a). Journal of Global Optimization. http://www.wkap.nl/journalhome.htm/0925-5001.

Pardalos, P. M. (1994-2002b). Nonconvex Optimization and its Applications. http://www.wkap.nl/series.htm/NOIA.

Parker, R. G. & Rardin, R. L. (1988). *Discrete optimization*, Computer Science and Scientific Computing, Academic Press, Boston, MA.

Phillips, A. T. & Rosen, J. B. (1988). A parallel algorithm for constrained concave quadratic global minimization, *Mathematical Programming* **42**: 421–448.

Phillips, A. T. & Rosen, J. B. (1992). Sufficient conditions for solving linearly constrained separable concave global minimization problems, *Journal of Global Optimization* **3**: 79–94.

Picard, J. & Queyranne, M. (1982). A network flow solution to some non-linear 0−1 programming problems with applications to graph theory, *Networks* **12**: 141–159.

Pretel, E. J., Lopez, P. A., Bottini, S. B. & Brignole, E. A. (1994). Computer-aided molecular design of solvents for separation processes, *AIChE J.* **40**: 1349–1360.

Quesada, I. & Grossmann, I. E. (1995a). A global optimization algorithm for linear fractional and bilinear programs, *Journal of Global Optimization* **6**: 39–76.

Quesada, I. & Grossmann, I. E. (1995b). Global optimization of bilinear process networks and multicomponent flows, *Computers & Chemical Engineering* **19**(12): 1219–1242.

Quesada, I. & Grossmann, I. E. (1996). Alternative bounding approximations for the global optimization of various engineering design problems, In I. E. Grossmann (ed.), *Global Optimization in Engineering Design*, Kluwer Academic Publishers, Boston, MA pp. 309–331.

Quist, A. J. (2000a). *Application of Mathematical Optimization Techniques to Nuclear Reactor Reload Pattern Design*, PhD thesis, Technische Universiteit, Delft, The Netherlands.

Quist, A. J. (2000b). On a nuclear reactor problem. Personal Communication.

Rao, M. R. (1971). Cluster analysis and mathematical programming, *Journal of the American Statistical Association* **66**: 622–626.

Rardin, R. L. & Choe, U. (1979). Tighter relaxations of fixed charge network flow problems, *Technical Report J-79-18, Industrial and Systems Engineering Report Series*, Georgia Institute of Technology, Atlanta, GA.

Reid, R. C., Prausnitz, J. M. & Poling, B. E. (1987). *The Properties of Gases and Liquids*, 4th edn, McGraw-Hill, New York, NY.

Rigby, B., Lasdon, L. S. & Waren, A. D. (1995). The evolution of Texaco's blending systems: From OMEGA to StarBlend, *Interfaces* **25**(5): 64–83.

Rikun, A. D. (1997). A convex envelope formula for multilinear functions, *Journal of Global Optimization* **10**: 425–437.

Rinnooy Kan, A. H. G. & Timmer, G. T. (1987a). Stochastic global optimization methods. I: Clustering methods, *Mathematical Programming* **39**: 27–56.

Rinnooy Kan, A. H. G. & Timmer, G. T. (1987b). Stochastic global optimization methods. II: Multilevel methods, *Mathematical Programming* **39**: 57–78.

Robillard, P. (1971). $(0,1)$ hyperbolic programming problems, *Naval Research Logistics Quarterly* **18**: 47–57.

Rockafellar, R. T. (1970). *Convex Analysis*, Princeton Mathematical Series, Princeton University Press.

Rockafellar, R. T. & Wets, R. J.-B. (1998). *Variational Analysis*, A Series of Comprehensive Studies in Mathematics, Springer, Berlin.

Rosen, J. B. & Pardalos, P. M. (1986). Global minimization of large-scale constrained concave quadratic problems by separable programming, *Mathematical Programming* **34**: 163–174.

Rosen, J. B. & van Vliet, M. (1987). A parallel stochastic method for the constrained concave global minimization problem, *Technical Report 87-31*, Computer Science Department, Institute of Technology, University of Minnesota, Minneapolis, MN.

Rote, G. (1992). The convergence rate of the sandwich algorithm for approximating convex functions, *Computing* **48**: 337–361.

Ryoo, H. S. (1999). *Global Optimization of Multiplicative Programs: Theory, Algorithms, and Applications*, PhD thesis, Department of Mechanical & Industrial Engineering, University of Illinois, Urbana, IL.

Ryoo, H. S. & Sahinidis, N. V. (1995). Global optimization of nonconvex NLPs and MINLPs with applications in process design, *Computers & Chemical Engineering* **19**: 551–566.

Ryoo, H. S. & Sahinidis, N. V. (1996). A branch-and-reduce approach to global optimization, *Journal of Global Optimization* **8**: 107–139.

Ryoo, H. S. & Sahinidis, N. V. (2001). Analysis of bounds for multilinear functions, *Journal Global Optimization* **19**: 403–424.

Ryoo, H. S. & Sahinidis, N. V. (2002). Global optimization of multiplicative programs. *Journal of Global Optimization*. Submitted.

Sahinidis, N. V. (1996). BARON: A general purpose global optimization software package, *Journal of Global Optimization* **8**: 201–205.

Sahinidis, N. V. (1999-2000). *BARON: Branch and Reduce Optimization Navigator, User's Manual, Version 4.0*. Available for download at http://archimedes.scs.uiuc.edu/baron.html.

Sahinidis, N. V. & Grossmann, I. E. (1991). Convergence properties of generalized Benders' decomposition, *Computers & Chemical Engineering* **15**: 481–491.

Sahinidis, N. V. & Tawarmalani, M. (2000). Applications of global optimization to process and molecular design, *Computers & Chemical Engineering* **24**: 2157–2169.

Sahinidis, N. V. & Tawarmalani, M. (2002). *GAMS/BARON 5.0: Global optimization of mixed-integer nonlinear programs*.

Sahinidis, N. V., Tawarmalani, M. & Yu, M. (2002). Design of alternative refrigerants via global optimization. *AIChE J.* Submitted.

Saipe, A. L. (1975). Solving a (0, 1) hyperbolic program by branch and bound, *Naval Research Logistics Quarterly* **22**: 497–515.

Sandgren, E. (1990). Nonlinear integer and discrete programming in mechanical design optimization, *Journal of Mechanical Design* **112**: 223–229.

Savelsbergh, M. W. P. (1994). Preprocessing and probing for mixed integer programming problems, *ORSA Journal on Computing* **6**: 445–454.

Schaible, S. (1995). Fractional programming with sums of ratios, *in* E. Castagnoli & J. Giorgi (eds), *Proceedings of the Workshop held in Milan on March 28, 1995*, Scalar and Vector Optimization in Economic and Financial Problems, pp. 163–175.

Schaible, S. & Sodini, C. (1995). Finite algorithm for generalized linear multiplicative programming, *Journal of Optimization Theory and Applications* **87**: 441–455.

Schoen, F. (1991). Stochastic techniques for global optimization: A survey of recent advances, *Journal of Global Optimization* **1**: 207–228.

Schrage, L. (1999). *Optimization Modeling with LINGO*, 3rd edn, LINDO Systems, Inc., Chicago, IL.

Schrijver, A. (1986). *Theory of Linear and Integer Programming*, Wiley-Interscience Series in Discrete Mathematics and Optimization, John Wiley & Sons, Chichester, Great Britain.

Schultz, R., Stougie, L. & van der Vlerk, M. H. (1996). Two-stage stochastic integer programming: A survey, *Statistica Neerlandica. Journal of the Netherlands Society for Statistics and Operations Research* **50**(3): 404–416.

Schultz, R., Stougie, L. & van der Vlerk, M. H. (1998). Solving stochastic programs with integer recourse by enumeration: A framework using Gröbner basis reductions, *Mathematical Programming* **83**: 229–252.

Shectman, J. P. (1999). *Finite Algorithms for Global Optimization of Concave Programs and General Quadratic Programs*, PhD thesis, Department of Mechanical & Industrial Engineering, University of Illinois, Urbana, IL.

Shectman, J. P. & Sahinidis, N. V. (1998). A finite algorithm for global minimization of separable concave programs, *Journal of Global Optimization* **12**: 1–36.

Sherali, H. D. (1997). Convex envelopes of multilinear functions over a unit hypercube and over special discrete sets, *Acta Mathematica Vietnamica* **22**: 245–270.

Sherali, H. D. & Adams, W. P. (1990). A hierarchy of relaxations between the continuous and convex hull representations for zero-one programming problems, *SIAM Journal of Discrete Mathematics* **3**: 411–430.

Sherali, H. D. & Adams, W. P. (1994). A hierarchy of relaxations and convex hull characterizations for mixed- integer zero-one programming problems, *Discrete Applied Mathematics* **52**(1): 83–106.

Sherali, H. D. & Adams, W. P. (1999). *A reformulation-linearization technique for solving discrete and continuous nonconvex problems*, Vol. 31 of *Nonconvex Optimization and its Applications*, Kluwer Academic Publishers, Dordrecht.

Sherali, H. D., Adams, W. P. & Driscoll, P. J. (1999). Exploiting special structures in constructing a hierarchy of relaxations for 0−1 mixed integer programs, *Operations Research* **46**: 396–405.

Sherali, H. D. & Alameddine, A. (1992). A new reformulation-linearization technique for bilinear programming problems, *Journal of Global Optimization* **2**: 379–410.

Sherali, H. D. & Tuncbilek, C. H. (1995). A reformulation-convexification approach for solving nonconvex quadratic programming problems, *Journal of Global Optimization* **7**: 1–31.

Sherali, H. D. & Wang, H. (2001). Global optimization of nonconvex factorable programming problems, *Mathematical Programming* **89**: 459–478.

Shin, D. K., Gurdal, Z. & Griffin, O. H. (1990). A penalty approach for nonlinear optimization with discrete design variables, *Engineering Optimization* **16**: 29–42.

Smith, E. M. B. & Pantelides, C. C. (1996). Global optimisation of general process models, In I. E. Grossmann (ed.), *Global Optimization in Engineering Design*, Kluwer Academic Publishers, Boston, MA pp. 355–386.

Soland, R. M. (1971). An algorithm for separable nonconvex programming problems II: Nonconvex constraints, *Management Science* **17**: 759–773.

Spaccamela, A. M., Rinnooy Kan, A. H. G. & Stougie, L. (1984). Hierarchical vehicle routing problems, *Networks* **14**: 571–586.

Speelpenning, B. (1980). *Compiling Fast Partial Derivatives of Functions Given by Algorithms*, PhD thesis, Department of Computer Science, University of Illinois, Urbana, IL.

Stancu-Minasian, I. M. (1997). *Fractional Programming*, Kluwer Academic Publishers, Netherlands.

Stephanopoulos, G. & Westerberg, A. W. (1975). The use of Hestenes' method of multipliers to resolve dual gaps in engineering system optimization, *Journal of Optimization Theory and Applications* **15**: 285–309.

Stoecker, W. F. (1971). *Design of Thermal Systems*, McGraw Hill Book Co., New York.

Stougie, L. & van der Vlerk, M. H. (1997). Stochastic integer programming, In *Annotated Bibliographies in Combinatorial Optimization*, M. Dell'Amico et al. (Eds), John Wiley & Sons, New York pp. 127–141.

Swaney, R. E. (1990). Global solution of algebraic nonlinear programs. *AIChE* Annual Meeting, Chicago, IL.

Sydney, J. B. (1975). Decomposition algorithm for single-machine sequencing with precedence relations and deferral costs, *Operations Research* **23**: 283–298.

Tawarmalani, M. (1997). *Multistage Network Optimization and Decomposition Algorithms*, Master's thesis, Department of Mechanical & Industrial Engineering, University of Illinois, Urbana, IL.

Tawarmalani, M. & Ahmed, S. (1997). Analysis of a machining process. Project Report on Design of Experiments.

Tawarmalani, M., Ahmed, S. & Sahinidis, N. V. (2002a). Global optimization of 0−1 hyperbolic programs. Accepted for publication in *Journal of Global Optimization*.

Tawarmalani, M., Ahmed, S. & Sahinidis, N. V. (2002b). Product disaggregation and relaxations of mixed-integer rational programs. Accepted for publication in *Optimization and Engineering*.

Tawarmalani, M. & Sahinidis, N. V. (1996). The time-dependent traveling salesman problem. Presented at INFORMS Spring Meeting, Washington D.C.

Tawarmalani, M. & Sahinidis, N. V. (1999a). BARON on the Web. http://archimedes.scs.uiuc.edu/cgi/run.pl.

Tawarmalani, M. & Sahinidis, N. V. (1999b). Global optimization of mixed-integer nonlinear programs: A theoretical and computational study. *Mathematical Programming*. Submitted.

Tawarmalani, M. & Sahinidis, N. V. (2001). Semidefinite relaxations of fractional programs via novel techniques for constructing convex envelopes of nonlinear functions, *Journal of Global Optimization* **20**: 137–158.

Tawarmalani, M. & Sahinidis, N. V. (2002a). Convex extensions and convex envelopes of l.s.c. functions. *Mathematical Programming*, DOI 10.1007/s10107-002-0308-z.

Tawarmalani, M. & Sahinidis, N. V. (2002b). Convexification and global optimization of the pooling problem. *Mathematical Programming*. Submitted.

Thakur, L. S. (1990). Domain contraction in nonlinear programming: Minimizing a quadratic concave function over a polyhedron, *Mathematics of Operations Research* **16**: 390–407.

Thoai, N. V. (1991). A global optimization approach for solving the convex multiplicative programming problems, *Journal of Global Optimization* **1**: 341–357.

Törn, A. & Zilinskas, A. (1989). *Global Optimization*, Lecture Notes in Computer Science, Vol. 350, Springer-Verlag, Berlin.

Tsai, L.-W. & Morgan, A. P. (1985). Solving the kinematics of the most general six- and five-degree-of-freedom manipulators by continuation methods, *ASME J. of Mechanisms, Transmissions and Automation in Design* **107**: 48–57.

Tuy, H. (1964). Concave programming under linear constraints, *Doklady Akademic Nauk* **159**: 32–35.

Tuy, H. (1985). Concave programming under linear constraints with special structure, *Optimization* **16**: 335–352.

Tuy, H. (1987). Global optimization of a difference of two convex functions, *Mathematical Programming Study* **30**: 150–182.

Tuy, H. & Horst, R. (1988). Convergence and restart in branch-and-bound algorithms for global optimization. Application to concave minimization and DC optimization problems, *Mathematical Programming* **41**: 161–183.

Tuy, H., Thieu, T. V. & Thai, N. Q. (1985). A conical algorithm for globally minimizing a concave function over a closed convex set, *Mathematics of Operations Research* **10**: 498–514.

Vaidyanathan, R. & El-Halwagi, M. (1996). Global optimization of nonconvex MINLP's by interval analysis, In I. E. Grossmann (ed.), *Global Optimization in Engineering Design*, Kluwer Academic Publishers, Boston, MA pp. 175–193.

van der Vlerk, M. H. (1995). *Stochastic Programming with Integer Recourse*, PhD thesis, University of Groningen, The Netherlands.

Van Hentenryck, P. (1999). *The OPL Optimization Programming Language*, MIT Press, Cambridge, MA. With contributions by I. Lustig, L. Michel, and J.-F. Puget.

Vandenberghe, L. & Boyd, S. (1996). Semidefinite programming, *SIAM Review* **38**: 49–95.

Venkatasubramanian, V., Chan, K. & Caruthers, J. M. (1994). Computer-aided molecular design using genetic algorithms, *Computers & Chemical Engineering* **18**: 833–844.

Venkatasubramanian, V., Chan, K. & Caruthers, J. M. (1995). Genetic algorithm approach for computer-aided molecular design, In C. Reynolds, K. Holloway, and H. Cox (eds.), *ACS Symposium Series Volume on Applications of Computer-Aided Molecular Design*, The American Chemical Society, Washington, D.C. pp. 396–414.

Venkatasubramanian, V., Sundaram, A., Chan, K. & Caruthers, J. M. (1996). Computer-aided molecular design using neural networks and genetic algorithms, In J. Devillers (ed.), *Genetic Algorithms in Molecular Design*, Academic Press pp. 271–302.

Verschelde, J. (1999). Algorithm 795: PHCpack: A general-purpose solver for polynomial systems by homotopy continuation, *ACM Transactions on Mathematical Software* **25**: 251–276.

Visweswaran, V. & Floudas, C. A. (1990). A global optimization algorithm (GOP) for certain classes of nonconvex NLPs—II. Applications of theory and test problems, *Computers & Chemical Engineering* **14**: 1419–1434.

Visweswaran, V. & Floudas, C. A. (1993). New properties and computational improvement of the GOP algorithm for problems with quadratic objective functions and constraints, *Journal of Global Optimization* **3**: 439–462.

Visweswaran, V. & Floudas, C. A. (1996). Computational results for an efficient implementation of the GOP algorithm and its variants, In I. E. Grossmann (ed.), *Global Optimization in Engineering Design*, Kluwer Academic Publishers, Boston, MA pp. 111–153.

West, D. B. (1996). *Introduction to Graph Theory*, Prentice Hall, New Jersey.

Westerberg, A. W. & Shah, J. V. (1978). Assuring a global optimum by the use of an upper bound on the lower (dual) bound, *Computers & Chemical Engineering* **2**: 83–92.

Williams, H. P. (1974). Experiments in the formulation of integer programming problems, *Mathematical Programming Study* **2**: 180–197.

Wise, S. M., Sommese, A. J. & Watson, L. T. (2000). Algorithm 801: POLSYS_PLP: A partitioned linear product homotopy code for solving polynomial systems of equations, *ACM Transactions on Mathematical Software* **26**: 176–200.

Wolfe, M. A. (1996). Interval methods for global optimization, *Appl. Math. Comput.* **75**: 179–206.

Wu, S.-J. & Chow, P.-T. (1995). Genetic algorithms for nonlinear mixed discrete-integer optimization problems via meta-genetic parameter optimization, *Engineering Optimization* **24**: 137–159.

Wu, T. (1997). A note on a global approach for general 0−1 fractional programming, *European Journal of Operational Research* **101**: 220–223.

Wuebbles, D. J. (1981). *The Relative Efficiency of a Number of Halocarbons for Destroying Stratospheric Ozone*, Report UCID-18924, Lawrence Livermore National Laboratory, Livermore, CA.

Wuebbles, D. J. & Edmonds, J. (1991). *A Primer on Green Gases*, Lewis Publishers, Chelsea, MI.

Xia, Q. (1996). A difficult problem for genetic algorithms. Available from http://solon.cma.univie.ac.at/~neum/glopt/xia.txt.

Yuan, X., Zhang, S., Pibouleau, L. & Domenech, S. (1988). Une méthode d'optimisation non linéaire en variables mixtes pour la conception de procédés, *Recherche Opérataionnelle/Operations Research* **22**: 331–346.

Zabinsky, Z. B. (1998). Stochastic methods for practical global optimization, *Journal of Global Optimization* **13**: 433–444.

Zamora, J. M. & Grossmann, I. E. (1998). MINLP model for heat exchanger networks, *Computers & Chemical Engineering* **22**: 367—384.

Zamora, J. M. & Grossmann, I. E. (1999). A branch and contract algorithm for problems with concave univariate, bilinear and linear fractional terms, *Journal of Global Optimization* **14**: 217–249.

Zhang, C. & Wang, H.-P. (1993). Mixed-discrete nonlinear optimization with simulated annealing, *Engineering Optimization* **21**: 277–291.

Index

adjoints, 220
affine hull, 29, 35
alkylation process design, 299
AMPL, 314
automotive, 22, 221, 230, 232, 233

BARON, xiv, xv, xvii, xviii, 16, 18, 22, 93, 96, 100, 112, 120–123, 194, 210, 211, 213–219, 221–225, 227, 228, 249, 280, 283, 285, 286, 293, 295, 298, 299, 311, 313–316, 321, 323–325, 330–333, 340, 347, 349–356, 363–365, 368–372, 375–382, 384, 385, 388, 389, 391, 394, 395, 397, 400, 401
benchmarks, xv, 23, 280, 285, 298–304, 313, 380
Benders cuts, 205
Benders decomposition, 222, 223
biconjugate, 148, 149
biconvex programming, 300
bilinear, 25, 28, 41, 48, 51, 56, 72, 75, 80–86, 93, 129, 184, 192, 194, 218, 256, 258, 260, 264, 266, 268, 269, 274–278, 395
bilinear programming, 21, 196, 255, 256, 299
binary, 36, 73, 74, 80, 104, 106, 107, 114–116, 198, 217–219, 239
binary expansion, 383
bisection, 10, 132, 135, 139, 144, 183, 185, 194, 223, 279
blending, xv, 22, 264
bounds contraction, *see* domain reduction
bounds reduction, *see* domain reduction
branch-and-bound, xiv, 3, 4, 6–8, 10–12, 14, 17, 18, 22, 26, 71, 73, 75, 88, 94, 96, 105, 108–112, 114, 160, 164, 182, 189, 190, 194, 196, 200, 205, 207, 211, 213, 214, 217, 218, 221–223, 253, 256, 276, 280, 286, 289, 293, 297, 299, 311, 330, 331, 333, 335, 336, 340, 341, 343, 347, 348, 350, 371, 384, 389
branching, 4, 6, 10, 17, 18, 20–22, 183, 187, 189–192, 194–197, 200, 202–205, 207, 210, 218, 222–224, 277–279, 331, 332, 334, 371, 375

chemical equilibrium, 299, 401
chlorofluorocarbons, 229, 230
collocation, 98, 99
complementarity, 177, 401

concave envelope, 25–29, 37, 40, 56, 57, 60, 62, 64, 72, 74–76, 80, 83, 84, 101, 127, 129, 130, 132, 258, 268, 277
concave extension, 28, 29, 33, 40, 76, 79, 80, 91, 92, 101–104
concave minimization, 214, 215, 285, 287, 288, 290, 300, 317, 379
concave quadratic programming, 215, 285, 287, 288, 290, 300, 317
concavoconvex, 130, 131
cone, 149, 150, 207
conic program, 61, 62, 65
conjugacy, 21, 150
conjugate, 147–150, 152, 165
CONOPT2, 348–352, 354, 355, 375, 377, 378
convergence, xv, 4, 6, 12, 14, 18, 20–22, 26, 75, 115, 135, 139, 141, 147, 183, 189, 191, 196, 218, 222, 223, 255, 278, 339, 371
convex envelope, xiv, 9, 20, 25–41, 43–47, 50, 51, 57, 59, 62–66, 68, 69, 76, 82–87, 131, 132, 268, 269, 274
convex extension, 20, 21, 25, 28, 29, 31–38, 40, 42, 43, 47, 49–52, 76, 79, 92, 101, 102, 104, 125, 151, 256, 266
convex hull, 10, 12, 29, 36, 39, 64, 75, 79, 82, 84, 253, 255, 269, 273
convex relaxation, 5, 9, 25, 35, 39, 79, 84, 88, 125, 178, 182, 269, 274

convexification, 28, 59, 63, 67, 69, 82, 253, 254, 256
convexoconcave, 130, 132
CPLEX, 75, 94, 114–116, 120–123, 209, 216, 320, 355, 389–393

data structures, 18, 22, 214–218, 224, 249, 325, 368
debugging, 224
design, 299, 300
design of experiments, 300
differentiation, 219, 220
direct pattern search, 307
disaggregation, xiv, 21, 71–74, 83, 93, 94, 97, 100, 205
discretization, 97–99
disjunction, 93
disjunctive, 93
disjunctive programming, 28, 44, 62, 253, 254, 256
domain contraction, *see* domain reduction
domain reduction, xiv, 4, 6, 15, 18, 20, 21, 95, 97, 142, 147, 153, 154, 159, 161, 163, 164, 177–179, 181–185, 196, 213, 214, 218, 222, 223, 277, 278, 280, 297, 333, 334, 353, 355, 367, 368, 381–383
dual, 21, 35, 135, 149, 152–155, 159, 160, 163, 164, 168, 169, 177–186, 194, 196, 222, 273–276, 278, 279
duality, 147, 153, 195, 206, 275, 333

economic equilibrium, 401
engineering design, xiii, xv, xvi, 1, 298, 299, 380

INDEX 465

enumeration, xv, 77, 198, 229, 231
enumerative, 73
envelope, *see* convex (concave) envelope
epigraph, 29, 30, 32, 33, 35–37, 39, 41, 44, 45, 63–65, 76, 79, 82, 83, 102, 148, 159, 163, 184
exhaustive, 22, 77, 183, 185, 194, 223, 278
exponential functions, 300
extension, *see* convex (concave) extension

facility location, xv, 21, 72, 73, 75, 110, 111, 115, 116
factorable programming, 4, 9, 10, 12, 21–23, 49, 62, 125–127, 132, 189, 190, 214, 215, 218, 285, 286, 298, 313, 315, 316
fathoming, 6, 21, 160, 183, 192, 203–205, 222
finite, xv, 6, 17, 20, 22, 189, 190, 196–198, 200, 205, 223
finiteness, 88, 189, 190, 205, 207, 210, 223
fixed-charge programming, 300
flowshop, 305
Fourier-Motzkin, 45, 180, 181, 247
fractional programming, 214, 215, 318

GAMS, xviii, 112, 224, 313, 314, 322, 323, 325, 332, 356, 357, 360, 368, 372, 388, 395, 399, 403, 405, 407, 409, 411, 413, 416, 418, 422, 425, 428, 429, 431

GAMS/CONOPT2, *see* CONOPT2
GAMS/DICOPT, 394
GAMS/MINOS, *see* MINOS
GAMS/SBB, 394
gamslib, 379, 385, 393
generating set, 25, 27, 37–39, 41, 43, 50, 56, 59, 62, 69, 268
globallib, 314, 376, 380, 395

heat exchanger network, 299
Hooke and Jeeves, 307
hyperbolic programming, 21, 29, 71, 72, 74, 75, 89–91, 100, 106–108, 110, 111, 114, 120–123
hypograph, 35, 37, 41, 76, 79, 83, 150

indefinite quadratic programming, 214, 215, 289, 291, 292, 319, 379
integer cuts, 384, 388–390, 392
interval arithmetic, 128, 218, 232, 277, 355, 381

just-in-time production, 305, 307, 313, 379

KKT conditions, 20, 35, 45, 46, 52, 177, 184, 193, 194

Lagrangian, 167, 175, 186
Lagrangian relaxation, 253
Lagrangian subproblem, 162, 164, 169, 175, 177–179, 181, 186, 275
Legendre-Fenchel transform, 147, 148
LeX, 215
linear programming, xiii, xiv, 1, 12, 14, 18, 20, 21, 91, 94, 180,

181, 196, 214, 216, 253, 275, 278, 279, 307
linear programming relaxation, xiv, 21, 71, 106, 111, 112, 114, 116, 125, 253, 256, 274, 280
linearization, 25, 28, 73, 75, 240, 384
LINGO, 314
LINPACK, 94, 209, 249, 280, 283
local search, 6, 10, 222, 278, 329–331, 334, 339, 342, 347–350, 352, 353, 355, 365, 367, 371, 375, 379–382
logarithmic, 129, 217, 218
LP, *see* linear programming

marginals, 15, 178, 179, 277, 278, 333
MILP, *see* mixed-integer linear programming
MINLP, *see* mixed-integer nonlinear programming
minlplib, 314, 376, 379, 380
minorant, 148
MINOS, 94, 112, 216, 320, 348–356, 375, 378
mixed-integer linear programming, xiv, 18, 22, 71, 74, 79, 83, 88, 207, 209, 214, 215, 320
mixed-integer nonlinear programming, xiii–xvi, 4, 5, 14, 16, 20, 22, 23, 71, 95, 190, 191, 214, 217, 221, 229, 231, 299, 305, 306, 311–316
modules, 22, 112, 214, 216–218
molecular design, 300, 313, 379, 391
monotonicity, 177, 204

multilinear functions, 28, 39–43, 74, 76, 77
multiplicative programming, 215, 293–297, 300, 313, 319, 325, 326, 330, 331, 333, 335, 336, 340–343, 346–354, 356, 357, 379
multiplier, 15, 35, 178–180, 182–185, 268, 333

nonlinear relaxation, 93, 132, 194

OPL, 314
OSL, 112, 216
outer-approximation, xiv, 3, 9, 26, 84, 132–135, 139, 141–144, 159, 163, 164, 222, 223, 253, 269, 342, 353
overestimator, 26, 126–128, 141
ozone, 229–231

parameter estimation, 300
parser, 224
partitioning, *see* branching
perturbation, 151, 152, 154, 163, 167, 200
polar, 150, 194
polyhedral, 39, 40, 43, 56, 75, 76, 84, 133, 144, 207, 253
polytope, 40, 41, 64, 110, 274
pooling problem, xv, 22, 23, 29, 253–256, 258, 261, 264, 273, 276, 278–280, 299, 313, 356, 360, 368, 372, 375, 380, 403, 405, 407, 409, 411, 413, 416, 418, 422, 425, 428, 429, 431
postponement, 22, 223
preprocessing, 18, 321, 323, 324, 330, 339, 342, 353, 355, 363–

INDEX

365, 367, 375, 381
priorities, 279
probing, 15, 181, 182, 276, 278, 279, 286, 289, 293, 297, 333, 334, 370
process synthesis, 299
projection, 29, 30, 32, 127, 149, 191, 193
pseudo-cost, 190, 191

quadratic programming, 299, 319

range contraction, *see* domain reduction
range reduction, *see* domain reduction
rational programming, 74
reaction, 97
reactor network, 299
reformulation-linearization technique, 79, 253, 256, 264
refrigerant, xv, 22, 221, 229–233, 235, 249–252
refrigerant design, *see* molecular design
refrigeration, 233
relaxation, xiv, 4–6, 9–18, 20–22, 25, 26, 29, 35, 46, 49, 57, 61, 62, 67, 69, 71–73, 75, 79, 81, 83, 84, 88, 93, 95, 100, 103–106, 108, 109, 111, 112, 114, 116, 125–129, 132, 133, 153, 155, 159, 161–163, 168, 169, 178–180, 182–186, 190–196, 200, 205, 206, 208, 213, 218, 224, 240, 243, 253–256, 258, 261, 262, 264, 266–270, 272–276, 278–280, 334

reliability, 300
RLT, *see* reformulation-linearization technique

sandwich algorithm, 21, 125, 135, 139, 142
Schur complement, 60, 62
SDP, *see* semidefinite programming
SDPA, 216
semidefinite programming, xvii, 20, 22, 29, 60–62, 67, 69, 129, 214–216
separable programming, 4
simplex, 262, 268, 274
SNOPT, 216, 320
special ordered sets, 4
stability, 401
stochastic programming, 22, 189, 190, 196–200, 202, 203, 207, 209, 223
subgradient, 33, 34, 133, 151, 157
systems of nonlinear equations, 394

truss design, 300

underestimator, 9, 10, 25, 26, 28, 29, 36, 37, 47–49, 51–53, 126–128, 194
univariate polynomial programming, 214, 215, 298, 318

violation, 10, 21, 189–195, 279

Weirstrass, 5

YACC, 215

Author Index

Achenie, L. E. K., 231, 232, 439–441
Adams, W. P., 4, 79, 253, 254, 256, 264, 266, 273, 293, 456
Adhya, N., 254, 255, 273, 274, 279, 280, 403, 435
Adjiman, C. S., 3, 435
Aggarwal, S. C., 73, 435
Ahmed, S., xvii, 22, 73, 114–116, 128, 190, 196–198, 223, 302, 435, 444, 446, 457, 458
Aho, A. V., 215, 435
Ahuja, R. K., 1, 435
Al-Khayyal, F. A., 29, 48, 56, 75, 129, 190, 196, 256, 260, 302, 435, 436
Alameddine, A., 3, 278, 286, 456
Alizadeh, F., 29, 436
Andersen, D. E., 18, 21, 147, 179, 436
Andersen, K. D., 18, 21, 147, 179, 436
Androulakis, I. P., 279, 325, 436, 449
Arnold, K. J., 302, 437
Arora, S. R., 73, 436
Ashrafi, N., 302, 436
Audet, C., 254, 255, 279, 280, 431, 436

Bagherpour, K., 231, 440
Balas, E., 28, 74, 93, 254, 436, 437
Bazaraa, M. S., 1, 20, 35, 42, 307, 437
Beale, E. M. L., 4, 437
Beck, J. V., 302, 437
Ben-Tal, A., 3, 254, 255, 261, 262, 275, 276, 279, 280, 411, 437
Bennett, K. P., 326, 437
Benson, H. P., 293, 295, 344, 437
Berger, A. J., 325, 449
Berman, O., 302, 436
Biegler, L. T., 98–100, 302, 437, 438, 449
Bienstock, D., 4, 198, 438
Bilde, O., 72, 447
Bitran, G. R., 198, 438
Blair, C. E., 207, 438
Blum, M., 91, 438
Boger, G. M., 293, 295, 344, 437
Boon, S., 397, 438
Borchers, B., 3, 4, 438
Borisov, W. W., 219, 451
Bottini, S. B., 231, 232, 438, 452
Boyd, A., 144, 438
Boyd, S., 29, 60, 61, 449, 459
Bracken, J., 301, 438
Brignole, E. A., 231, 232, 438, 452

Brimberg, J., 254, 255, 279, 280, 431, 436
Brook, A., 112, 224, 314, 439
Burkard, R. E., 21, 135, 139, 439, 441
Bussieck, M. R., 314, 439

Caruthers, J. M., 232, 459, 460
Carøe, C. C., 198, 205, 208, 209, 211, 439
Ceria, S., 28, 254, 436, 439
Chan, K., 232, 459, 460
Charnes, A., 91, 106, 439
Cheung, B. K.-S., 3, 439
Choe, U., 72, 453
Chou, C.-T., 3, 302, 448
Chow, P.-T., 3, 461
Churi, N., 231, 232, 439, 440
Ciric, A. R., 301, 441
Colville, A. R., 302, 440
Constantinou, L., 231, 440
Cooper, L. G., 116, 451
Cooper, W. W., 91, 106, 439
Cornuèjols, G., 254, 436
Craig, S., 73, 115, 442
Crama, Y., 28, 31, 76, 128, 440

Dakin, R. J., 6, 440
Dallwig, S., 3, 435
Dantzig, G. B., 1, 440
Delmaire, H., 3, 439
Dempster, M. A. H., 198, 440
Devillers, J., 232, 440
Dixon, L. C. W., 302, 440
Doig, A. G., 6, 448
Domenech, S., 301, 461
Dorneich, M. C., 326, 440
Driscoll, P. J., 273, 456

Duran, M. A., 3, 4, 440
Duvedi, A. P., 231, 232, 441

Edmonds, J., 231, 461
Eiger, G., 3, 254, 255, 261, 262, 275, 276, 279, 280, 411, 437
El-Halwagi, M., 3, 459
Epperly, T. G. W., 3, 441

Falk, J. E., 3, 6, 29, 48, 56, 75, 129, 256, 260, 297, 302, 435, 441
Floudas, C. A., 3, 4, 254, 279, 280, 283, 286, 289, 301, 322, 325, 435, 436, 441, 449, 460
Floyd, R. W., 91, 438
Forrest, J. J. H., 4, 437
Foulds, L. R., 254, 279, 280, 416, 441
Fourer, R., 314, 441
Fournier, R. L., 231, 451
Fredenslund, A., 231, 232, 442
Fruhwirth, B., 21, 139, 441
Fujie, T., 62, 441
Fukushima, M., 73, 444

GAMS Development Corporation, 314, 442
Gani, R., 231, 232, 438, 440, 442
Gay, D. M., 314, 441
Gershovitz, V., 3, 254, 255, 261, 262, 275, 276, 279, 280, 411, 437
Ghildyal, V., 213, 442
Ghosh, A., 73, 115, 116, 442
Gill, P. E., 216, 320, 442
Gilmore, P. C., 73, 442
Glover, F., 74, 76, 80, 293, 442
Gomory, R. E., 73, 442
Goyal, S. K., 305, 307, 308, 443

Granot, D., 73, 442, 443
Granot, F., 73, 442, 443
Greenberg, H. J., 254, 443
Griewank, A., 219, 443
Griffin, O. H., 302, 456
Grossmann, I. E., 3, 4, 21, 28, 47, 57, 83, 147, 179, 254, 264, 301, 302, 326, 437, 440, 446, 452, 454, 461
Gruber, P. M., 133, 443
Grunspan, M., 73, 443
Gunasekaran, A., 305, 307, 308, 443
Gunn, D. J., 97, 443
Gupta, O. K., 3, 4, 311, 312, 443
Gurdal, Z., 302, 456
Gutierrez, R. A., xvii, 305, 307, 308, 443

Haas, E. A., 198, 438
Hamacher, H., 21, 135, 439
Hamed, A. S. E., 21, 147, 443
Hammer, P. L., 73, 76, 443
Hansen, E. R., 3, 444
Hansen, P., 3, 21, 73, 74, 91, 147, 254, 255, 279, 280, 293, 298, 431, 436, 444
Haque, M. A., 114–116, 444
Hashizume, S., 73, 444
Haugland, D., 254, 279, 280, 416, 441
Haverly, C. A., 254–256, 258, 279, 301, 428, 444
Heller, W. R., 326, 449
Henderson, J. M., 325, 444
Hillestad, R. J., 3, 444
Hiriart-Urruty, J.-B., 147, 444, 445
Hock, W., 94, 302, 445
Hoffman, K., 3, 445

Hooker, J., 16, 445
Hopcroft, J. E., 215, 445
Horst, R., 3, 4, 6, 29, 194, 286, 445, 459
Horvath, A. L., 230, 231, 445

Ibaraki, T., 73, 444
IBM, 112, 216, 445
ILOG, 75, 94, 114, 216, 320, 445
Isbelle, J. R., 73, 445

Jacobsen, S. E., 3, 444
Jaumard, B., 3, 21, 74, 91, 147, 293, 298, 444
Jeroslow, R. G., 207, 254, 438, 445
Joback, K. G., 229, 231, 232, 239, 247, 248, 446
Jorjani, S., 209, 446
Jornsten, K., 254, 279, 280, 416, 441

Kabadi, S. N., 286, 451
Kalantari, B., 286, 302, 446
Katoh, N., 73, 444
Kearfott, R. B., 3, 218, 446
Keeney, R. L., 326, 446
Kenderov, P., 133, 443
Kendrick, D., 112, 224, 314, 439
Kernighan, B. W., 314, 441
Kim, Y. S., 302, 446
King, A. J., 198, 446
Klein, J. A., 231, 440
Klein Haneveld, W. K., 198, 446
Kocis, G. R., 301, 446
Kojima, M., 62, 441, 447
Kokossis, A. C., 232, 449
Konno, H., 293–295, 297, 302, 325, 326, 447
Krarup, J., 72, 447

Kuno, T., 293–295, 297, 302, 326, 447

Lamar, B. W., 21, 147, 179, 447
Land, A. H., 6, 448
Langevin, A., 3, 439
Laporte, G., 198, 448
Lasdon, L. S., 254, 301, 448, 453
Lawler, E. L., 73, 448
Lazimy, R., 4, 448
Lebret, H., 29, 61, 449
Lee, E. K., 3, 4, 448
Lemaréchal, C., 147, 444, 445
Li, H.-L., 3, 74, 92, 93, 101, 105, 106, 112, 302, 448
Liebman, J., 301, 448
Liu, M. L., 194, 286, 448
Liu, X. J., 293, 295, 344, 448
Lobo, M. S., 29, 61, 449
Logsdon, J. S., 98, 99, 449
Lokketangen, A., 209, 449
Lopez, P. A., 231, 232, 452
Louveaux, F. V., 198, 448
Lovàsz, L., 254, 449
Lu, S.-H., 3, 21, 147, 444

Macchietto, S., 231, 239, 244, 247, 248, 449, 451
Magnanti, T. L., 1, 435
Maling, K., 326, 449
Mangasarian, O. L., 21, 147, 177, 326, 437, 449
Manousiouthakis, V., 301, 449
Maranas, C. D., 325, 449
Marcoulaki, E. C., 232, 449
Marin, R., 302, 449
Markowitz, H. M., 326, 450
Marlow, W. H., 73, 445

Martelli, G., 21, 450
Martikainen, T., 305, 307, 308, 443
Martin, R. K., 72, 450
Mathon, V., 74, 91, 293, 444
Matsui, T., 293, 447, 450
Matsuo, H., 198, 438
Mazzola, J. B., 74, 437
McBride, R. D., 4, 450
McCormick, G. P., 3, 9, 12, 21, 28, 48, 72, 75, 125–127, 129, 147, 184, 260, 301, 315, 438, 443, 450
McLafferty, S., 73, 115, 116, 442
McLinden, L., 21, 147, 177, 449
Meeraus, A., 112, 224, 314, 439, 450
Megiddo, N., 73, 88, 89, 450
Minoux, M., 1, 20, 450
Mitchell, J. E., 3, 4, 438, 448
Mittelmann, H. D., 314, 450
Mladenovic, N., 254, 255, 279, 280, 431, 436
Morgan, A. P., 395, 459
Mueller, S. H., 326, 449
Mulvey, J. M., 325, 326, 449, 450
Murray, W., 216, 320, 442
Murtagh, B. A., 94, 112, 216, 320, 450
Murty, K. G., 286, 451

Nabar, S. V., 3, 451
Nakanishi, M., 116, 451
Nanda, G., 232, 239, 244, 245, 451
Naser, S. F., 231, 451
Nemhauser, G. L., 4, 72, 83, 93, 179, 451
Neumaier, A., 3, 435
Nielsen, B., 231, 232, 442

AUTHOR INDEX

Odele, O., 231, 239, 244, 247, 248, 449, 451
Omatsone, O., 231, 449
Orlin, J. B., 1, 435
Ostrovskii, G. M., 219, 451
Ourique, J. E., 232, 451

Pantelides, C. C., 3, 4, 457
Pardalos, P. M., 3, 286, 322, 441, 451–453
Parker, R. G., 4, 72, 83, 93, 452
Phillips, A. T., 286, 289, 452
Pibouleau, L., 301, 461
Picard, J., 73, 452
Poggi de Aragao, M. V., 73, 444
Poling, B. E., 231, 453
Polocsay, S. W., 297, 441
Pratt, V., 91, 438
Prausnitz, J. M., 231, 453
Pretel, E. J., 231, 232, 452
Puri, M. C., 73, 436
Putavy, C., 232, 440

Quandt, R. E., 325, 444
Quesada, I., 83, 254, 264, 326, 452
Queyranne, M., 73, 452
Quist, A. J., 95, 96, 452

Raiffa, H., 326, 446
Rao, M. R., 73, 453
Rardin, R. L., 4, 72, 83, 93, 452, 453
Ravindran, A., 3, 4, 311, 312, 443
Reid, R. C., 231, 232, 446, 453
Ribeiro, C. C., 73, 444
Rigby, B., 254, 453
Rikun, A. D., 28, 39, 40, 69, 128, 254, 274, 453

Rinnooy Kan, A. H. G., 3, 198, 453, 457
Rivest, R. L., 91, 438
Robillard, P., 73, 453
Rockafellar, R. T., 28–30, 33, 44, 63, 80, 147, 149–153, 156, 161, 254, 453
Rosen, J. B., 286, 289, 302, 446, 452, 453
Rote, G., 21, 125, 135, 139, 439, 441, 454
Rudeanu, S., 73, 76, 443
Ruszczyński, A., 198, 439
Ryoo, H. S., 3, 4, 14, 16, 17, 21, 128, 147, 178, 179, 181, 182, 293, 297, 299, 326, 336, 344, 454

Sahinidis, N. V., xvii, 3, 4, 14, 16, 17, 21, 22, 73, 112, 128, 147, 178, 179, 181, 182, 190, 194, 196, 197, 213, 223–225, 239, 244, 245, 254–256, 266, 273, 274, 279, 280, 286, 299, 301, 302, 305, 307, 308, 314, 325, 326, 336, 344, 403, 435, 440, 442, 443, 448, 451, 454, 456, 458
Saipe, A. L., 73, 91, 107, 110, 111, 455
Sandgren, E., 302, 455
Saunders, M. A., 94, 112, 216, 320, 442, 450
Savelsbergh, M. W. P., 21, 147, 179, 181, 455
Schaible, S., 73, 297, 455
Schittkowski, K., 94, 302, 445
Schoen, F., 4, 455

Schrage, L., 3, 301, 314, 448, 451, 455
Schrijver, A., 4, 72, 83, 93, 254, 449, 455
Schultz, R., 198, 199, 205, 208, 209, 439, 455
Scott, C. H., 209, 446
Sethi, R., 215, 435
Shah, J. V., 301, 460
Shapiro, J. F., 198, 438
Shectman, J. P., 17, 21, 147, 179, 190, 194, 196, 223, 286, 289, 448, 455, 456
Sherali, H. D., 1, 3, 4, 20, 28, 35, 42, 79, 128, 190, 196, 253, 254, 256, 264, 266, 273, 278, 286, 293, 307, 436, 437, 456
Shetty, C. M., 1, 20, 35, 42, 307, 437
Shin, D. K., 302, 456
Shirakawa, H., 325, 326, 447
Smith, E. M. B., 3, 4, 457
Soares, J., 28, 439
Sodini, C., 297, 455
Soland, R. M., 3, 6, 301, 302, 441, 457
Sommese, A. J., 394, 460
Sourlas, D., 301, 449
Spaccamela, A. M., 198, 457
Speelpenning, B., 219, 457
Stancu-Minasian, I. M., 73, 89, 107, 457
Stephanopoulos, G., 229, 231, 232, 239, 247, 248, 301, 446, 457
Stoecker, W. F., 301, 457
Stougie, L., 198, 199, 209, 446, 455, 457
Sundaram, A., 232, 460

Swaney, R. E., 3, 301, 441, 457
Swarup, K., 73, 436
Sydney, J. B., 73, 457
Szegoe, G. P., 302, 440

Takriti, S., 198, 446
Tarjan, R. E., 91, 438
Tawarmalani, M., xvii, 3, 22, 73, 128, 190, 196, 197, 213, 223–225, 254–256, 266, 273, 274, 279, 280, 302, 305, 314, 325, 403, 435, 454, 457, 458
Telles, A. S., 232, 451
Thai, N. Q., 3, 286, 459
Thakur, L. S., 21, 147, 179, 302, 458
Thieu, T. V., 3, 286, 459
Thoai, N. V., 3, 293–295, 302, 336, 445, 458
Thomas, M. E., 73, 443
Thomas, W. J., 97, 443
Timmer, G. T., 3, 453
Tind, J., 198, 439
Tjoa, I. B., 99, 100, 438
Tomlin, J. A., 4, 437
Tsai, L.-W., 395, 459
Tuncbilek, C. H., 286, 456
Tunçel, L., 62, 447
Tuy, H., 3, 4, 6, 29, 194, 286, 445, 459
Törn, A., 4, 458

Ullman, J. D., 215, 435, 445
Umegaki, T., 293, 295, 344, 448

Vaidyanathan, R., 3, 459
Vandenberghe, L., 29, 60, 61, 449, 459
Vanderbei, R. J., 326, 450

van der Vlerk, M. H., 198, 199, 209, 446, 455, 457, 459
Van Hentenryck, P., 314, 459
van Vliet, M., 286, 453
Venkatasubramanian, V., 232, 459, 460
Verschelde, J., 394, 397, 460
Visweswaran, V., 3, 254, 279, 280, 283, 286, 289, 301, 436, 460
Volin, Y. M., 219, 451

Wang, H., 4, 456
Wang, H.-P., 3, 461
Waren, A. D., 254, 301, 448, 453
Watson, L. T., 394, 460
West, D. B., 248, 460
Westerberg, A. W., 301, 302, 437, 457, 460
Wets, R. J.-B., 147, 149, 152, 453
Williams, H. P., 73, 74, 93, 460
Wise, S. M., 394, 460
Wolfe, M. A., 3, 461
Wolsey, L. A., 4, 72, 83, 93, 179, 451
Woodruff, D. L., 209, 446, 449
Woolsey, E., 74, 76, 80, 293, 442
Wu, D. T., 231, 440
Wu, S.-J., 3, 461
Wu, T., 74, 92, 93, 101, 105, 461
Wuebbles, D. J., 231, 461

Xia, Q., 302, 461
Xiong, J., 298, 444

Yajima, Y., 293, 294, 297, 326, 447
Yamamoto, Y., 293, 295, 344, 448
Yamazaki, H., 325, 326, 447
Yli-Olli, P., 305, 307, 308, 443
Yormark, J. S., 4, 450

Yu, M., xvii, 22, 302, 454
Yuan, X., 301, 461

Zabinsky, Z. B., 3, 461
Zamora, J. M., 21, 28, 47, 57, 147, 179, 461
Zenios, S. A., 326, 450
Zhang, C., 3, 461
Zhang, S., 301, 461
Zilinskas, A., 4, 458

Nonconvex Optimization and Its Applications

22. H. Tuy: *Convex Analysis and Global Optimization.* 1998   ISBN 0-7923-4818-4
23. D. Cieslik: *Steiner Minimal Trees.* 1998   ISBN 0-7923-4983-0
24. N.Z. Shor: *Nondifferentiable Optimization and Polynomial Problems.* 1998
    ISBN 0-7923-4997-0
25. R. Reemtsen and J.-J. Rückmann (eds.): *Semi-Infinite Programming.* 1998
    ISBN 0-7923-5054-5
26. B. Ricceri and S. Simons (eds.): *Minimax Theory and Applications.* 1998
    ISBN 0-7923-5064-2
27. J.-P. Crouzeix, J.-E. Martinez-Legaz and M. Volle (eds.): *Generalized Convexitiy, Generalized Monotonicity: Recent Results.* 1998   ISBN 0-7923-5088-X
28. J. Outrata, M. Kočvara and J. Zowe: *Nonsmooth Approach to Optimization Problems with Equilibrium Constraints.* 1998   ISBN 0-7923-5170-3
29. D. Motreanu and P.D. Panagiotopoulos: *Minimax Theorems and Qualitative Properties of the Solutions of Hemivariational Inequalities.* 1999   ISBN 0-7923-5456-7
30. J.F. Bard: *Practical Bilevel Optimization.* Algorithms and Applications. 1999
    ISBN 0-7923-5458-3
31. H.D. Sherali and W.P. Adams: *A Reformulation-Linearization Technique for Solving Discrete and Continuous Nonconvex Problems.* 1999   ISBN 0-7923-5487-7
32. F. Forgó, J. Szép and F. Szidarovszky: *Introduction to the Theory of Games.* Concepts, Methods, Applications. 1999   ISBN 0-7923-5775-2
33. C.A. Floudas and P.M. Pardalos (eds.): *Handbook of Test Problems in Local and Global Optimization.* 1999   ISBN 0-7923-5801-5
34. T. Stoilov and K. Stoilova: *Noniterative Coordination in Multilevel Systems.* 1999
    ISBN 0-7923-5879-1
35. J. Haslinger, M. Miettinen and P.D. Panagiotopoulos: *Finite Element Method for Hemivariational Inequalities.* Theory, Methods and Applications. 1999
    ISBN 0-7923-5951-8
36. V. Korotkich: *A Mathematical Structure of Emergent Computation.* 1999
    ISBN 0-7923-6010-9
37. C.A. Floudas: *Deterministic Global Optimization: Theory, Methods and Applications.* 2000   ISBN 0-7923-6014-1
38. F. Giannessi (ed.): *Vector Variational Inequalities and Vector Equilibria.* Mathematical Theories. 1999   ISBN 0-7923-6026-5
39. D.Y. Gao: *Duality Principles in Nonconvex Systems.* Theory, Methods and Applications. 2000   ISBN 0-7923-6145-3
40. C.A. Floudas and P.M. Pardalos (eds.): *Optimization in Computational Chemistry and Molecular Biology.* Local and Global Approaches. 2000   ISBN 0-7923-6155-5
41. G. Isac: *Topological Methods in Complementarity Theory.* 2000 ISBN 0-7923-6274-8
42. P.M. Pardalos (ed.): *Approximation and Complexity in Numerical Optimization: Concrete and Discrete Problems.* 2000   ISBN 0-7923-6275-6
43. V. Demyanov and A. Rubinov (eds.): *Quasidifferentiability and Related Topics.* 2000
    ISBN 0-7923-6284-5

Nonconvex Optimization and Its Applications

44. A. Rubinov: *Abstract Convexity and Global Optimization.* 2000
    ISBN 0-7923-6323-X
45. R.G. Strongin and Y.D. Sergeyev: *Global Optimization with Non-Convex Constraints.* 2000   ISBN 0-7923-6490-2
46. X.-S. Zhang: *Neural Networks in Optimization.* 2000   ISBN 0-7923-6515-1
47. H. Jongen, P. Jonker and F. Twilt: *Nonlinear Optimization in Finite Dimensions. Morse Theory, Chebyshev Approximation, Transversability, Flows, Parametric Aspects.* 2000   ISBN 0-7923-6561-5
48. R. Horst, P.M. Pardalos and N.V. Thoai: *Introduction to Global Optimization.* 2nd Edition. 2000   ISBN 0-7923-6574-7
49. S.P. Uryasev (ed.): *Probabilistic Constrained Optimization. Methodology and Applications.* 2000   ISBN 0-7923-6644-1
50. D.Y. Gao, R.W. Ogden and G.E. Stavroulakis (eds.): *Nonsmooth/Nonconvex Mechanics. Modeling, Analysis and Numerical Methods.* 2001   ISBN 0-7923-6786-3
51. A. Atkinson, B. Bogacka and A. Zhigljavsky (eds.): *Optimum Design 2000.* 2001
    ISBN 0-7923-6798-7
52. M. do Rosário Grossinho and S.A. Tersian: *An Introduction to Minimax Theorems and Their Applications to Differential Equations.* 2001   ISBN 0-7923-6832-0
53. A. Migdalas, P.M. Pardalos and P. Värbrand (eds.): *From Local to Global Optimization.* 2001   ISBN 0-7923-6883-5
54. N. Hadjisavvas and P.M. Pardalos (eds.): *Advances in Convex Analysis and Global Optimization. Honoring the Memory of C. Caratheodory (1873-1950).* 2001
    ISBN 0-7923-6942-4
55. R.P. Gilbert, P.D. Panagiotopoulos[†] and P.M. Pardalos (eds.): *From Convexity to Nonconvexity.* 2001   ISBN 0-7923-7144-5
56. D.-Z. Du, P.M. Pardalos and W. Wu: *Mathematical Theory of Optimization.* 2001
    ISBN 1-4020-0015-4
57. M.A. Goberna and M.A. López (eds.): *Semi-Infinite Programming. Recent Advances.* 2001   ISBN 1-4020-0032-4
58. F. Giannessi, A. Maugeri and P.M. Pardalos (eds.): *Equilibrium Problems: Nonsmooth Optimization and Variational Inequality Models.* 2001   ISBN 1-4020-0161-4
59. G. Dzemyda, V. Šaltenis and A. Žilinskas (eds.): *Stochastic and Global Optimization.* 2002   ISBN 1-4020-0484-2
60. D. Klatte and B. Kummer: *Nonsmooth Equations in Optimization. Regularity, Calculus, Methods and Applications.* 2002   ISBN 1-4020-0550-4
61. S. Dempe: *Foundations of Bilevel Programming.* 2002   ISBN 1-4020-0631-4
62. P.M. Pardalos and H.E. Romeijn (eds.): *Handbook of Global Optimization, Volume 2.* 2002   ISBN 1-4020-0632-2
63. G. Isac, V.A. Bulavsky and V.V. Kalashnikov: *Complementarity, Equilibrium, Efficiency and Economics.* 2002   ISBN 1-4020-0688-8
64. H.-F. Chen: *Stochastic Approximation and Its Applications.* 2002
    ISBN 1-4020-0806-6